Deconvolution

With Applications in Spectroscopy

Deconvolution

With Applications in Spectroscopy

Edited by

PETER A. JANSSON

E. I. du Pont de Nemours and Company (Inc.)
Engineering Physics Laboratory
Wilmington, Delaware

1984

ACADEMIC PRESS, INC.

(Harcourt Brace Jovanovich, Publishers)

Orlando   San Diego   San Francisco   New York   London
Toronto   Montreal   Sydney   Tokyo

ACADEMIC PRESS, INC.
Orlando, Florida 32887

United Kingdom Edition published by
ACADEMIC PRESS, INC. (LONDON) LTD.
24/28 Oval Road, London NW1 7DX

Library of Congress Cataloging in Publication Data

Main entry under title:

Deconvolution : with applications in spectroscopy.

   Includes index.
   1. Spectrum analysis--Deconvolution. I. Jansson,
Peter A.
QC451 6.D45 1984   535.8'4   83-15647
ISBN 0-12-380220-2

PRINTED IN THE UNITED STATES OF AMERICA

84 85 86 87   9 8 7 6 5 4 3 2 1

To Muriel, Karen, and Jonathan

In Memory of
Earle K. Plyler
Educator, Counselor, Physicist

Contents

Contributors   xi
Preface   xiii

1. Convolution and Related Concepts
Peter A. Jansson

    I. Introduction   3
   II. Definition of Convolution   4
  III. Properties   7
  IV. Fourier Transforms   11
   V. The Problem of Deconvolution   28
       References   34

2. Distortion of Optical Spectra
Peter A. Jansson

    I. Inherent Line Breadth   38
   II. Contribution of Dispersive Spectrometer   44
  III. Resolution Criteria   61
       References   64

3. Traditional Linear Deconvolution Methods
Peter A. Jansson

    I. Direct Approach   69
   II. Van Cittert's Method   71
  III. Matrix Inversion   73
  IV. Inverse Filters   80
   V. Other Linear Methods   87
  VI. Overview of Linear Methods   89
       References   91

4. Modern Constrained Nonlinear Methods
Peter A. Jansson

    I. Introduction   96
   II. Meaning of Constraints   97
  III. The Promise of Analytic Continuation   97

IV. Early Constrained Methods   99
V. Greater Realization of Constraint Benefits   102
VI. Other Methods   128
VII. Concluding Remarks   130
    References   132

5. Application to Electron Spectroscopy for Chemical Analysis

Peter A. Jansson

I. Introduction   136
II. Distortions Present in ESCA Spectra   138
III. Correction of Distortions   141
IV. Applications   145
V. Concluding Remarks   150
    References   151

6. Instrumental Considerations

William E. Blass and George W. Halsey

I. Introduction   155
II. Resolution-Acquisition-Time Trade-Offs   156
III. Acquiring the Data: Dispersive Spectrophotometers   157
IV. Details of Spectroscopic Experiments   173
V. Preparing to Deconvolve a Data Set   179
VI. Deconvolving the Data: Constrained Deconvolution   182
    References   185

7. Deconvolution Examples

George W. Halsey and William E. Blass

I. Introduction   188
II. Examining the Deconvolution Process   188
III. Deconvolution, Noise, and Smoothing   195
IV. Relaxation Methods   201
V. Operating Considerations   206
VI. Deconvolution of Other Types of Spectra   211
VII. Deconvolution Examples   215
    References   225

8. Maximum-Likelihood Estimates of Spectra

B. Roy Frieden

I. Introduction 229
II. Orientation 230
III. Physical Model for the Object 232
IV. Most Probable Object: Definition 235
V. Most Probable Object in the Presence of Data 237
VI. Object-Class Law 238
VII. Case of a "White" Object; Maximum Entropy 239
VIII. Scenarios for Knowing $p(q_1, \ldots, q_M)$ 240
IX. Estimators 246
X. How to Handle Noise 250
XI. Test Cases 252
XII. Conclusion 258
References 258

9. Fourier Spectrum Continuation

Samuel J. Howard

I. Introduction 262
II. Advantages of Fourier Analysis 264
III. Restoration and Enhancement of Experimental Data 264
IV. Discrete Fourier Transform and Discrete Function Continuation 271
V. Constraint of Finite Extent 278
VI. Concluding Remarks 285
References 286

10. Minimum-Negativity-Constrained Fourier Spectrum Continuation

Samuel J. Howard

I. Theory 290
II. Application 295
III. Concluding Remarks 323
References 330

Index 333

Contributors

Numbers in parentheses indicate the pages on which the authors' contributions begin.

William E. Blass (153, 187), Department of Physics and Astronomy, The University of Tennessee, Knoxville, Tennessee 37996

B. Roy Frieden (227), Optical Sciences Center, The University of Arizona, Tucson, Arizona 85721

George W. Halsey (153, 187), Department of Physics and Astronomy, The University of Tennessee, Knoxville, Tennessee 37996

Samuel J. Howard (261, 289), Physics Department, The Florida State University, Tallahassee, Florida 32306

Peter A. Jansson (1, 35, 67, 93, 135), E. I. du Pont de Nemours and Company (Inc.), Engineering Physics Laboratory, Wilmington, Delaware 19898

Preface

The literature on deconvolution is rich with the contributions of many investigators. These contributions are, however, scattered among journals devoted to numerous specialties. No single volume has been available that provides both an overview and the detail needed by a newcomer to this field. When a specific need arises, a recent journal article or the advice of a colleague often initiates a considerable, but not always successful, search. The lack of a suitable volume has fostered this understandable approach. Although it may be sorely needed, deconvolution is a distraction for the researcher in pursuit of personal scientific goals. The present work conveys an understanding of the field and presents under one cover a selection of the most effective, practical techniques.

The growing interest in deconvolution is timely for at least two reasons: (1) the inexorable advance of computer technology has given us hardware for the computation-intensive programs needed, and (2) the advantages of modern nonlinear constrained methods have made deconvolution really worth the effort. Two such advantages—reduced noise sensitivity and superresolving capability—produce effects that are sometimes astonishing when judged by the standards of traditional linear methods. The clear message in this work is that reexamination of dogmatic "truths" can sometimes yield surprises. For years we were led to believe that frequencies beyond the cutoff of an observing instrument were not recoverable. How could they be? The information was not apparent in the data. Now, we have learned to look elsewhere for the additional information required to restore those frequencies. We found it in seemingly unimportant physical-realizability constraints. All of the contributors to the present volume have extensive experience with one or more of the constrained nonlinear methods.

The reader should have a background in physical science, engineering, mathematics, or statistics. A working knowledge of calculus is assumed. Previous experience with convolutions or Fourier transforms would be helpful but is not absolutely necessary, because the required material is developed. In this respect, *Deconvolution* is reasonably self-contained. Throughout, we have emphasized methods that work; methods of purely theoretical interest are described or referenced only to round out the picture. Because it is not possible to be totally comprehensive in a book of this size and scope, we have covered areas pivotal to the evolution of the most practical and effective methods. Where details are omitted, references are made to pertinent literature.

The ten chapters are organized into three main sections. The first four chapters (Jansson) introduce the reader to basic concepts and progress through a survey of both traditional linear and modern nonlinear methods. Chapters 5 (Jansson), 6 (Blass and Halsey), and 7 (Halsey and Blass) detail specific applications of a proven method to the fields of electron spectroscopy for chemical analysis (ESCA) and high-resolution infrared spectroscopy via three different instrumental techniques. Also included are brief examples of applications to nuclear and Raman spectroscopy. The final section, Chapters 8 (Frieden), 9 (Howard), and 10 (Howard), illustrates recent work and reveals some directions for potential future research.

Much of this volume has generally applicability. Certainly Chapters 1, 3, and 4 are not in any way confined to spectroscopy applications. Significant portions of Chapters 8-10 are easily generalized. The remaining chapters, although dealing with the specifics of spectroscopy, can function as a guide to new areas of application.

As a whole, the volume provides an introduction to deconvolution for the scientist who needs it but does not know how to get started. The volume should also serve as a good beginning for the scientist or student preparing for research aimed at improved deconvolution methods. It provides a guide to the literature that will enable the researcher to identify quickly many of the important contributions. Although specific examples are chosen from spectroscopy, deconvolution's general applicability should make the volume useful in diverse fields that deal in both single and multidimensional data. It should, for example, find application in digital image processing, signal analysis, and statistics, to name but a few.

Acknowledgments

I shall begin by thanking my family for their forbearance during the writing and editing of this volume. Full-time occupation with my regular duties in optics and digital image processing necessarily diverted the present effort to my time at home. The support of my wife, Muriel, has been deeply appreciated. Her contributions extend to enlightened criticism and extensive checking of the drafts.

For technical criticisms and suggestions, I am indebted to my fellow authors, especially my friend Roy Frieden, who has taught me much over the years. I am grateful to my Du Pont Colleagues Barry Rubin and Jeffrey G. Yorker for their critical reading of the manuscript. I also thank my friend and former mentor, Robert H. Hunt of Florida State University, for his suggestions.

For their collaboration during the research reported in Chapter 5, I am indebted to my colleagues in Du Pont's ESCA program, who are separately acknowledged at the end of that chapter.

Valuable suggestions on style and editorial assistance have come from our division technical editor, Dorothy M. Welsh. This help has significantly improved my contribution and I am duly grateful. I have appreciated the efforts of Maria T. Carter, Mary June Peterson, Claire T. Campbell, Barbara A. Sama, and Frances R. Johansen, who have borne the brunt of the word processing chores, and members of Du Pont's Central Research & Development Department's graphics group for skilled preparation of the figures. My work with the literature was greatly aided by the competence and willingness of Du Pont's Lavosier library staff. I wish especially to thank Sara L. Hamlin and Marie P. Rule for their help with the references.

Last, but not least, I wish to acknowledge E. I. du Pont de Nemours and Company (Inc.) for providing the services essential to getting the job done.

CHAPTER **1**

Convolution and Related Concepts

Peter A. Jansson

Engineering Physics Laboratory, E. I. du Pont de Nemours and Company (Inc.)
Wilmington, Delaware

I.	Introduction	3
II.	Definition of Convolution	4
	A. Discrete Case	4
	B. Continuous Case	6
III.	Properties	7
	A. Integration and Differentiation	7
	B. Central-Limit Theorem	8
	C. Voigt Function	10
IV.	Fourier Transforms	11
	A. Special Symbols and Useful Functions	12
	B. Some Properties and Relationships	18
	C. Sampling and the Discrete Fourier Transform	24
V.	The Problem of Deconvolution	28
	A. Defining the Problem	28
	B. Difficulties	29
	C. Alternatives to Deconvolution	30
	References	34

LIST OF SYMBOLS

a, b, g	functions $a(x), b(x), g(x)$, with no explicit dependences shown
a', b', g'	first derivatives of functions a, b, g
a_n, b_n, g_n	sampled values of $a(x), b(x), g(x)$
$a(x), b(x), g(x)$	functions used to illustrate properties of convolution
A, B, G	functions $A(\omega), B(\omega), G(\omega)$, with no explicit dependences shown
$A(\omega), B(\omega), G(\omega)$	Fourier transforms of $a(x), b(x), g(x)$
c	constant
C	constant
$f(x), F(\omega)$	function and Fourier transform, respectively
$F'(\omega), F''(\omega)$	first and second derivatives of $F(\omega)$, respectively
$G(\zeta)$	Fourier transform of $g(x)$ given by alternative convention
$H(x)$	1 when $x > 0$, 0 when $x \le 0$
$i(x), i$	"image" data that incorporate smearing by $s(x)$
$\hat{i}_M(\mathbf{v}, x)$	model representing idealized image data
j	imaginary operator such that $j^2 = -1$

1

DECONVOLUTION:
WITH APPLICATIONS IN SPECTROSCOPY

N_a, N_b, N_g	number of samples available for functions a, b, g
$o(x), o$	"object" or function sought by deconvolution, usually the true spectrum, but also the instrument function when this is sought by deconvolution
$\hat{o}(x)$	estimate of $o(x)$
$\hat{o}_M(\upsilon, x)$	model of true spectrum or true object
q	independent variable given in scaled units of Gaussian half-widths $(q = x\sqrt{\ln 2}/\Delta x_G)$
$\text{rect}(x)$	rectangle function having half-width $\frac{1}{2}$
$\text{sgn}(x)$	1 when $x > 0$, -1 when $x \le 0$
$\text{sinc}(x)$	$(\sin \pi x)/\pi x$
$\text{Si}(x)$	$\int_0^x \left[(\sin x')/x'\right] dx'$
$s(x), s$	spread function, usually the instrument function, but also spreading due to other causes
$s(x, x')$	general integral equation kernel; shift-variant spread function
x, x'	generalized independent variables and arguments of various functions
$\langle x \rangle, \langle x^2 \rangle$	first and second moments of a distribution of x
α	fractional increase in instrument response-function breadth due to convolution with narrow spectral line
β	parameter specifying influence of sharpness or smoothness criteria
$\otimes$	convolution operation
$\delta(x), \delta$	Dirac δ function or impulse
$\delta'(x), \delta'$	first derivative of $\delta(x)$
Δx	half-width at half maximum (HWHM)
$\Delta x_G, \Delta x_C$	Gaussian and Cauchy half-widths at half maximum
Δx_N	Nyquist interval
ζ	conjugate of x in alternative Fourier transform system; Fourier frequency in cycles per units of x; variable of integration
$\theta(x), \Theta(\omega)$	spurious part of solution $\hat{o}(x)$ and its Fourier transform $\hat{O}(\omega)$
$\Lambda(x)$	triangle function of unit height and half-width $\frac{1}{2}$
μ	scaled ratio of Cauchy to Gaussian half-widths, $\Delta x_C \sqrt{\ln 2}/\Delta x_G$
σ	standard deviation
σ^2	variance
$\sigma_a^2, \sigma_b^2, \sigma_g^2$	variances of a, b, g
$\sigma_A^2, \sigma_B^2, \sigma_G^2$	variances of A, B, C
$\tau(\omega), \tau$	$\tau(\omega)$, Fourier transform of $s(x)$
υ	vector having components υ_l
υ_l	parameters of a model comprising multiple peaks
$\Phi(\upsilon)$	objective function to be minimized
ω	conjugate of x; Fourier frequency in radians per units of x
Ω	cutoff frequency such that $\tau(\omega) = 0$ for $|\omega| > \Omega$
$\text{п}(x)$	positive-impulse pair $\frac{1}{2}\delta(x + \frac{1}{2}) + \frac{1}{2}\delta(x - \frac{1}{2})$
$^\text{I}\text{п}(x)$	impulse pair with positive and negative components, $\frac{1}{2}\delta(x + \frac{1}{2}) - \frac{1}{2}\delta(x - \frac{1}{2})$
$\text{III}(x)$	Dirac "comb" $\sum_{n=-\infty}^{\infty} \delta(x - n)$

I. INTRODUCTION

Our daily experience abounds with phenomena that can be described by the mathematical process of convolution. Spreading, blurring, and mixing are qualitative terms frequently used to describe these phenomena. Sometimes the spreading is caused by physical occurrences unrelated to our mechanisms of perception; sometimes our sensory inputs are directly involved. The blurred visual image is an example that comes to mind. The blur may exist in the image that the eye views, or it may result from a physiological defect. Biological sensory perception has parallels in the technology of instrumentation. Like the human eye, most instruments cannot discern the finest detail. Instruments are frequently designed to determine some observable quantity while an independent parameter is varied. An otherwise isolated measurement is often corrupted by undesired contributions that should rightfully have been confined to neighboring measurements. When such contributions add up linearly in a certain way, the distortion may be described by the mathematics of convolution.

Spectroscopy is profoundly affected by these spreading and blurring phenomena. The recovery of a spectrum as it would be observed by a hypothetical, perfectly resolving instrument is an exciting goal and, with recent advances, one that has stimulated the development of restoration methods that merit serious consideration. Through these methods, the spectroscopist has access to information that would otherwise remain unavailable. The advent of the new methods has even changed the way we regard experimental apparatus. Furthermore, these methods cannot fail to have a marked influence on the progress of science outside the disciplines of spectroscopy.

Before we can confront the problem of undoing the damage inflicted by spreading phenomena, we need to develop background material on the mathematics of convolution (the function of this chapter) and on the nature of spreading in a typical instrument, the optical spectrometer (see Chapter 2). In this chapter we introduce the fundamental concepts of convolution and review the properties of Fourier transforms, with emphasis on elements that should help the reader to develop an understanding of deconvolution basics. We go on to state the problem of deconvolution and its difficulties.

The most important symbols introduced in this chapter are used throughout the volume; that we occasionally deviate from this notation is testimony to the diversity of applications and to the limited number of clear and convenient notational possibilities. We have avoided mathematical rigor in favor of developing an intuitive grasp of the fundamentals. The reader is referred to the outstanding and readable text by Bracewell (1978) for added depth.

Although the present work is concerned primarily with spectra, its applicability does not lie only in that area. The term "spectra" should be understood to include one-dimensional data from experiments that do not explicitly involve optical phenomena. Data from fields as diverse as radio astronomy, statistics, separation science, and communications are suitable candidates for treatment by the methods described here. Confusion arises when we discuss Fourier transforms of these quantities, which may also be called spectra. To avoid this confusion, we adopt the convention of referring to the latter spectra as Fourier spectra. When this term is used without the qualifier, the data space (nontransformed regime) is intended.

Also, application of these methods is not limited to one-dimensional problems. Most of the concepts may readily be extended to multiple dimensions. Digital image processing, which usually deals with two independent variables, has been the object of considerable investigation and the source of many advances in our understanding of the deconvolution problem.

II. DEFINITION OF CONVOLUTION

A. Discrete Case

Convolution, in its simplest form, is usually considered to be a blurring or smoothing operation. A "rough" or "bumpy" function is convolved with a smoothing function to yield a smoother output. Typically, each output value is identified with a corresponding input value. It is obtained, however, by processing that value *and* some of its neighbors. One simple way of smoothing the fluctuations in a sequence of numbers is to perform a moving average. Below, in row a, we see a smoothed sequence partially computed for the noisy values in row g:

g		4	3	9	8	2	1	6	5	4	2	3
a				5.2	4.6	5.2	4.4	3.6				

In this particular case, we chose to average five adjacent g numbers to produce each value in row a. We wrote each element in row a opposite the center of the averaging interval. Giving each element of rows a and g indices n and m, respectively, we write the moving average

$$a_n = (g_{n-2} + g_{n-1} + g_n + g_{n+1} + g_{n+2})/5 = \sum_{m=n-2}^{n+2} g_m/5. \qquad (1)$$

It could suit our purpose to apply a weight to each of the g elements before summing:

g	4	3	9	8	2	1	6	5	4	2	3
					×	×	×	×			
b					0.1	0.3	0.5	0.1	→		
					0.2	0.3	3.0	0.5			
a	···		6.6	7.2	4.4	2.5	4.0				

In this example the sum of the b weights is 1. Therefore, we may say that b is *normalized*. We see that each element of row a is a *linear combination* of a corresponding subset of the g elements. Sequential values of a are computed by sliding the set of b factors along row g. It is usually convenient to write each a value opposite the largest weighting coefficient in row b:

g_1	g_2	g_3	g_4	g_5	g_6	g_7	g_8	g_9	g_{10}	g_{11}
				×	×	×	×	×		
				b_{+2}	b_{+1}	b_0	b_{-1}	b_{-2}	→ slide	
a_1	a_2	a_3	a_4	a_5	a_6	a_7				

Thus,

$$a_7 = b_2 g_5 + b_1 g_6 + b_0 g_7 + b_{-1} g_8 + b_{-2} g_9. \tag{2}$$

We generalize this procedure in the following equation:

$$a_n = \sum_{m=n-2}^{n+2} b_{n-m} g_m. \tag{3}$$

For the present case, b vanishes everywhere except over the interval -2 to $+2$. We also can define b as an infinite sequence of numbers. In this case we may write

$$a_n = \sum_{m=-\infty}^{\infty} b_{n-m} g_m. \tag{4}$$

This is sometimes called the discrete convolution or serial product. The values of b and g, however, may just be samples of continuous functions.

Note here that because each value of a_n requires values of g_{n-2} through g_{n+2}, N values of a can never be computed from N values of g. If N_a is the number of a values desired, and N_g the number of g values available, we must always have $N_g > N_a$, except in the trivial case where b has only one non-zero value. The representation of real-world situations usually involves some accommodation of this effect, possibly by filling the ends of arrays with zeros or by treating the data values as if they were cyclic and "wrapped around" on themselves. Often, sections of spectra can be chosen so that they begin and end in regions of zero base line. Alternatively, one may choose the range of interest far from the ends of the data so that end effects can be ignored.

B. Continuous Case

We may choose the sample interval to be as fine as we please, thus extending the concept of summation into the continuous regime. With a, b, and g now written as functions of the continuous variables x and x', we may write the convolution integral

$$a(x) = \int_{-\infty}^{\infty} b(x - x')g(x')\, dx'. \tag{5}$$

Each value of a is a weighted integral of the g function, the function b supplying the required weight and being slid along g according to the displacement specified by x'.

If the area under $b(x)$ is unity,

$$\int_{-\infty}^{\infty} b(x)\, dx = 1, \tag{6}$$

we may say that $b(x)$ is normalized. Equation (5) then represents a moving weighted average. In the study of instrumental resolution, we often think of the convolution integral in this way.

We may denote convolution by the symbol $\otimes$, and rewrite Eq. (5):

$$a = b \otimes g. \tag{7}$$

The presence of the minus sign in the argument of b in Eq. (5) gives the convolution integral the highly useful property of commutativity, that is,

$$b \otimes g = g \otimes b. \tag{8}$$

This is easily shown by redefining the variable of integration. Convolution also obeys associativity,

$$(d \otimes b) \otimes g = d \otimes (b \otimes g), \tag{9}$$

and is distributive with respect to addition,

$$d \otimes (b + g) = (d \otimes b) + (d \otimes g). \tag{10}$$

These properties carry back to the discrete formulation. We shall use both discrete and continuous formulations in this volume, changing back and forth as needs require. The continuous regime allows us to avoid consideration of sampling effects when such consideration is not of immediate concern. Deconvolution algorithms, on the other hand, are numerically implemented on sampled data, and we find the discrete representation indispensable in such cases.

III. PROPERTIES

In addition to the relations just presented, we shall state without proof some other properties of convolutions. Most of these are quite easily proved.

A. Integration and Differentiation

If b and g are peaked functions (such as in a spectral line), the area under their convolution product is the product of their individual areas. Thus, if b represents instrumental spreading, the area under the spectral line is preserved through the convolution operation. In spectroscopy, we know this phenomenon as the invariance of the equivalent width of a spectral line when it is subjected to instrumental distortion. This property is again referred to in Section II.F of Chapter 2 and used in our discussion of a method to determine the instrument response function (Chapter 2, Section II.G).

Convolution has an interesting property with respect to differentiation. The first derivative of the convolution product of two functions may be given by the convolution of either function with the derivative of the other. Thus, if

$$a = b \otimes g, \tag{11}$$

then

$$a' = b' \otimes g = b \otimes g', \tag{12}$$

where the primes denote differentiation.

This property follows from the associative and commutative properties if we allow the concept of a differentiation operator δ' that performs its function by convolution. We see that

$$a' = \delta' \otimes a = (\delta' \otimes b) \otimes g = b \otimes (\delta' \otimes g). \tag{13}$$

Such an operator is indeed the first derivative of the familiar impulse or Dirac δ function. It can, like the δ function, be represented as the limiting form of a variety of functions. In the case of δ', we have adjacent positive- and negative-going impulses. It can be shown to have the following properties:

$$\int_{-\infty}^{\infty} \delta'(x)\,dx = 0, \tag{14}$$

$$\delta'(0) = 0, \tag{15}$$

$$\int_{-\infty}^{\infty} x\delta'(x)\,dx = -1, \tag{16}$$

$$x\delta'(x) = -\delta(x). \tag{17}$$

B. Central-Limit Theorem

From experience, we know that a curve obtained by integrating a function is smoother than the function itself. This characteristic also applies to the convolution integral. We know that an instrumentally smeared spectrum is smoother than the original as it would be observed by a perfect instrument.

If we convolve two single-peaked functions, the result is smoother than either component. Two rectangle functions convolved yield a triangle; two triangle functions convolved (four rectangles convolved) produce a result that is astonishingly close to a gaussian (Fig. 1).

A Gaussian function has the form

$$f(x) = \exp(-x^2). \tag{18}$$

Here the height in the center is unity and is $1/e$ when $x = 1$.

There are various ways of normalizing the gaussian, in both abscissa and ordinate. In statistics, we often deal with

$$f(x) = [1/(\sigma\sqrt{2\pi})]\exp(-x^2/2\sigma^2), \tag{19}$$

where σ is the standard deviation, and the factor before the exponential guarantees $f(x)$ to have unit area. The central ordinate is no longer 1. The variance is given by the square of σ.

A form familiar to the spectroscopist is

$$f(x) = \exp[-\ln 2(x/\Delta x)^2]. \tag{20}$$

Here the central ordinate is again unity, and Δx represents the half-width at half maximum (HWHM), so that $f(|\Delta x|) = \frac{1}{2}$.

We have seen that convolving rectangles gives a gaussianlike function. A gaussian is, in fact, the exact result of an infinite number of convolutions provided that the functions convolved obey certain conditions. The rigorous

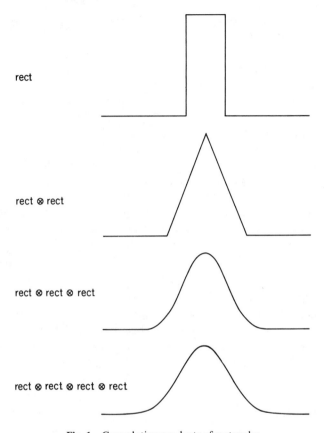

rect

rect ⊗ rect

rect ⊗ rect ⊗ rect

rect ⊗ rect ⊗ rect ⊗ rect

Fig. 1 Convolution products of rectangles

statement of this property is the central-limit theorem. *The central-limit theorem states that when a function $f(x)$ is convolved with itself n times, in the limit $n \rightarrow \infty$, the convolution product is Gaussian with variance n times the variance of $f(x)$, provided that the area, mean, and variance of $f(x)$ are finite.*

The conditions may be stated as

$$\int_{-\infty}^{\infty} f(x)\,dx < \infty, \tag{21}$$

$$\int_{-\infty}^{\infty} xf(x)\,dx < \infty, \tag{22}$$

and

$$\int_{-\infty}^{\infty} x^2 f(x)\,dx < \infty. \tag{23}$$

If nonidentical functions are convolved, the conditions are somewhat more complex (Bracewell, 1978).

If two Gaussian functions are convolved, the result is a gaussian with variance equal to the sum of the variances of the components. Even when two functions are not Gaussian, their convolution product will have variance equal to the sum of the variances of the component functions. Furthermore, the second moment of the convolution product is given by the sum of the second moments of the components. The horizontal displacement of the centroid is given by the sum of the component centroid displacements. Kendall and Stuart (1963) and Martin (1971) provide helpful additional discussions of the central-limit theorem and attendant considerations.

Most peaklike functions become more gaussianlike when convolved with one another. One notable exception of interest to spectroscopists is the Cauchy function, which is the familiar Lorentzian shape assumed by lines in the spectra of gases subject to pressure broadening:

$$f(x) = \frac{1}{1 + (x/\Delta x)^2}. \tag{24}$$

If two Cauchy distributions having half-widths Δx_1 and Δx_2 are convolved, the result is a similar distribution that has a half-width of $\Delta x_1 + \Delta x_2$. If n Cauchy distributions having half-width Δx_C are convolved, the result is a Cauchy distribution of half-width $n \, \Delta x_C$.

C. Voigt Function

One might well ask what the result would be if the Cauchy distribution were convolved with the Gaussian distribution. The result may be obtained from Eqs. (20) and (24):

$$\frac{1}{1 + (x/\Delta x_C)^2} \otimes \exp[-\ln 2(x/\Delta x_G)^2]. \tag{25}$$

Redefining the independent variable in terms of the Gaussian half-width, $q = x\sqrt{\ln 2}/\Delta x_G$, we may write this convolution as

$$\mu \, \Delta x_C \int_{-\infty}^{\infty} \frac{\exp(-\zeta^2)}{\mu^2 + (q - \zeta)^2} \, d\zeta, \tag{26}$$

where μ is proportional to the ratio of Cauchy to Gaussian half-widths:

$$\mu = (\Delta x_C/\Delta x_G)\sqrt{\ln 2}. \tag{27}$$

If we require this function to have unit area when plotted as a function of q, we must modify the scale of the ordinate. The resulting expression is

$$\frac{\mu}{\pi^{3/2}} \int_{-\infty}^{\infty} \frac{\exp(-\zeta^2)}{\mu^2 + (q - \zeta)^2} \, d\zeta. \tag{28}$$

Multiplying this expression by π yields the Voigt function that occurs in the description of spectral-line shapes resulting from combined Doppler and pressure broadening. We elaborate on these phenomena in Section I of Chapter 2.

IV. FOURIER TRANSFORMS

Although a number of effective deconvolution algorithms do not use Fourier methods, these methods shed considerable light on the performance of the algorithms. For this reason, we introduce the Fourier transform and outline some of its most-useful properties. Only a brief treatment is given here. For additional detail, we again refer the reader to the excellent practical text on this subject by Bracewell (1978).

Scale factors can be used in various ways to define Fourier transform pairs. We adopt the symmetrical convention

$$F(\omega) = \frac{1}{\sqrt{2\pi}} \int_{-\infty}^{\infty} f(x)e^{-j\omega x} \, dx \tag{29}$$

and

$$f(x) = \frac{1}{\sqrt{2\pi}} \int_{-\infty}^{\infty} F(\omega)e^{j\omega x} \, d\omega \tag{30}$$

for use in most parts of this volume. In these expressions, x is the spatial coordinate or independent variable of the measurement space (wavelength, wave number, mass, grating angle, etc.) and ω the independent variable of the Fourier space given in radians per units of x. We usually denote functions of x by lowercase letters, such as the f used here, and their transforms by the corresponding capital letters.

An alternative symmetrical convention that has gained popularity specifies the Fourier transform pair $g(x)$ and $G(\zeta)$ to be related by

$$G(\zeta) = \int_{-\infty}^{\infty} g(x)e^{-j2\pi\zeta x} \, dx$$

and

$$g(x) = \int_{-\infty}^{\infty} G(\zeta)e^{j2\pi\zeta x} \, d\zeta,$$

where ζ is frequency in cycles per units of x. Conversion from one system to the other is easy. If x is the same for both systems, the frequency variable may be related by $\omega = 2\pi\zeta$. On the other hand, if a transform pair is known in the ω system of Eqs. (29) and (30), multiplying both arguments x and ω by $\sqrt{2\pi}$ and substituting ζ for ω yields a corresponding pair in the alternative system. Scale changes by variable substitution and similarity theorem (Section IV.B.4) then tailor the pair to the situation at hand.

A. Special Symbols and Useful Functions

It is convenient to introduce some special symbols and functions. These simple representations, when combined with the various Fourier transform properties and a little practice, will enable the reader to gain a deeper understanding of, and intuition for, convolution and deconvolution.

1. Rectangle, Sinc, and Si

Consider a function, centered at the origin, defined as unity out to $\frac{1}{2}$ in both positive and negative directions and vanishing beyond $\frac{1}{2}$. More precisely, we define

$$\text{rect}(x) = \begin{cases} 0, & |x| > \frac{1}{2}, \\ \frac{1}{2}, & |x| = \frac{1}{2}, \\ 1, & |x| < \frac{1}{2}. \end{cases} \tag{31}$$

This function may be illustrated by a rectangle. A scaled version of it, $\text{rect}(x/\sqrt{2\pi})$, is shown in Fig. 2, along with its transform and a number of other functions that follow in this section. Its transform $F(\omega)$ can be computed in a straightforward way by replacing the infinite limits of integration with $\pm\sqrt{2\pi}/2$ and setting $f(x)$ equal to unity:

$$\begin{aligned} F(\omega) &= \frac{1}{\sqrt{2\pi}} \int_{-\infty}^{\infty} \text{rect}\left(\frac{x}{\sqrt{2\pi}}\right) e^{-j\omega x}\, dx \\ &= \frac{1}{\sqrt{2\pi}} \int_{-\sqrt{2\pi}/2}^{\sqrt{2\pi}/2} e^{-j\omega x}\, dx \\ &= \frac{\sin(\omega\sqrt{2\pi}/2)}{\omega\sqrt{2\pi}/2} \\ &= \text{sinc}\left(\frac{\omega}{\sqrt{2\pi}}\right), \end{aligned} \tag{32}$$

where we have used the opportunity to define the sinc function

$$\text{sinc}(x) = \frac{\sin \pi x}{\pi x}.$$ (33)

This function is used extensively in optics and elsewhere, as well as its integral

$$\text{Si}(x) = \pi \int_0^{x/\pi} \text{sinc}(x')\, dx' = \int_0^x \frac{\sin x'}{x'}\, dx'.$$ (34)

Some authors define sinc without the π factors, so beware of this variation when comparing works.

2. Triangle and Sinc Squared

By simple geometrical arguments and the definition of convolution, we can verify that the convolution of $\text{rect}(x)$ with itself yields $\Lambda(x)$, which we define as

$$\Lambda(x) = \begin{cases} 0, & |x| \geq 1, \\ 1 - |x|, & |x| < 1. \end{cases}$$ (35)

Its transform is just $(1/\sqrt{2\pi})\, \text{sinc}^2(\omega/2\pi)$. This may easily be shown with the aid of the convolution theorem discussed in Section IV.B.10.

3. Gaussian Function

The Gaussian function is its own Fourier transform; that is, when we define

$$f(x) = \exp(-x^2/2),$$ (36)

we obtain

$$F(\omega) = \exp(-\omega^2/2).$$ (37)

Allowing for the case of arbitrary amplitude and width, we obtain the transform pair

$$C \exp\left(-\frac{cx^2}{2}\right) \leftrightarrow \frac{C}{c} \exp\left(-\frac{\omega^2}{2c}\right).$$ (38)

Here and henceforth the symbol $\leftrightarrow$ indicates correspondence between the two domains. The scale factor c demonstrates the inverse relationship between the widths of the gaussians in the two domains. The wider a gaussian, the narrower is the gaussian to which it transforms.

$$f(x) = \frac{1}{\sqrt{2\pi}} \int_{-\infty}^{\infty} F(\omega) e^{j\omega x} d\omega \qquad F(\omega) = \frac{1}{\sqrt{2\pi}} \int_{-\infty}^{\infty} f(x) e^{-j\omega x} dx$$

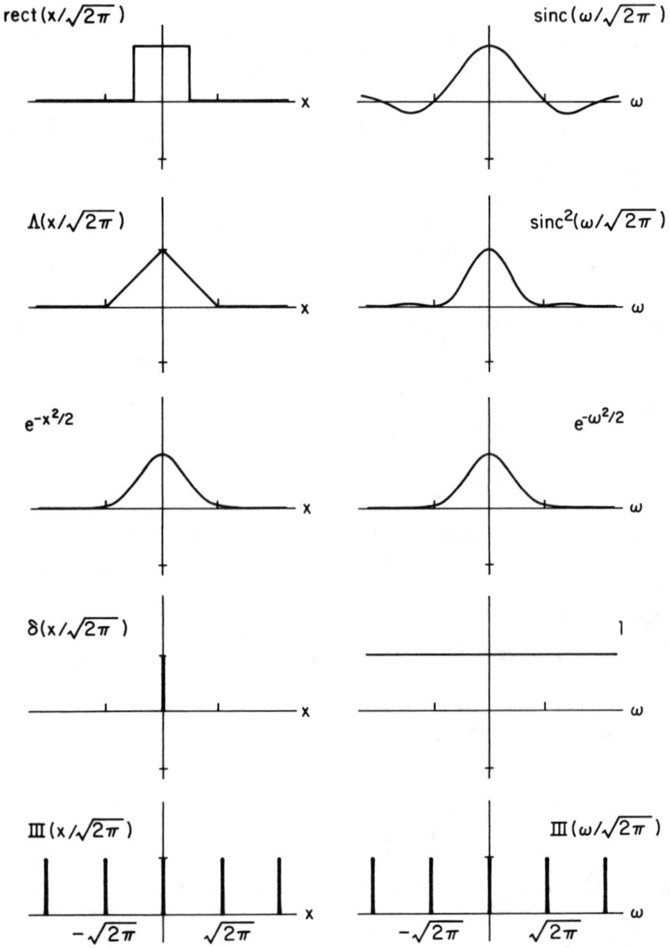

Fig. 2 Fourier transform directory. The Dirac δ function having area $\sqrt{2\pi}$ is shown as a bar of unit height. Imaginary components are shown dashed. The vertical-axis tick mark is at 1, the horizontal-axis tick mark at $\sqrt{2\pi}$.

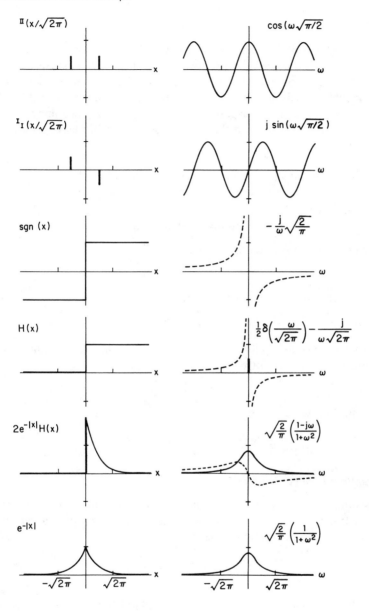

Fig. 2 *continued.*

4. *The δ Function—Alone and in Pairs*

The transform of $\delta(x)$ is the constant $1/\sqrt{2\pi}$. The transform of $\delta(x/\sqrt{2\pi})$ is unity owing to the scaling property $\delta(cx) = (1/|c|)\delta(x)$. Displaced from the origin, this function transforms to sinusoids having real and imaginary components. We can use two positive impulses, equally displaced from the origin in the positive and negative directions, to create two imaginary sine components in the transform that cancel each other but leave a real cosine term. If we set

$$f(x) = \delta(x/\sqrt{2\pi} + \tfrac{1}{2}) + \delta(x/\sqrt{2\pi} - \tfrac{1}{2}), \tag{39}$$

we obtain

$$F(\omega) = \cos(\omega\sqrt{\pi/2}). \tag{40}$$

Let us define the positive-impulse pair

$$\amalg(x) = \tfrac{1}{2}\delta(x + \tfrac{1}{2}) + \tfrac{1}{2}\delta(x - \tfrac{1}{2}). \tag{41}$$

Then we can rewrite the transform pair

$$\amalg(x/\sqrt{2\pi}) \leftrightarrow \cos(\omega\sqrt{\pi/2}). \tag{42}$$

Similarly, an odd-impulse pair may be used to generate an imaginary sine component in the transform space. Let us define

$$^{\mathrm{I}}\!\amalg(x) = \tfrac{1}{2}\delta(x + \tfrac{1}{2}) - \tfrac{1}{2}\delta(x - \tfrac{1}{2}). \tag{43}$$

This definition yields the pair

$$^{\mathrm{I}}\!\amalg(x/\sqrt{2\pi}) \leftrightarrow j\,\sin(\omega\sqrt{\pi/2}). \tag{44}$$

5. *Infinitely Replicated Impulses*

The infinite string of δ functions causes certain difficulties in rigor when its Fourier transform is considered. Nevertheless, the concept is a very useful one. It is well worth defining a special symbol

$$\mathrm{III}(x) = \sum_{n=-\infty}^{\infty} \delta(x - n). \tag{45}$$

Many of its properties follow readily from the properties of $\delta(x)$. Of greatest interest, however, is the fact that its Fourier transform is a similar function that has reciprocal spacing between its impulses:

$$\mathrm{III}\!\left(\frac{cx}{\sqrt{2\pi}}\right) \leftrightarrow \mathrm{III}\!\left(\frac{\omega}{c\sqrt{2\pi}}\right). \tag{46}$$

The effect of sampling a function $f(x)$ can be simulated by multiplying it by $III(x)$. Likewise, convolving $f(x)$ with $III(x)$ replicates $f(x)$ infinitely in both directions.

6. Sign and Heaviside Step Functions

We may define the sign function

$$\text{sgn}(x) = \begin{cases} 1, & x > 0, \\ -1, & x \le 0, \end{cases} \tag{47}$$

and note that its Fourier transform is given by

$$-\frac{j}{\omega}\sqrt{\frac{2}{\pi}}. \tag{48}$$

The Heaviside step, although similar, is not purely an odd function:

$$H(x) = \begin{cases} 1, & x > 0, \\ 0, & x \le 0. \end{cases} \tag{49}$$

We see that the $H(x)$ step is half as high as the sgn, and is raised up by $\frac{1}{2}$. The required added constant is responsible for an impulse in its Fourier transform,

$$\frac{1}{\sqrt{2\pi}}\left[\pi\delta(\omega) - \frac{j}{\omega}\right]. \tag{50}$$

Note that we obtain a very nice ramp function through multiplying $H(x)$ by a straight line of unit slope, and that the same result can be obtained by the self-convolution of $H(x)$:

$$\text{ramp} = xH(x) = H(x) \otimes H(x). \tag{51}$$

One obtains an impulse by differentiating $H(x)$:

$$dH(x)/dx = \delta(x). \tag{52}$$

7. Truncated Exponentials and Resonance Contours

When simple electrical RC filters are treated, the truncated exponential $e^{-|x|}H(x)$ is indispensable. Its transform is given by $(2\pi)^{-1/2}(1 - j\omega)/(1 + \omega^2)$. If the truncated exponential is reflected about the origin, eliminating $H(x)$ and leaving $e^{-|x|}$, the imaginary part of the transform disappears. We obtain the transform $(2/\pi)^{1/2}/(1 + \omega^2)$. This is the resonance contour, Cauchy distribution, or Lorentzian shape encountered previously in Section III.B.

B. Some Properties and Relationships

1. *Superposition*

A fundamental property of the Fourier transform is that of superposition. The usefulness of the Fourier method lies in the fact that one can separate a function into additive components, treat each one separately, and then build up the full result by summing the individual results. It is a beautiful and explicit example of the stepwise refinement of complex problems. In stepwise refinement, one successfully tackles the most difficult tasks and solves problems far beyond the mind's momentary grasp by dividing the problem into its ultimately simple pieces. The full solution is then obtained by reassembling the solved pieces.

In linear superposition, the method is literally that of adding components. When treating the optics of coherent light, for example, the instantaneous values of the field vectors are superimposed. Incoherent light, on the other hand, requires us to deal with the time-averaged square of the field. In nonlinear optics, superposition breaks down as it does in other nonlinear systems. Even when it does not hold exactly, however, superposition is often useful as a first-order approximation.

We may use the concept of orthogonal functions to identify components. In the present case, our component functions are sinusoids, and we find that the Fourier transform of the sum is the sum of the Fourier transforms:

$$f_1(x) + f_2(x) \leftrightarrow F_1(\omega) + F_2(\omega). \tag{53}$$

2. *Oddness and Evenness*

When we say that $f(x)$ is an even function, we mean that $f(-x) = f(x)$; if it is odd then $f(-x) = -f(x)$. It may easily be verified that a real even function has a real even transform, and an imaginary even function an imaginary even transform. Real odd functions have imaginary odd transforms. Any function may be decomposed into odd and even parts. For linear operations by superposition, they may be treated separately and the results reassembled. Both real and imaginary parts of a function may have odd and even components. In this case superposition is again useful.

3. *Fourier's Integral Theorem*

If we transform and then inverse transform a function, we should get back to where we started. By substituting Eq. (29) into Eq. (30), we obtain

$$f(x) = \int_{-\infty}^{\infty} f(x') \left\{ \frac{1}{2\pi} \int_{-\infty}^{\infty} \exp[j\omega(x - x')] \, d\omega \right\} dx'. \tag{54}$$

The quantity in braces must have the property of a Dirac δ function, so we write

$$\delta(x - x') = \frac{1}{2\pi} \int_{-\infty}^{\infty} \exp[j\omega(x - x')] \, d\omega. \tag{55}$$

4. *Sign and Scale*

From the concept of linearity it follows that multiplication of a function by a constant scale factor c results in a corresponding increase in its transform:

$$cf(x) \leftrightarrow cF(\omega). \tag{56}$$

A change of scale in x or ω, however, results in a reciprocal change in the transform variable and an amplitude change as well:

$$f(cx) \leftrightarrow \frac{1}{|c|} F\left(\frac{\omega}{c}\right). \tag{57}$$

This is sometimes called the similarity theorem. The horizontal stretching of a function in one domain results in horizontal contraction and amplitude growth in the other. In fact, these changes occur in such a way that the area under the curve in the other domain remains constant.

5. *Power Theorem and Rayleigh's Theorem*

The previous section dealt with a horizontal scale change in one domain resulting in both horizontal and vertical changes in the other domain. Rayleigh's theorem, on the other hand, makes a statement about the area under the squared modulus. This integral, in fact, has the same value in both domains:

$$\int_{-\infty}^{\infty} |f(x)|^2 \, dx = \int_{-\infty}^{\infty} |F(\omega)|^2 \, d\omega. \tag{58}$$

In its more general form, we have the power theorem

$$\int_{-\infty}^{\infty} f(x)g^*(x) \, dx = \int_{-\infty}^{\infty} F(\omega)G^*(\omega) \, d\omega. \tag{59}$$

6. *Swapping Domains*

We can foresee a situation in which we already know a transform pair but require a corresponding pair with reversed domains. Specifically, suppose we know that $\alpha(x)$ has a transform $\beta(\omega)$, but our problem presents us with $\alpha(\omega)$, and we desire its x-domain transform:

$$\alpha(x) \leftrightarrow \beta(\omega), \tag{60}$$

$$? \;\leftrightarrow \alpha(\omega). \tag{61}$$

By using the properties of even and odd functions, it is easy to show that $\beta(x)$ is the unknown function if $\alpha(\omega)$ is an even function. On the other hand, if $\alpha(\omega)$ is an odd function, then the unknown function is $-\beta(x)$. These relationships hold for both real and imaginary components. By superposition, any function can be broken into odd and even components and treated separately. The results may then be combined.

7. Shift Theorem

Our known transform pairs are usually centered at the origin. To apply these known transforms to real situations, we must know what happens to them as they are shifted along the abscissa. The result of such a shift is a simple change in phase:

$$f(x) \leftrightarrow F(\omega), \tag{62}$$

$$f(x - c) \leftrightarrow F(\omega) \exp(-jc\omega), \tag{63}$$

$$f(x) \exp(jcx) \leftrightarrow F(\omega - c). \tag{64}$$

No change in amplitude occurs.

8. Differentiation

By using the definition of differentiation, we can show that

$$df(x)/dx \leftrightarrow j\omega F(\omega) \tag{65}$$

and

$$-jxf(x) \leftrightarrow dF(\omega)/d\omega. \tag{66}$$

An important consequence of these relationships is that differentiation increases the amplitude of high-frequency components. It is well known that data-differentiation procedures often yield unsatisfactory results when applied to noisy data. In spectroscopy, peak positions are sometimes sought by looking for a vanishing first derivative. The spectral peaks contain predominantly low and middle frequencies, but the noise often contains high frequencies as well. A possible remedy to the problem is to fit a polynomial to the data over a limited domain centered on each point at which a derivative is required. The first derivative of the polynomial is found to be a simple linear combination of the neighboring data ordinates. A whole curve may be quickly differentiated in this way. The process is one of simple convolution. This method of differentiation contains its own low-pass filter. It has been described by Savitzky and Golay (1964). Some numerical errors that appear in their paper have been corrected by Steinier *et al.* (1972).

9. Area, Moments, and Variances

Let us write the Fourier transform of $f(x)$:

$$F(\omega) = \frac{1}{\sqrt{2\pi}} \int_{-\infty}^{\infty} f(x) \exp(-j\omega x)\, dx. \tag{67}$$

The transform evaluated at the origin is proportional to the area under $f(x)$:

$$F(0) = \frac{1}{\sqrt{2\pi}} \int_{-\infty}^{\infty} f(x) \exp(0)\, dx = \frac{1}{\sqrt{2\pi}} \int_{-\infty}^{\infty} f(x)\, dx \tag{68}$$

and

$$\int_{-\infty}^{\infty} f(x)\, dx = \sqrt{2\pi} F(0). \tag{69}$$

Similarly, we have

$$\int_{-\infty}^{\infty} F(\omega)\, d\omega = \sqrt{2\pi} f(0). \tag{70}$$

The first moment can be shown to be proportional to the slope of the transform at the origin. Precisely, we have

$$\int_{-\infty}^{\infty} xf(x)\, dx = j\sqrt{2\pi} F'(0). \tag{71}$$

This may be proved by exercising the differentiation property previously discussed.

The mean value of x, where $f(x)$ is a distribution, may also be called the centroid and is defined

$$\langle x \rangle = \int_{-\infty}^{\infty} xf(x)\, dx \bigg/ \int_{-\infty}^{\infty} f(x)\, dx. \tag{72}$$

Substitutions from Eqs. (69) and (71) yield

$$\langle x \rangle = jF'(0)/F(0). \tag{73}$$

The second moment is given by

$$\int_{-\infty}^{\infty} x^2 f(x)\, dx = -\sqrt{2\pi} F''(0). \tag{74}$$

Here we twice differentiated $F(\omega)$. The mean value of x^2 is given by

$$\langle x^2 \rangle = \int_{-\infty}^{\infty} x^2 f(x)\, dx \Big/ \int_{-\infty}^{\infty} f(x)\, dx = -F''(0)/F(0). \tag{75}$$

The variance of a distribution $f(x)$ is given by

$$\sigma^2 = \langle (x - \langle x \rangle)^2 \rangle = \left[\frac{F'(0)}{F(0)}\right]^2 - \frac{F''(0)}{F(0)}. \tag{76}$$

10. Convolution Theorem

We shall now present a theorem that is fundamentally important in Fourier analysis for understanding data acquisition and the performance of spectrometers. It is essential to the thought process required for both the qualitative understanding of concepts and precise mathematical analysis. Under certain circumstances, it can substantially reduce computation.

Easily proved from the definition of the Fourier transform, this theorem states that convolving two functions is equivalent to finding the product of their Fourier transforms. Specifically, if $a(x)$, $b(x)$, and $g(x)$ have transforms $A(\omega)$, $B(\omega)$, and $G(\omega)$, then

$$a(x) = b(x) \otimes g(x) \tag{77}$$

is equivalent to

$$A(\omega) = B(\omega)G(\omega). \tag{78}$$

The theorem works equally well whether we are using plus or minus transforms, so we may write the pair of relationships

$$a \otimes g \leftrightarrow AG, \tag{79}$$

$$ag \leftrightarrow A \otimes G. \tag{80}$$

Superposition may be invoked to determine the behavior of the theorem when the functions are subjected to changes in the sign of real or imaginary, odd or even, components.

We have seen in Section III.B that when two functions are convolved, the variance of the convolution product is the sum of the variances of the individual functions. That is, if

$$a = b \otimes g \tag{81}$$

holds true, then we may write

$$\sigma_a^2 = \sigma_b^2 + \sigma_g^2. \tag{82}$$

We see that when two gaussians of equal breadth are convolved, the result is a gaussian $\sqrt{2}$ times broader.

We know that the variance of a gaussian is the reciprocal of the variance of its transform. We apply the convolution theorem to obtain

$$\sigma_A^2 = \frac{\sigma_B^2 \sigma_G^2}{(\sigma_B^2 + \sigma_G^2)}. \tag{83}$$

The transform of our convolved equal-breadth gaussians is narrower than either individual transformed gaussian. In the present case $1/\sqrt{2}$ is the appropriate scaling.

Normally, discrete convolution involves shifting, adding, and multiplying —a laborious and time-consuming process, even in a large digital computer. The convolution theorem presents us with an alternative. It reveals the possibility of computing in the Fourier domain. What are the trade-offs between the two methods?

Conventionally, if the numbers a_i are the N_a sampled values of the function $a(x)$ over its domain of nonvanishing values, and b_i are the N_b sampled values of the function $b(x)$ over its domain of nonvanishing values, then the discrete convolution of a and b involves computing $N_a N_b$ sums and $N_a N_b$ products, or $2N_a N_b$ arithmetic operations all together. This result is demonstrated by a visualization similar to that in Section II.A. In this example, all nonvanishing values of the product are computed.

The fast Fourier transform (FFT) requires $5N \log N$ (Bergland, 1969) elementary arithmetic operations for an array of N samples. Computation of the convolution product requires three such transforms plus $4N$ elementary operations (N complex products) in the transform domain.

Precise comparison of the two methods of computing a convolution requires careful attention to details such as whether aliasing, computing the "ends" of the function, matching array lengths to powers of 2, or whatever other FFT base is employed. It is apparent, however, that when $N_a = N_b$, the FFT method is superior. When $N_a \ll N_b$, the FFT method involves considerable unnecessary computation. In instrumental resolution studies, one of the two functions typically has a considerably smaller extent than the other; that is, the response function is usually narrow relative to the extent of the data. In this case, it is usually more efficient to perform convolution directly, without transformation. The FFT, however, does have an important place. To use the FFT effectively, we need to know a little more about the influence of sampling. This is covered in Section IV.C.

The convolution theorem plays a valuable role in both exact and approximate descriptions of functions useful for analyzing resolution distortion and in helping us understand the effects of these functions in Fourier space. Functions of interest and their transforms can be constructed from our directory in Fig. 2 by forming their sums, products, and convolutions. This technique adds immeasurably to our intuitive grasp of resolution limitations imposed by instrumentation.

C. Sampling and the Discrete Fourier Transform

We cannot manipulate a continuous function in a digital computer. If we wish to deal with such a function—call it $f(x)$—we must employ its equivalent discretely sampled approximation. In this case $f(x)$ is usually sampled and encoded into numerical values at evenly spaced steps of its independent variable x. These values can be obtained only to a limited precision because they are subject to roundoff or quantization effects. In modern digital computers it is usually possible to attain adequate encoding precision for $f(x)$, so we shall not continue further along this line.

Let us instead turn our attention to the consequences of sampling the function at evenly spaced intervals of x. Consider the Λ function and its transform, a sinc function squared, shown in Fig. 3. Suppose that we wish to compute that transform numerically. First, let us replicate the Λ by convolving it with a low-frequency III function. Now multiply it by a high-frequency III function to simulate sampling. We see a periodically replicated and sampled Λ. The value of each sample is represented as the scaled area under a Dirac δ function.

By applying the convolution theorem, we see that replication in the x domain has produced a sampling effect in the frequency domain. The wider the replication interval, the finer is the frequency sampling. Sampling in the x domain, on the other hand, appears in Fourier space as replication. Fine sampling in x produces wide spacing between cycles in ω. The area under each scaled Dirac function of ω may be taken as the numerical value of a sample.

The discrete Fourier transform gives us the ability to compute the numbers representing one cycle of $F(\omega)$ from the numbers representing one cycle of $f(x)$ and vice versa. We may adjust scale factors, number of samples, sample density, and so on in this representation to elucidate many features of importance in discrete–Fourier-transform applications.

For example, suppose that we had sampled $f(x)$ too coarsely. The result would be closely spaced replicas in the Fourier domain. The high-frequency lobes of the sinc-squared cycles would overlap significantly, and high frequencies from an adjacent replica would appear superimposed on the lower frequencies. This phenomenon is known as frequency aliasing. When such aliasing might cause difficulty, it is necessary to band-limit the original by filtering it before it is sampled. If this is not done and the signal is sampled, there will be no way to separate the true from the aliased components based on the information in the sampled data alone. Both intuitively, then, and explicitly via the foregoing, fine sampling is required for adequate representation of an arbitrary function $f(x)$. The function can, in fact, be perfectly reconstructed from its evenly spaced samples provided that *at least two samples are taken for each cycle of the highest frequency that it contains*. This statement comes from the famous Whittaker–Shannon sampling theorem. The minimum sampling interval required for perfect reconstruction is known as the Nyquist interval. If $F(\omega)$ is contained entirely within frequency bounds $\pm\Omega$, evenly spaced samples of $f(x)$ must be no farther apart than the Nyquist interval $\Delta x_N = \pi/\Omega$. The sinc function provides a means of performing the reconstruction for all x:

$$f(x) = \sum_{n=-\infty}^{\infty} f\left(\frac{n\pi}{\Omega}\right) \operatorname{sinc}\left(\frac{\Omega x}{\pi} - n\right). \tag{84}$$

The reader may wish to extend the type of analysis shown in Fig. 3 by considering the effects of "boxcar" averaging. In this process, $f(x)$ is convolved with a rect(x) function before sampling. Each sample then represents a uniformly weighted average. It is instructive to consider the effect of varying the number of samples used for the transform. In Chapter 9, Howard introduces the mathematics needed for practical application of the discrete Fourier transform.

Previously, we referred to the FFT. Its development was a giant step that enabled the efficient computation of discrete Fourier transforms. The FFT algorithm and its variations have revolutionized signal analysis and made interferometric infrared spectroscopy practical. Both NMR spectra and mass spectra are also now computed from data that are acquired in their Fourier transform domain. The rediscovery of this algorithm by Cooley and Tukey (1965) is responsible for its current widespread use. Summaries of its properties and pitfalls are provided by Bergland (1969), Brigham (1974), and Bracewell (1978).

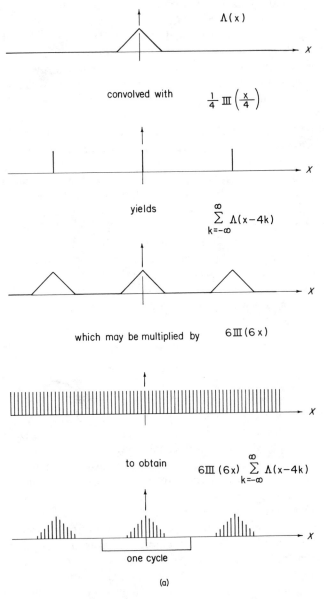

$\Lambda(x)$

convolved with $\frac{1}{4}\, \text{III}\left(\frac{x}{4}\right)$

yields $\sum\limits_{k=-\infty}^{\infty} \Lambda(x-4k)$

which may be multiplied by $6\,\text{III}\,(6\,x)$

to obtain $6\,\text{III}\,(6x)\sum\limits_{k=-\infty}^{\infty}\Lambda(x-4k)$

one cycle

(a)

Fig. 3 Sampling and replication in (a) the x domain and (b) the ω domain.

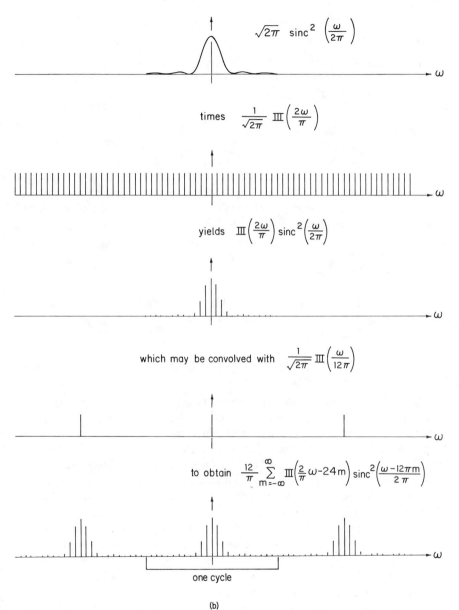

$$\sqrt{2\pi} \ \ \text{sinc}^2 \left(\frac{\omega}{2\pi}\right)$$

times $\dfrac{1}{\sqrt{2\pi}} \ \text{III}\left(\dfrac{2\omega}{\pi}\right)$

yields $\text{III}\left(\dfrac{2\omega}{\pi}\right) \text{sinc}^2\left(\dfrac{\omega}{2\pi}\right)$

which may be convolved with $\dfrac{1}{\sqrt{2\pi}} \ \text{III}\left(\dfrac{\omega}{12\pi}\right)$

to obtain $\dfrac{12}{\pi} \displaystyle\sum_{m=-\infty}^{\infty} \text{III}\left(\dfrac{2}{\pi}\omega - 24m\right) \text{sinc}^2\left(\dfrac{\omega - 12\pi m}{2\pi}\right)$

one cycle

(b)

Fig. 3 *continued.*

V. THE PROBLEM OF DECONVOLUTION

In Section II we developed the concept of the convolution integral and its discrete approximation. No conceptual difficulty is therefore encountered in computing A from B and G:

$$A_n = \sum_{m=-\infty}^{\infty} B_{n-m}G_m. \tag{85}$$

Even considering our brief treatment of this subject thus far, we may have enough information to foresee that difficulties could arise when we attempt to compute values of G from given values of A and B. Such difficulties do indeed occur. The problem of deconvolution has therefore been the subject of a vast literature spanning the numerous special fields of science.

A. Defining the Problem

Let us begin our study of the problem by reformulating it in the continuous representation, with definitions of variables that have enjoyed some popularity:

$$i(x) = \int_{-\infty}^{\infty} s(x - x')o(x')\, dx'. \tag{86}$$

In optics, we seek the object function $o(x)$ when we are given the image data $i(x)$ and the known spread function $s(x)$ of the intervening optical system. This notation will be used throughout this volume. In our adaptation, the object o is usually the true spectrum as it would be observed by the ideal, perfectly resolving spectrometer. The spread function s may incorporate both optical blurring and electronic filtering. The image i may be the spectrum as it normally would appear on a strip-chart recorder. Regardless of which blurring effects we chose to include in s, and even in the case of nonoptical spectroscopies (e.g., electron spectroscopy for chemical analysis; see Chapter 5), in this work the quantities o, s, and i always have the same meaning in terms of the deconvolution problem: i *represents the known data; s the blurring phenomenon, the effects of which are to be eliminated; and o the function that we seek, free of spreading due to s.*

If we are observing a spectrum $o(x')$ with the aid of an instrument having a characteristic response function $s(x - x')$, then i represents the data acquired. If we have a perfectly resolving instrument, then $s(x - x')$ is a Dirac function, and our data $i(x)$ directly represent the true spectrum, that is, $o(x)$. In this case we have no need for deconvolution.

Conversely, we may observe an exceedingly narrow spectral line, so that $o(x')$ is approximated by $\delta(x')$. Now the data $i(x)$ represent the response function. This principle can, in fact, be used to determine the response function of a spectrometer. The laser, for example, is a tempting source of monochromatic radiation for measuring the response function of an optical spectrometer. Coherence effects, however, complicate the issue. We present further detail in Section II of Chapter 2.

Even if perfectly narrow spectral lines are not available, we may take a clue from this approach and measure a spectral line of known shape. Deconvolution should then yield the instrument function. This technique does indeed work (Chapter 2, Section II.G.3). Some of the information needed to generate the known line shape can even be obtained directly from the observed data.

Let us again consider the convolution integral. Equation (86) is an example of a Fredholm integral equation of the first kind. In such equations the kernel can be expressed as a more-general function of both x and x':

$$i(x) = \int_{-\infty}^{\infty} s(x, x')o(x')\,dx'. \tag{87}$$

This equation describes situations where the functional form of s varies with its argument. Even in such so-called shift-variant cases, the true convolution integral may be an adequate representation if the range of x is not too large.

Some techniques for solving deconvolution problems also adapt readily to the more-general Fredholm case. The relaxation methods of, for example, Van Cittert (1931) and Jansson can be so adapted (Jansson, 1968, 1970; Jansson, Hunt and Plyler, 1968, 1970).

Methods that have been developed for the solution of the Fredholm equation sometimes rely on continuous functional representations of $s(x - x')$ and $i(x)$. These methods are limited in usefulness for the experimentalist, who wishes to apply deconvolution techniques by computer to digitized spectra.

B. Difficulties

Difficulties in solving the deconvolution problem are revealed if we examine it in the continuous representation given by Eq. (86). Suppose that the solution for $o(x)$ is to some degree uncertain, and that it may be written as the sum of a desired solution $o(x')$ and a spurious part $\theta(x')$:

$$\hat{o}(x') = o(x') + \theta(x'). \tag{88}$$

We may then ask what the character of $\theta(x')$ must be so that Eq. (86) holds true. Certainly any $\theta(x')$ for which

$$\int_{-\infty}^{\infty} s(x - x')\theta(x') = 0 \tag{89}$$

can be summed with $o(x')$ to form a valid solution. We see the problem as one of uniqueness: how do we know that deconvolution is giving us the $o(x)$ solution that we want, free of the spurious components $\theta(x)$? The Fourier transform version of Eq. (89) gives us a clue. By the convolution theorem we have

$$\tau(\omega)\Theta(\omega) = 0, \tag{90}$$

where τ and Θ are the transforms of s and θ, respectively. Usually, instrumental distortion is characterized by the passage of low frequencies and the attenuation or complete loss of high frequencies. The transfer function $\tau(\omega)$ must be large for ω small, and small or vanishing for ω large. When ω is small, clearly $\Theta(\omega) = 0$. For ω large, when $\tau(\omega) = 0$, $\Theta(\omega)$ could take on any finite value whatsoever. The spurious component of the solution, $\Theta(\omega)$, has only high-frequency content. Perhaps smoothing the high frequencies will help us. These matters will assume additional significance in Chapter 3, and indeed the rest of the volume.

The convolution theorem aided us in understanding a property of the spurious component $\theta(x)$. It also hints at a method of deconvolution when applied to Eq. (86). This method is developed in Section IV of Chapter 3.

C. Alternatives to Deconvolution

Much of this volume is devoted to the concept that the best solutions result from incorporating the maximum amount of prior knowledge. Assume, for example, that we know the object to have precisely four Gaussian components. Various fitting methods may be applied to determine the positions, widths, and amplitudes of these components. This approach has great value, where applicable, and is sometimes called deconvolution, although this usage is erroneous. Because these fitting methods are viable alternatives to deconvolution in many cases, we discuss them further.

Occasionally, useful information may be gleaned from the observed spectrum of an isolated line without deconvolution, even though the instrument response function is wider than the line itself. We see this in the application of the method of equivalent widths to the determination of line strengths (Chapter 2, Sections II.F and II.G). When more complete knowledge is sought, we can often achieve the desired end by employing fewer degrees of freedom than a true deconvolution process utilizes.

1. *Spectral Models*

When we are solving for an unknown spectrum, each data point contains some information about the component spectral lines. A data point far from the center of a given line would be expected to contain very little information about that line. A hypothetical model of the spectrum could incorporate widely varying estimates of the amplitude of that line without influencing the fit to the data point in question. Three data points moderately distributed near the line center—say, spanning the interval between the half-maximum points—affix the parameters of the line more reliably. Instead of taking equally spaced samples of a trial solution as independent unknowns (the deconvolution approach), we can express the sought-after true spectrum in terms of its spectral-line parameters—amplitudes, half-widths, positions, and so on—provided that we can assume such a model with some confidence.

It is possible to estimate the influence of the spectrometer on linewidth measurements by convolving the response function with a mathematical model of the line. The resulting curve may be compared with the observed line and the true width inferred. Hunt *et al.* (1968) have employed corrections of this type supplied by the author in their determination of carbon monoxide self-broadened linewidths.

2. *Objective Functions*

The central-limit theorem (Section III.B) suggests that when a measurement is subject to many simultaneous error processes, the composite error is often additive and Gaussian distributed with zero mean. In this case, the least-squares criterion is an appropriate measure of goodness of fit. The least-squares criterion is even appropriate in many cases where the error is not Gaussian distributed (Kendall and Stuart, 1961). We may thus construct an objective function that can be minimized to obtain a best estimate. Suppose that our data $i(x)$ represent the measurements of a spectral segment containing spectral-line components that are specified by the N parameters $v_1, \ldots, v_N$, which may be considered as the components of a vector $\mathbf{v}$. The objective function may be written

$$\Phi(\mathbf{v}) = \int [i(x) - \hat{i}_M(\mathbf{v}, x)]^2 \, dx, \tag{91}$$

where $\hat{i}_M(\mathbf{v}, x)$ is the model the parameters of which we wish to establish. We seek the parameters that minimize $\Phi(\mathbf{v})$, thereby ensuring the best fit. If we know that instrumental spreading is described by convolution of the true spectrum with a spread function $s(x)$, we may write

$$\hat{i}_M(\mathbf{v}, x) = \int s(x - x')\hat{o}_M(\mathbf{v}, x') \, dx', \tag{92}$$

where $\hat{o}_M(\upsilon, x)$ is the model of the true spectrum based on the parameters υ_l. We may then substitute Eq. (92) in Eq. (91). Typically the independent variable must be considered as discrete, so summations are employed instead of integrals.

Only in the simplest cases—a single Gaussian component, for example— may conventional linear least-squares method be employed to solve for υ. More commonly, either approximate linearized methods or nonlinear methods are employed.

3. Linearized Methods

One way of linearizing the problem is to use the method of least squares in an iterative linear differential correction technique (McCalla, 1967). This approach has been used by Taylor et al. (1980) to solve the problem of modeling two-dimensional electrophoresis gel separations of protein mixtures. One may also treat the components—in the present case spectral lines—one at a time, approximating each by a linear least-squares fit. Once fitted, a component may be subtracted from the data, the next component fitted, and so forth. To refine the overall fit, individual components may be added separately back to the data, refitted, and again removed. This approach is the basis of the CLEAN algorithm that is employed to remove antenna-pattern sidelobes in radio-astronomy imagery (Högbom, 1974) and is also the basis of a method that may be used to deal with other two-dimensional problems (Lutin et al., 1978; Jansson et al., 1983).

4. Nonlinear Optimization

When these methods are unsuitable, nonlinear methods may be applied. The function $\Phi(\upsilon)$ may be considered as a surface in an $(N + 1)$-dimensional hyperspace. We seek the components of υ that correspond to the minimum value of $\Phi(\upsilon)$, the bottom of a dishlike region of that surface. The chief difficulties in solving this problem are avoidance trapping in local minima and overall computational efficiency. The function $\Phi(\upsilon)$ is often expensive to compute, so maximum advantage must accrue from each evaluation of it. To this end, numerous methods have been developed. Optimization is a field of ongoing research. No one single method is best for all types of problem. Where $\Phi(\upsilon)$ is a sum of squares, as we have expressed it, and where derivatives $d\Phi/d\upsilon_l$ are available, the method of Marquardt (1963) and its variants are perhaps best. Other methods may be desirable where constraints are to be applied to the υ_l, or where $\Phi(\upsilon)$ cannot be formulated as a sum of squares. One method of treating the sum-of-squares problem (Lawson and Hanson, 1974) has potentially special value because it can employ constraints such as positivity (see Chapter 4). When noise is not additive but signal dependent, least squares no longer leads to a correct formulation, and $\Phi(\upsilon)$ takes on a different form.

Martin (1971) gives a brief tutorial on the subject of optimization. Fletcher (1971) gives an excellent review and presents criteria for the selection of an appropriate method. Rigler and Pegis (1980) discuss the various methods from the viewpoint of optical applications. Pitha and Jones (1966) compare several methods applied to fitting infrared band envelopes.

5. A Curve-Resolving Instrument

An interesting method of fitting was presented with the introduction, some years ago, of the model 310 curve resolver by E. I. du Pont de Nemours and Company. With this equipment, the operator chose between super-positions of Gaussian and Cauchy functions electronically generated and visually superimposed on the data record. The operator had freedom to adjust the component parameters and seek a visual best match to the data. The curve resolver provided an excellent graphic demonstration of the ambiguities that can result when any method is employed to resolve curves, whether the fit is visually based or firmly rooted in rigorous least squares. The operator of the model 310 soon discovered that, when data comprise two closely spaced peaks, "acceptable" fits can be obtained with more than one choice of parameters. The closer the blended peaks, the wider was the choice of parameters. The part played by noise also became rapidly apparent. The noisy data trace allowed the operator additional freedom of choice, when he considered the "error bar" that is implicit at each data point.

6. Constraints and True Deconvolution

Although useful, and even preferable in limited circumstances, none of the aforementioned methods may properly be called deconvolution. We address ourselves in this volume to the case in which our prior knowledge of $o(x)$ is less complete and in which we seek a more general solution. In Chapter 4, we shall see that certain modest elements of prior knowledge may be handled within the context of deconvolution, and that this knowledge exerts a powerful beneficial effect. It is valuable to know a little more about the distortion of spectra before tackling the deconvolution problem. Accordingly, in the next chapter we examine the details of spectral-line broadening.

REFERENCES

Bergland, G. D. (1969). *IEEE SPECTRUM* **6**, 41-52.

Bracewell, R. (1978). "The Fourier Transform and Its Applications," (2nd ed). McGraw-Hill, New York.

Brigham, E. O. (1974). "The Fast Fourier Transform." Prentice-Hall, Englewood Cliffs, New Jersey.

Cooley, J. W., and Tukey, J. W. (1965). *Math. Comput.* **19**, 297-301.

Fletcher, R. (1971). Proceedings of the Symposium on Computer-Aided Engineering, Univ. Waterloo, Waterloo, Ontario, pp. 123-155.

Högbom, J. A. (1974). *Astron. Astrophys. Suppl. Ser.* **15**, 417-426.

Hunt, R. H., Toth, R. A., and Plyler, E. K. (1968). *J. Chem. Phys.* **49**, 3909-3912.

Jansson, P. A. (1968). Ph.D. Dissertation, Florida State Univ., Tallahassee.

Jansson, P. A. (1970). *J. Opt. Soc. Am.* **60**, 184-191.

Jansson, P. A., Hunt, R. H., and Plyler, E. K. (1968). *J. Opt. Soc. Am.* **58**, 1665-1666.

Jansson, P. A., Hunt, R. H., and Plyler, E. K. (1970). *J. Opt. Soc. Am.* **60**, 596-599.

Jansson, P. A., Grim, L. B., Elias, J. G., Bagley, E. A., and Lonberg-Holm, K. K. (1983). *Electrophoresis* **4**, 82-91.

Kendall, M. G., and Stuart, A. (1961). "The Advanced Theory of Statistics," Vol. 2, Inference and Relationship, 3-Vol. ed. Hafner, New York.

Kendall, M. G., and Stuart, A. (1963). "The Advanced Theory of Statistics," Vol. 1, Distribution Theory, 3-Vol. ed. Hafner, New York.

Lawson, C. L., and Hanson, R. J. (1974). "Solving Least Squares Problems." Prentice-Hall, Englewood Cliffs, New Jersey.

Lutin, W. A., Kyle, C. F., and Freeman, J. A. (1978). *In* "Electrophoresis '78" (N. Catsimpoolas, ed.), "Dev. Biochem.", Vol. 2, pp. 93-106. Elsevier, New York.

McCalla, T. R. (1967). "Introduction to Numerical Methods and FORTRAN Programming." Wiley, New York.

Marquardt, D. W. (1963). *J. Soc. Ind. Appl. Math.* **11**, 431-441.

Martin, B. R. (1971). "Statistics for Physicists." Academic Press, New York.

Pitha, J., and Jones, R. N. (1966). *Can. J. Chem.* **44**, 3031-3050.

Rigler, A. K., and Pegis, R. J. (1980). *In* "Top. Appl. Phys.: The Computer in Optical Research: Methods and Applications," Vol. 41 (B. R. Frieden, ed.), pp. 211-268. Springer-Verlag, New York.

Savitzky, A., and Golay, M. J. E. (1964). *Anal. Chem.* **36**, 1627-1639.

Steinier, J., Termonia, Y., and Deltour, J. (1972). *Anal. Chem.* **44**, 1906-1909.

Taylor, J., Anderson, N. L., Coulter, B. P., Scandora, A. E. Jr., and Anderson, N. G. (1980). *In* "Electrophoresis '79" (B. J. Radola, ed.), pp. 329-339.

Van Cittert, P. H. (1931). *Z. Phys.* **69**, 298-308.

CHAPTER 2

Distortion of Optical Spectra

Peter A. Jansson

Engineering Physics Laboratory, E. I. du Pont de Nemours and Company (Inc.)
Wilmington, Delaware

I.	Inherent Line Breadth	38
	A. Natural Line Broadening	38
	B. Collision Broadening	39
	C. Doppler Broadening	40
	D. Combined Effects	40
	E. Absorption Spectroscopy	41
	F. Infrared Spectra of Liquids	44
II.	Contribution of Dispersive Spectrometer	44
	A. Coherent Irradiation	45
	B. Incoherent Irradiation	49
	C. Real Spectrometers	51
	D. Electrical Filtering	51
	E. Measuring Transmittance and Absorptance	54
	F. Measurements Independent of Spectrometer Broadening	56
	G. Determining the Spectrometer Response	58
III.	Resolution Criteria	61
	A. Two-Point Criteria	61
	B. Conclusion	64
	References	64

LIST OF SYMBOLS

A_L	measured absorptance amplitude of narrow rectangular line as observed by Gaussian response function
$A_M(x)$, $A_T(x)$	measured and "true" absorptances given by $1 - T_M(x)$ and $1 - T_T(x)$, respectively
B_L	maximum absorptance amplitude of narrow-line measurement
$B_M(x)$	flux in absence of sample as measured by a spectrometer of finite resolution
c	velocity of light
C	electrical capacitance
D	linear dimension of aperture of spectrometer grating or prism
$D_M(x)$	additive offset due to stray flux or electrical errors as might be recorded by a spectrometer

35

DECONVOLUTION:
WITH APPLICATIONS IN SPECTROSCOPY

$\mathscr{E}(t)$, $\mathscr{E}_1(t)$, $\mathscr{E}_2(t)$	quantities proportional to electric field, time dependence explicitly shown
$E(z)$	quantity proportional to net field at distance z from image center
f	collimator focal length
f'	telescope focal length
$f(x)$	function
h	Planck's constant
$i(t)$	electrical current
$i(x)$, i	"image" data that incorporate smearing by $s(x)$
J_N, J_P, J_D, J_V	probabilities per unit frequency of a transition between energy states for spectral lines showing natural, pressure (combined natural and collision), Doppler, and Voigt (combined) broadening
k	Boltzmann's constant
m	molecular mass
$o(x)$, o	"object" or function sought by deconvolution, usually the true spectrum, but also the instrument function when this is sought by deconvolution
P'	$S\sqrt{\ln 2/\pi}/\Delta x_D$
P_N, P_P, P_D, P_V	spectral absorption coefficients in cases of natural, pressure-broadened, Doppler-broadened, and combined Doppler- and pressure-broadened profiles, respectively
$P(x)$, P	spectral absorption coefficient, where $x = v/c$ and is given in cm^{-1} (c is velocity of light)
$q(t)$	electrical charge
R	electrical resistance
$\text{rect}(x)$	rectangle function having half-width $\frac{1}{2}$
$r(x)$	normalized spectrometer response function
$r_{coh}(z)$, $r_{inc}(z)$	unnormalized coherent and incoherent optical spectrometer response functions, respectively
S	line strength
$\text{sinc}(x)$	$(\sin \pi x)/\pi x$
$\text{Si}(x)$	$\int_0^x [(\sin x')/x'] \, dx'$
$s(x)$, s	spread function, usually the instrument function, but also spreading due to other causes
t, t'	time
T	absolute temperature
$T_M(x)$	transmittance as measured $[U_M(x) - D_M(x)]/[B_M(x) - D_M(x)]$
$T_T(x)$	"true" transmittance $[U_T(x) - D_T(x)]/[B_T(x) - D_T(x)]$
U, U_1, U_2	irradiances given by time averages $\overline{[\mathscr{E}(t)]^2}$, $\overline{[\mathscr{E}_1(t)]^2}$, $\overline{[\mathscr{E}_2(t)]^2}$, respectively
U_s	scale factor accounting for entrance-slit irradiance and optical-system efficiency
$U(x)$, $U_T(x)$, U	spectral flux transmitted through sample as it would be recorded by a perfectly resolving spectrometer
$U_0(x)$, U_0	spectral flux incident on sample
$U_M(x)$	flux passed through sample as measured by a spectrometer of finite resolution

$U_T(x), B_T(x), D_T(x)$ — "true" versions of functions $U_M(x), B_M(x), D_M(x)$ as would be recorded by a perfectly resolving spectrometer

$U_{coh}(z), U_{inc}(z)$ — irradiances in slit image plane when slit is coherently and incoherently irradiated, respectively

$U_{obs}(z)$ — electrically recorded spectrum

$v(t), v_C(t)$ — voltages applied to circuit and across capacitor, respectively

$V(\omega), V_C(\omega)$ — Fourier transforms of voltages $v(t), v_C(t)$

w — entrance-slit width

w' — detector width

W — equivalent width

W_M — measured equivalent width

$W(SX)$ — equivalent width expressed as a series

x, x' — generalized independent variables and arguments of various functions, in places given specifically in cm^{-1}

x — location of line center in cm^{-1}

X — pressure of sample gas multiplied by optical path length

z — distance from center of line image expressed in Rayleigh widths

z_0 — half-width of geometrical image of entrance slit expressed in Rayleigh widths

Z — collision frequency

α — fractional increase in instrument response-function breadth due to convolution with narrow spectral line

γ_C — $2Z$

γ_N — sum of reciprocals of natural upper- and lower-state lifetimes

γ_P — $\gamma_N + \gamma_C$

$\delta(x), \delta$ — Dirac delta function or impulse

$\Delta E, \Delta t$ — energy and time uncertainties, respectively

$\Delta\lambda$ — Rayleigh resolution in wavelength units

Δx_G — Gaussian half-width at half maximum

$\Delta x_L, \sigma_L^2$ — half-width and variance of narrow rectangular line, respectively

$\Delta x_N, \Delta x_P, \Delta x_D, \Delta x_V$ — line-profile half-widths in cm^{-1} for natural, pressure-broadened, Doppler-broadened, and combined Doppler- and pressure-broadened cases, respectively; generally $\Delta x = \Delta v/c$, where c is velocity of light

$\Delta x_R, \sigma_R^2$ — half-width and variance of instrument response function, respectively

$\Delta\bar{v}$ — Rayleigh resolution in wave-number units

Δv_C — half-width $\gamma_C/4\pi$ observed when collision broadening dominates (given in hertz)

Δv_D — Doppler half-width in hertz

Δv_N — natural half-width $1/4\pi\gamma_N$

Δv_P — pressure-broadened half-width $1/4\pi\gamma_P$

ζ — variable of integration

λ — wavelength of radiant flux

$\Lambda(x)$ — triangle function of unit height and half-width $\frac{1}{2}$

μ — scaled ratio $(\Delta v_N + \Delta v_C)\sqrt{\ln 2}/\Delta v_D$

ν	frequency of electromagnetic wave in hertz
$\bar{\nu}$	wave number, usually in cm^{-1}
ν_0	center frequency of transition
ξ	distance from line center in scaled units of Doppler half-width
$\xi(t)$	response function of single-stage RC circuit
$\Xi(\omega)$	transfer function of single-stage RC circuit
σ^2	variance
$\tau(\omega), \tau$	Fourier transform of $s(x)$
ψ	Gaussian line-broadening function
$\psi(SX)$	$W_M - W(SX)$
ω	conjugate of x; Fourier frequency in radians per units of x
Ω	cutoff frequency such that $\tau(\omega) = 0$ for $|\omega| > \Omega$

Because the need for deconvolution arises in many disciplines, the details of the blurring mechanisms involved are varied. It is neither possible nor desirable to cover all circumstances here. Although the content of the present chapter may diverge from the reader's own specialty, it serves to illustrate considerations needed in understanding spreading phenomena. Optical spectrum broadening originates from two principal sources. One is inherent in the phenomenon being observed; the other is caused by the spectrometer. We begin by describing inherent broadening.

I. INHERENT LINE BREADTH

Restricting ourselves to the optical spectra of gases, we now describe phenomena that cause a spectral line to have an inherent breadth. Even observations by a perfectly resolving spectrometer would show this broadening. Can the broadening be removed? And if so, under what circumstances? To answer these questions, we must first discuss the nature of the broadening.

A. Natural Line Broadening

The Heisenberg principle tells us that the energy uncertainty of a given state is inversely proportional to the uncertainty in the time during which the corresponding energy level is occupied:

$$\Delta E \, \Delta t \approx h/2\pi, \tag{1}$$

where ΔE and Δt are the energy and time uncertainties, respectively, and h is Planck's constant. The quantity Δt may be identified with the lifetime of an energy state. There are two such states, upper and lower, associated with a simple transition involving absorption or emission of a photon. The energy uncertainties combine as an overall "fuzziness" in our knowledge of the energy converted in a transition between the two states. A distribution is therefore required to characterize the probability of seeing a transition involving light of a given frequency. For such natural line broadening, we state that the probability per unit frequency of a transition yielding frequency v is given by

$$J_N = \frac{\gamma_N}{4\pi^2(v - v_0)^2 + \gamma_N^2/4},\tag{2}$$

where γ_N is the sum of the reciprocals of the natural upper- and lower-state lifetimes, and v_0 the center frequency of the emission. Equation (2) represents the Lorentzian spectral-line shape, and we recognise it as having the Cauchy mathematical form encountered in Sections III and IV.A.7 of Chapter 1.

B. Collision Broadening

To consider gas molecules as isolated from interactions with their neighbors is often a useless approximation. When the gas has finite pressure, the molecules do in fact collide. When natural and collision broadening effects are combined, the line shape that results is also a lorentzian, but with a modified half-width at half maximum (HWHM). Twice the reciprocal of the mean time between collisions must be added to the sum of the natural lifetime reciprocals to obtain the new half-width. We may summarize by writing the probability per unit frequency of a transition at a frequency v for the combined natural and collision broadening of spectral lines for a gas under pressure:

$$J_P = \frac{\gamma_P}{4\pi^2(v - v_0)^2 + \gamma_P^2/4}.\tag{3}$$

In this expression the line breadth is dependent on $\gamma_P = \gamma_N + \gamma_C$. The collision contribution is given by $\gamma_C = 2Z$, where Z is the collision frequency. The resulting half-width is $\Delta v_P = \gamma_P/4\pi$. At low pressure γ_N will dominate, and γ_C will dominate at high pressure. Half-widths for these limiting line profiles are given by $\Delta v_N = \gamma_N/4\pi$ and $\Delta v_C = \gamma_C/4\pi$, respectively.

Quantum-theoretical calculations also predict a shift in line frequency under the latter condition. References and additional detail on all of the material in this section are given by Penner (1959).

C. Doppler Broadening

From statistical mechanics we know that molecular velocities obey the Maxwell distribution law. This distribution is Gaussian in form. The Doppler effect influences light emitted from the moving molecules. The light from those in motion toward the observer is shifted upwards in frequency at the observation point, and the light from those moving away is shifted downwards. When this is taken into account, it can be determined that the probability per unit frequency of observing light at a frequency v is given by

$$J_D = \frac{1}{\Delta v_D} \sqrt{\frac{\ln 2}{\pi}} \exp\left[\frac{-\ln 2(v - v_0)^2}{\Delta v_D^2}\right], \tag{4}$$

where the Doppler half-width is given by

$$\Delta v_D = v_0 \sqrt{(2kT \ln 2)/mc^2}. \tag{5}$$

In this expression, m is the molecular mass, k is Boltzmann's constant, T is the absolute temperature of the gas, and c is the velocity of light.

D. Combined Effects

The effects of Doppler and collision broadening are independent. Let us consider the case of emission, which can be generalized to the case of absorption as well. Each ensemble of molecules moving with a *given specific velocity* relative to the observer emits flux according to the spectral-distribution formula for combined natural and collision broadening. The flux observed from each such ensemble appears Doppler shifted according to its velocity relative to the observer. We combine by integration the contributions from all the molecules. The same result can also be obtained by considering the individual contributions of all molecules emitting at a given frequency in the Lorentzian profile and realizing that these molecules have Gaussian-distributed velocities. We would expect the two approaches to give the same result because the convolution operation is associative and commutative.

The function that we obtain is the Voigt function encountered in Section III.C of Chapter 1. With definition of variables and scaling appropriate to the case of combined Doppler, natural, and collision broadening, the Voigt function appears as the probability per unit frequency of observing light at a frequency v:

$$J_V = \frac{1}{\Delta v_D} \sqrt{\frac{\ln 2}{\pi}} \frac{\mu}{\pi} \int_{-\infty}^{\infty} \frac{\exp(-\zeta^2)}{\mu^2 + (\xi - \zeta)^2} d\zeta, \tag{6}$$

where

$$\xi = \frac{v - v_0}{\Delta v_D} \sqrt{\ln 2} \tag{7}$$

and

$$\mu = \frac{\Delta v_P}{\Delta v_D} \sqrt{\ln 2}. \tag{8}$$

Although this function has been extensively investigated, it has not been evaluated in closed form. A program has been developed by Armstrong (1967) for its evaluation. References to many approximations and expansions are given by Jansson and Korb (1968). Kielkopf (1973), Puerta and Martin (1981), and others have also developed useful approximations.

When the molecular mean free path is small compared with wavelength c/v_0, the mean Doppler shift is smaller than that for a free molecule. Under these circumstances, Eq. (4) is no longer valid, even when collision and natural broadening are neglected. This phenomenon of collisional narrowing was first described by Dicke (1953). A line-shape function given by Galatry (1961) is standardized in dimensionless form by Herbert (1974), who also develops and summarizes methods of evaluating it. Rodgers (1976) describes its effect on the equivalent width and gives criteria for judging when collisional narrowing must be taken into account. Pine (1980) has tested various line-shape models by observing the fundamental vibration of hydrogen fluoride gas in the infrared with the aid of a tunable-laser difference-frequency spectrometer.

E. Absorption Spectroscopy

Until now we have developed equations most directly adaptable to a particular case of spectral emission. In this case, the radiant flux, once emitted, suffers no further interactions on its way to the spectrometer. The intensity of the emitted flux is directly proportional to the probability distributions J_N, J_P, J_D, and J_V.

Many cases of real spectral emission, however, are not so simple, and the blackbody law must be invoked, along with attendant considerations of the relationships between transmissivity, reflectivity, emissivity, and absorptivity. We refer the reader to Penner (1959) for development of these concepts.

1. Absorption Lines

For the present, let us turn our attention to absorption phenomena. As the flux passes through the absorbing gas, each successive lamellar slice of the gas absorbs a fixed fraction of the light incident upon it from the previous neighboring slice. The total flux absorbed by the gas is proportional to the flux incident upon it, and is given by a single integration. We recognize the result as the Bouger–Lambert law. The transmitted spectral flux is given by

$$U = U_0 \exp(-PX), \tag{9}$$

where U_0 is the spectral flux incident upon the sample, X the pressure times the optical path length, and P the spectral absorption coefficient. The probability distribution J that we have already discussed may be adapted to this situation in the manner described by Penner (1959). At this point, we shall also begin dealing with the independent variable x in reciprocal centimeters (wave numbers), instead of the frequency v in hertz employed until now. With this in mind, we may write

$$P(x) = cSJ(v), \tag{10}$$

where c is the velocity of light and S the line strength. Bearing also in mind that $v = cx$, we may now write the spectral absorption-coefficient profile. For the profile with purely natural broadening, we find

$$P_N(x) = \frac{S}{\pi} \frac{\Delta x_N}{(x - x_0)^2 + \Delta x_N^2}, \tag{11}$$

where $\Delta x_N = \gamma_N/4\pi c$. The quantity x_0 is the location of the line center in reciprocal centimeters. In the pressure-broadened case, where both natural and collision broadening are present, we obtain

$$P_P(x) = \frac{S}{\pi} \frac{\Delta x_P}{(x - x_0)^2 + \Delta x_P^2}. \tag{12}$$

In this profile we define $\Delta x_P = \gamma_P/4\pi c$. Doppler broadening yields

$$P_D(x) = \frac{S}{\Delta x_D} \sqrt{\frac{\ln 2}{\pi}} \exp\left[\frac{-\ln 2(x - x_0)^2}{\Delta x_D^2}\right], \tag{13}$$

where the Doppler half-width (HWHM) is given by

$$\Delta x_D = x_0 \sqrt{(2kT \ln 2)/mc^2}. \tag{14}$$

Expressing the distance from the line center in terms of Doppler half-widths, we identify

$$\xi = (v - v_0)\sqrt{\ln 2}/\Delta v_D = (x - x_0)\sqrt{\ln 2}/\Delta x_D. \tag{15}$$

Defining $P' = P_D(0) = (S/\Delta x_D)\sqrt{(\ln 2)/\pi}$, we write the Doppler profile in a much simpler form,

$$P_D = P' \exp(-\xi^2).$$ (16)

By using the same definitions, the profile for combined Doppler, natural, and collision broadening may be expressed in terms of the Voigt function:

$$P_V = P'\frac{\mu}{\pi}\int_{-\infty}^{\infty}\frac{\exp(-\zeta^2)}{\mu^2 - (\xi - \zeta)^2}\,d\zeta.$$ (17)

2. Deconvolution of Inherent Broadening

It may have occurred to the reader that inherent line broadening could be removed by deconvolution. This appears especially attractive as a means of separating the various contributions. In the particular case of absorption spectra, however, it is not as straightforward as it at first appears. Referring to Eq. (9), we see that the various broadening-profile convolutions appear in the exponential:

$$U = U_0 \exp(-\text{natural} \otimes \text{collision} \otimes \text{Doppler} \otimes \text{spectrum}).$$ (18)

The data that we acquire are usually in the form of U/U_0. Only if there is no significant instrumental broadening present, or if it has already been corrected, may we remove the inherent broadening. In this case, we operate on the logarithm of the data or corrected data:

$$K(x) = \text{natural} \otimes \text{collision} \otimes \text{Doppler} \otimes \text{spectrum} = \ln[U_0(x)/U(x)].$$ (19)

There is another difficulty, however. Additive instrumental noise is also logarithmically converted by the operation. This in turn may affect the validity of the noise assumptions employed in developing certain deconvolution algorithms. When absorptions are very small, we may approximate

$$XP(x) \approx 1 - U(x)/U_0(x),$$ (20)

and the inherent broadening convolutions are done in the same regime as those for instrumental broadening. When absorptions are very small, however, the signal-to-noise ratio (or, more properly, peak-height-to-noise ratio) is usually poor. Deconvolution requires a good signal-to-noise ratio, so that this case would not appear to be practical.

It may be surprising, then, that Pliva et al. (1980) have been successful in partially removing inherent broadening by operating directly on absorption data $U(x)/U_0(x)$. Their success appears to be due in part to the stability enforced by physical-realizability constraints.

Measurements of the total integrated absorption $S = \int P(x)\,dx$ may be made by using the method of equivalent widths. Such measurements are independent of instrumental broadening and are considered further in Sections II.F and II.G.

F. Infrared Spectra of Liquids

In liquids the interactions between neighboring molecules are considerably more complicated than in gases. The resultant broadening obliterates the fine line structure seen in gas spectra, leaving only broad band profiles. There are many possible contributors to this broadening. In some cases, adequate approximation is obtained by assuming that the band contour is established by collisions. Ramsay (1952) has noted that substitution of appropriate molecular density and collision diameter numbers in the collision broadening formula results in realistic band widths for certain liquid-phase systems. In such systems, the bands typically show an approximately Lorentzian profile. Approximate deconvolution of inherently broadened liquid-phase spectra may therefore be obtained on the basis of the assumption of Lorentzian shape (Kauppinen et al., 1981).

It is common, however, for liquid-phase systems to include many specific absorbing species. Such species could include isotopic variations, conformational isomers, and solvent–solute interactions resulting in varied-lifetime transient associations between molecules. Distributions resulting from these effects give the Voigt profile utility in studying liquid spectra. We must understand, however, that the functions introduced here are only rough approximations when applied to the spectra of liquids because of the complexities just mentioned and others beyond the scope of this work.

II. CONTRIBUTION OF DISPERSIVE SPECTROMETER

We now consider broadening caused by the spectrophotometric system itself. This broadening is of special interest to us in this work because it is not part of the phenomenon under observation. Instead, it represents our inability to observe this phenomenon accurately. Elimination of spectrometer broadening is the usual goal of deconvolution.

The discussion in the present section concerns only dispersive optical spectroscopy. We shall not treat the resolution characteristic of a Fourier interferometric spectrometer, which is determined by the optical path difference scanned and the apodizing function used. Nevertheless, these considerations are pertinent to Howard's work in extending Fourier interferograms. (See Chapters 9 and 10, which contain appropriate references.) The other deconvolution methods described in this volume may, with suitable adaptation, also be applied to Fourier spectroscopy.

Broadening caused by the dispersive optical spectrometer can be divided into electrical and predominantly optical parts. When photographic detection is employed, other considerations apply as well, but we shall not discuss them here. As for optics, the idealized model of a perfect diffraction pattern integrated over the slit openings is complicated by the need to consider slit irradiation coherence. Furthermore, optical aberrations and grating imperfections can contribute significantly. These considerations might lead us to believe that empirical determination of the spectrometer response function is the best. This belief is often well justified in spite of the obstacles that also lie along this route.

A brief treatment of the major contributors to optical spreading is certainly in order because it will clarify some of the considerations in designing and adjusting the spectrometer. In some cases it will also allow us to compute a useful approximation to the optical spread function. In the same vein, we later describe the behavior of the most common, useful, and easily described form of electrical noise smoothing: the single-stage RC filter.

A. Coherent Irradiation

Consider a spectrometer used for measuring either emission or absorption spectra. The flux from the sample is typically focused onto an entrance slit. The spectral distribution of this flux is the result of an emission process, possibly also followed by absorption. It may be a line spectrum affected by broadening of the type previously discussed. In one possible optical arrangement (Fig. 1), the flux emerging from the slit is incident upon a collimating mirror, which we here take to have focal length f. The flux may then be dispersed by a prism or grating having an aperture of width D. A telescope mirror, which we take to have focal length f', is then employed to focus the dispersed flux to an image of the entrance slit.

1. Monochromatic Case

Assume first that we have closed the entrance slit to the smallest possible opening so that a perfect geometrical line is approximated. From elementary physical optics we know that monochromatic illumination of the idealized slit in this system gives rise to a diffraction phenomenon owing to spatial truncation of the wave fronts by the aperture of width D. Thus, an irradiance

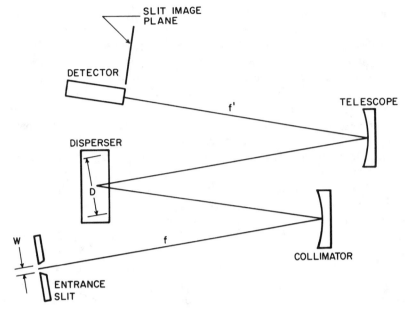

Fig. 1 Optical schematic of a typical dispersive spectrometer.

pattern is formed in the image plane having $\operatorname{sinc}^2(z)$ dependence, where z is the distance from the center of the line image as measured in Rayleigh widths. The Rayleigh width is given by $f'\lambda/D$, where λ is the wavelength of the monochromatic light.

Let us see what happens as the entrance slit is opened to a width w. It can now be thought of as comprising a large number of narrow slits placed next to one another. If this entrance slit is uniformly filled with monochromatic light, then the image plane will show the superposition of a large number of diffraction patterns, a sinc function as shown in Fig. 2 for the field owing to each narrow elemental entrance-slit component.

Before we state this mathematically, let us consider the implication of perfectly monochromatic illumination. We shall assume that the slit elements are illuminated by light of *exactly* the same wavelength and frequency. Each given pair of points in the entrance slit will then have its own fixed phase relationship. Neither point can gain or lose phase relative to the other because the fields have *exactly* the same frequency. The fields at these two points are perfectly *coherent*. Because this is true for any given pair of points in the slit, all the points have a fixed phase relationship. We call this the case of *coherent irradiation*.

From the physics of electricity and magnetism we know that electric fields obey the superposition principle. To find the net field in the image plane, we

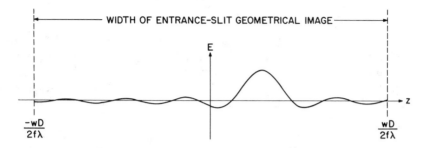

Fig. 2 Diffraction image of an entrance-slit element. Abscissa z is distance in Rayleigh widths; ordinate E is proportional to electric-field amplitude.

must sum the field components, not the irradiances. For simplicity, let us consider only one plane of polarization. In the limit of contributions from infinitesimal slit elements, we obtain

$$E(z) = \sqrt{U_s} \int_{-wD/2f\lambda}^{wD/2f\lambda} \text{sinc}(z - z') \, dz', \qquad (21)$$

where $E(z)$ is proportional to the net field at a distance z from the center of the image. We introduced the sinc function in Section IV.A.1 of Chapter 1. The constant U_s is a scale factor accounting for the entrance-slit irradiance and optical-system efficiency. The limits of integration represent the breadth of the slit image when the optical-system magnification is taken into account and are given in Rayleigh widths.

We can also express Eq. (21) as the convolution of a sinc with a rectangle,

$$E(z) = \sqrt{U_s} \, \text{sinc}(z) \otimes \text{rect}(fz/f'w)$$

or explicitly

$$E(z) = \sqrt{U_s} \int_{-wD/2f\lambda}^{wD/2f\lambda} \frac{\sin[\pi(z - z')]}{\pi(z - z')} \, dz' = \sqrt{U_s} \int_{z-z_0}^{z+z_0} \frac{\sin \pi z'}{\pi z'} \, dz'. \quad (22)$$

We have defined $z_0 = wD/2f\lambda$ as the half-width of the geometrical image of the entrance slit, as measured in Rayleigh widths. Referring to the Si function defined in Section IV.A.1 of Chapter 1 we now may write

$$E(z) = (1/\pi)\sqrt{U_s} \, \{\text{Si}[\pi(z + z_0)] - \text{Si}[\pi(z - z_0)]\}. \qquad (23)$$

The irradiance in the image plane is proportional to the square of the field:

$$U_{coh}(z) = U_s \pi^{-2} \{\text{Si}[\pi(z + z_0)] - \text{Si}[\pi(z - z_0)]\}^2. \qquad (24)$$

2. General Case

The irradiance of the entrance slit just described was perfectly mono-chromatic and coherent. If we now irradiate the entrance slit with flux having a spectrum of finite breadth, but arrange the optics so as to guarantee continued perfect spatial coherence in the entrance-slit plane, the resulting irradiance in the image plane is given by the convolution of $U_{coh}(z)$ with the spectrum of the source.

Now assume that we have placed a flat-surfaced detector of width w' in the image plane, and that each point on the surface of the detector yields an electrical-signal contribution proportional to the irradiance (not the field). Assume further that the response across the detector is uniform, and that contributions from all parts of the detector combine by superposition. The resulting signal is then proportional to the direct superposition of the irradiance contributions. If we sweep this slit detector across the spectrally dispersed irradiance, the resulting signal, as a function of time, is given by the convolution of the spectrum with $U_{coh}(z)$ and $\text{rect}(z/w')$, the rectangle function this time representing the exit slit. The unnormalized optical response function of the spectrometer is thus given by

$$\text{rect}(z/w') \otimes [\text{rect}(fz/f'w) \otimes \text{sinc}(z)]^2. \qquad (25)$$

Note that the inner convolution results from the superposition of coherent fields and the outer convolution from the superposition of electrical-signal contributions.

Suppose, instead, that we had placed a lens near the plane of the exit slit, in the path of the flux emerging from that slit, and imaged the spectrometer pupil onto the detector. Then the field at each point in the exit slit would contribute equally to the field at any particular point on the detector. The net field at that point on the detector is given by linear superposition of the fields. The electrical contribution, as before, goes as the square of this field. Sweeping the spectrum again, we observe the signal given by convolving the spectrum with a different unnormalized optical response function $r_{coh}(z)$:

$$U_{obs}(z) = r_{coh}(z) \otimes U(z), \qquad (26)$$

where $U_{obs}(z)$ is the electrically recorded spectrum and $U(z)$ the spectrum as it would be recorded by a perfectly resolving spectrometer. The response function is now given by

$$r_{coh}(z) = [\text{rect}(z/w') \otimes \text{rect}(fz/f'w) \otimes \text{sinc}(z)]^2. \qquad (27)$$

In the common case where the exit-slit width is given by $w' = f'w/f$ (a typical situation is $f' = f$ and $w' = w$), we obtain

$$r_{\text{coh}}(z) = [\Lambda(fz/f'w) \otimes \text{sinc}(z)]^2, \tag{28}$$

where, as stated earlier, f and f' are the focal lengths of the collimating and telescope elements, respectively, w and w' are the entrance- and exit-slit widths, respectively, and z is the distance in the direction of dispersion in the exit-slit plane as measured in Rayleigh widths.

B. Incoherent Irradiation

1. *Monochromatic Case*

Now let us assume that a monochromatic source of flux is placed in the plane of the entrance slit so that there is no constant phase relationship between the fields at any two given points in the slit. This, in itself, is a contradiction, because a perfect source monochromaticity implies both spatial and temporal coherence. By definition of coherence, a constant phase relationship would result. To eliminate the possibility of such a relationship, we must require the source spectrum to have finite breadth. Let us modify the assumption accordingly but specify the source spectrum breadth narrow enough so that its spatial extent when dispersed is negligible compared with the breadth of the slits, diffraction pattern, and so on. Whenever time integrals are required to obtain observable signals from superimposed fields, we evaluate them over time periods that are long compared with the reciprocal of the frequency difference between the fields. We shall call the assumed source a quasi-monochromatic source.

Let us consider the flux from the quasi-monochromatic incoherent slit source imaged through the spectrometer as in the previous example. Slit width and focal lengths are defined as before. Again the exit-slit–plane image will consist of the superimposed fields from the sinc(z) diffraction pattern due to each vertical element. The instantaneous fields obey superposition. We may write the field resulting from two contributions, for example, as simply proportional to $\mathscr{E}_1(t) + \mathscr{E}_2(t)$. The resulting irradiance is given by the time average,

$$U = \overline{[\mathscr{E}_1(t) + \mathscr{E}_2(t)]^2} = \overline{\mathscr{E}_1^2(t)} + \overline{\mathscr{E}_2^2(t)} + \overline{2\mathscr{E}_1(t)\mathscr{E}_2(t)}. \tag{29}$$

The irradiance U is seen to be the sum of the irradiances due to each component plus a cross term $\overline{2\mathscr{E}_1(t)\mathscr{E}_2(t)}$. But the assumption of complete incoherence and quasi-monochromaticity guarantees that no two components $\mathscr{E}_1(t)$ and $\mathscr{E}_2(t)$ could have a constant phase difference and still yield a product having a finite time integral. (Remember that the integration time must be long compared with the reciprocal of the frequency difference.) For perfectly incoherent slit irradiation, then, the resulting exit-slit–plane irradiance is the linear superposition of the diffraction-pattern irradiances, and we find that

$$U_{inc}(z) = U_s \int_{-wD/2f\lambda}^{wD/2f\lambda} \text{sinc}^2(z - z') \, dz'. \tag{30}$$

As before, we may express an integral of this type as a convolution:

$$U_{inc}(z) = U_s \, \text{sinc}^2(z) \otimes \text{rect}(fz/f'w). \tag{31}$$

This time the convolution is linear in the irradiances, not the fields. We may rewrite Eq. (30) as

$$U_{inc}(z) = U_s \int_{z-z_0}^{z+z_0} \left(\frac{\sin \pi z'}{\pi z'}\right)^2 dz'. \tag{32}$$

Integration by parts yields

$$U_{inc}(z) = \frac{1}{\pi} U_s \left\{ \text{Si}[2\pi(z + z_0)] - \text{Si}[2\pi(z - z_0)] \right.$$

$$\left. - \frac{\sin^2[(z + z_0)/\pi]}{(z + z_0)/\pi} + \frac{\sin^2[(z - z_0)/\pi]}{(z - z_0)/\pi} \right\}, \tag{33}$$

where z and z_0 are defined as before.

2. General Case

Let us assume now that, instead of looking at a quasi-monochromatic source, we are filling the entrance slit with a truly polychromatic source. The irradiance components in the image plane add by incoherent superposition, and we find that the dispersed spectrum in the image plane is obtained by convolving $U_{inc}(z)$ with the spectrum of the source. A detector of width w that obeys the superposition criteria established previously may be placed in the image plane and swept across the spectrum. Its contribution to the spectral broadening will be a linear superposition of irradiances over its width, resulting in convolution with $\text{rect}(z/w')$. Consequently, the electrically recorded spectrum is given by

$$U_{obs}(z) = r_{inc}(z) \otimes U(z), \tag{34}$$

where $U(z)$ is the undistorted spectrum as it would be recorded by a perfectly resolving spectrometer, and the unnormalized incoherent optical response function is given by

$$r_{inc}(z) = \text{rect}(z/w') \otimes \text{rect}(fz/f'w) \otimes \text{sinc}^2(z). \tag{35}$$

When $w' = f'w/f$, by analogy with the previous derivation, we find

$$r_{inc}(z) = \Lambda(fz/f'z) \otimes \text{sinc}^2(z). \tag{36}$$

We also obtain this result if we use a lens or mirror to image the exit slit onto a detector, provided that any blur produced by this imaging element is negligible compared with the breadth of $r_{inc}(z)$.

C. Real Spectrometers

Unlike the situation just discussed we are usually not concerned with measuring the spectrum of a highly coherent source like a laser. It is also not usual to place an incoherent source directly in the entrance slit. A more common situation than either of these extremes is one in which the source is imaged onto the entrance slit by foreoptics.

To deal properly and completely with this general case, we must consider the effects of partial coherence. Discussion of this subject is beyond the scope of the present work. Mielenz (1967) has derived appropriate response-function formulas and gives references to the literature.

D. Electrical Filtering

The spectrum is usually observed by passing it across the detector, thereby effectively converting it to a function of time. Wavelength or wave number may (or may not) be a linear function of time, depending on the drive mechanism employed. Over limited spectral regions, it is often adequate to assume that either wavelength or wave number is proportional to time.

Whatever its source, the electrical signal usually contains noise components having frequencies far higher than those found in the spectrum itself. Electrical filtering is then called for to remove these high-frequency components.

1. Single-Stage RC Filter

The design of optimum electrical filters is a highly developed discipline that we shall not address here. It is worthwhile, however, to discuss briefly the simple single-stage RC filter that is widely used. When all factors are considered, the performance advantages of elaborate designs are often outwieghed by simplicity.

We shall not rely on the electrical engineer's formalism to deal with this problem. Instead, because it is instructive, we appeal to basic physics for a description of this filter's performance.

Figure 3 shows a simple single-section RC low-pass filter network. Kirchhoff's law tells us that

$$v(t') - i(t')R - v_C(t') = 0, \tag{37}$$

where $v(t')$ is the input signal voltage, $i(t')$ the current in the loop, R the resistance, $v_C(t')$ the voltage across the capacitor, and t' time. Now $v_C(t') = q(t')/C$, where $q(t')$ is the charge on the capacitor with capacitance C. Furthermore, $i(t') = dq(t')/dt'$. Therefore, from Eq. (37), we find

$$\frac{d}{dt'}q(t') + \frac{1}{RC}q(t') - \frac{1}{R}v(t') = 0. \tag{38}$$

Multiplying by the integrating factor $\exp(t'/RC)$ and integrating with respect to time from $-\infty$ to t, we obtain

$$\frac{1}{C}q(t)\exp\left(\frac{t}{RC}\right) = \frac{1}{RC}\int_{-\infty}^{t} v(t')\exp\left(\frac{t'}{RC}\right)dt'. \tag{39}$$

Noting that $q(t)/C = v_C(t)$, and moving $\exp(t/RC)$ into the integrand, we have

$$v_C(t) = \frac{1}{RC}\int_{-\infty}^{t} v(t')\exp\left[-\frac{(t-t')}{RC}\right]dt'. \tag{40}$$

If we let

$$\xi(t) = (1/RC)H(t)\exp(-t/RC), \tag{41}$$

where $H(t)$ is the Heaviside step function, we obtain

$$v_C(t) = \xi(t) \otimes v(t). \tag{42}$$

The output voltage v_C is obtained from the input voltage by a simple convolution. Referring to the directory of Fourier transforms (Chapter 1, Fig. 2), we may write

$$V_C(\omega) = \Xi(\omega)V(\omega). \tag{43}$$

Here the frequency spectra of the output and input signals are given by $V_C(\omega)$ and $V(\omega)$, respectively, and the complex filter transfer function is given by

$$\Xi(\omega) = \frac{1}{\sqrt{2\pi}}\frac{1 - j\omega RC}{1 + (\omega RC)^2}. \tag{44}$$

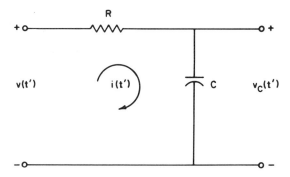

Fig. 3 Single-section RC low-pass filter.

We see that the amplitude-transfer characteristic is given by $\{2\pi[1 + (\omega RC)^2]\}^{-1/2}$. The power-transfer characteristic is given by the square of this quantity. It has the form of a Cauchy function and attenuates high frequencies. Brodersen (1953) and Stewart (1967) have analyzed in detail the performance of other linear electrical filters applied in spectroscopy.

The spectrum distortion introduced by all such filters limits the spectrometer's scanning rate. The subtleties of the scan rate's dependence on electrical filtering and slit width for constant signal-to-noise ratio are explored by Blass and Halsey (1981).

2. *Applications*

The single-section RC filter is often used after phase-sensitive detection. In this method of observing spectra, the radiant flux is typically modulated by a mechanical chopper. The resulting signal is usually demodulated synchronously by switching. This operation can be considered as multiplication of the waveform by a bipolar square wave. Sometimes a narrow-band filter is used to reduce high-frequency noise and permit additional gain without amplifier saturation by noise before demodulation. The result of using such a narrow-band filter is improved dynamic range. By such a demodulation scheme, the fundamental frequency in the signal is shifted to the origin. The modulated information is available after demodulation as a slowly varying (nearly dc) signal. We see that the effect of the RC filter is, in frequency space, equivalent to applying a square-root Cauchy amplitude-transfer characteristic centered at the modulation frequency.

The entire analysis of synchronous detection, or lock-in amplification as it is sometimes called, can be conveniently analyzed by straightforward application of the Fourier transform techniques, transform directory, and convolution theorem developed in Section IV of Chapter 1.

In concluding this section, we point out that the effect of any electrical filter composed of purely linear elements, whether they be passive like resistors, capacitors, and inductors or active like linear amplifiers, can be represented as a convolution. The various other spreading phenomena that are described by convolution in the same domain may therefore be lumped together with the electrical contribution and comprehensively called the spectrometer response function. Even inherent line broadening may be included, provided that the convolution does not appear in an exponent, as in the case of absorption spectra.

E. Measuring Transmittance and Absorptance

We have shown that the radiant flux spectrum, as recorded by the spectrometer, is given by the convolution of the "true" radiant flux spectrum (as it would be recorded by a perfect instrument) with the spectrometer response function. In absorption spectroscopy, absorption lines typically appear superimposed upon a spectral background that is determined by the emission spectrum of the source, the spectral response of the detector, and other effects. Because we are interested in the properties of the absorbing molecules, it is necessary to correct for this background, or baseline as it is sometimes called. Furthermore, we shall see that the valuable physical-realizability constraints presented in Chapter 4 are easiest to apply when the data have this form.

1. Transmittance

In double-beam spectrometers, the correction is made before the spectrum is recorded by dividing:

$$T_M(x) = [U_M(x) - D_M(x)]/[B_M(x) - D_M(x)]. \qquad (45)$$

In this expression, x is the independent variable of measurement, whether it be wavelength, wave number, or other parameter. The quantity $U_M(x)$ is the flux, as measured by the spectrometer, after it has passed through the absorbing sample. The quantity $B_M(x)$ represents the flux that has passed over an identical path but in the absence of an absorbing sample. The term $D_M(x)$ represents any additive offsets that might be introduced by "stray" flux, errors in electronic amplification and digital conversion, or other causes. We call the resulting measured transmittance $T_M(x)$.

If we have a single-beam spectrometer, we may separately record spectra $B_M(x)$, $U_M(x)$, and $D_M(x)$ and apply Eq. (45) later in the computer. With special rapid-scanning spectrometers this approach may be practical, but source and detector drift usually preclude use of this procedure for high-accuracy, slow-scanning spectroscopy.

More typically, one seeks $U_M(x)$ regions of negligible absorption and connects them with a fitted curve (often a straight line is adequate over a limited region) to approximate $B_M(x)$. If $D_M(x)$ represents mainly offsets in the electronics, a shutter placed in the beam facilitates its determination.

2. Effect of Instrumental Spreading

In Eq. (45), T_M represents the "true" transmittance only if instrumental spreading is negligible. In previous sections, however, we learned that instrumental spreading may be described by the convolution product of the flux and the spectrometer response function, here called $r(x)$, that incorporates all the instrumental spreading phenomena:

$$U_M(x) = r(x) \otimes U_T(x). \tag{46}$$

Here $U_T(x)$ represents the "true" flux.

Both $B_M(x)$ and $D_M(x)$ are similarly distorted versions of true background and offset functions that we may identify, respectively, as $B_T(x)$ and $D_T(x)$. Thus we may write the measured transmittance as

$$T_M(x) = \frac{U_M(x) - D_M(x)}{B_M(x) - D_M(x)} = \frac{r(x) \otimes U_T(x) - r(x) \otimes D_T(x)}{r(x) \otimes B_T(x) - r(x) \otimes D_T(x)}$$

$$= \frac{\displaystyle\int r(x - x')[U_T(x') - D_T(x')] \, dx'}{\displaystyle\int r(x - x')B_T(x') \, dx' - \int r(x - x')D_T(x') \, dx'}, \tag{47}$$

where the limits cover a spectral region of interest.

Let us now assume that $B(x)$ and $D(x)$ are either constant or slowly varying compared with $r(x)$, so that we may take $B(x') = B(x)$ and $D(x') = D(x)$ in the denominator. The normalization of $r(x)$ then permits us to write

$$T_M(x) = \frac{1}{B_T(x) - D_T(x)} \int r(x - x')[U_T(x') - D_T(x')] \, dx'. \tag{48}$$

Again taking advantage of the slow variation of $B_T(x)$ and $D_T(x)$, we may include them in the integrand:

$$T_M(x) = \int r(x - x') \frac{U_T(x') - D_T(x')}{B_T(x') - D_T(x')} \, dx' = r(x) \otimes T_T(x). \tag{49}$$

We see that transmittance obeys the convolution relation for instrumental spreading provided that $B_T(x)$ and $D_T(x)$ are either constant or slowly varying compared with $r(x)$.

3. Absorptance

We define measured and true absorptance by the pair of relations

$$A_M(x) = 1 - T_M(x) \tag{50}$$

and

$$A_T(x) = 1 - T_T(x). \tag{51}$$

From Eqs. (49) through (51) we may obtain

$$A_M(x) = 1 - r(x) \otimes T_T(x) = r(x) \otimes [1 - T_T(x)] = r(x) \otimes A_T(x). \tag{52}$$

We see that absorptance obeys the convolution relation for instrumental spreading under the same conditions of slowly varying $B_T(x)$ and $D_T(x)$.

We note here the similarity between the terms absorptance and absorbance. The latter is frequently employed in chemical infrared spectroscopy to denote $\log_{10}[1/T(x)]$, and may also be called optical density. As discussed in Section I.E.2, absorbance may not be directly employed in the convolution relation for instrumental spreading.

F. Measurements Independent of Spectrometer Broadening

In many applications, it is convenient to define the equivalent width of a function $f(x)$ as the width of a rectangle having the same area and maximum height as $f(x)$. Following this definition for a function $f(x)$ having its maximum at $x = 0$, we have for the equivalent width

$$\left[\int_{-\infty}^{\infty} f(x)\,dx \right] \Big/ f(0). \tag{53}$$

In absorption spectroscopy it is more useful to define the equivalent width of a spectral line as the width of an idealized, totally absorbing rectangular line having the same area (Fig. 4). The equivalent width of an isolated absorption line is therefore given by

$$W = \int A_T(x)\,dx = \int [1 - T_T(x)]\,dx, \tag{54}$$

where the limits of integration are understood to extend on either side of the line until the absorption encountered is negligible. Because the area under the

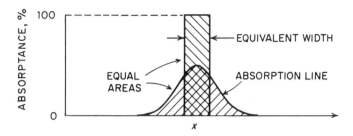

Fig. 4   Equivalent width of an absorption line.

instrument response function is normalized to unity, we may insert that integral under the integrand of Eq. (54):

$$W = \int A_T(x') \left[\int r(x - x') \, dx \right] dx'. \tag{55}$$

Reversing the order of integration and recognizing that $A_M(x) = r(x) \otimes A_T(x)$, we obtain

$$W = \int \int r(x - x') A_T(x') \, dx' \, dx = \int A_M(x) \, dx = \int [1 - T_M(x)] \, dx. \tag{56}$$

Therefore, we have shown that

$$W = \int A_T(x) \, dx = \int A_M(x) \, dx. \tag{57}$$

The equivalent width W of an isolated absorption line is independent of spectrometer broadening effects.

Noise causes excursions above and below the desired trace when a scan is made across such an absorption line. The small positive and negative contributions made by these excursions tend to cancel in the computation of area. Such areas can thus be measured to relatively high precision. Furthermore, when lines exhibiting predominantly Doppler broadening are measured, the wings fall rapidly to zero at a moderate distance from the line center. This behavior facilitates determination of the base line and contributes to the accuracy of measurement. Accurate measurements of line strength can therefore be made by these methods (Korb et al., 1968) because the equivalent width may be expressed as a function of the line strength S times the pressure path-length product X. The quantity W is given by different series for Doppler- and pressure-broadened cases. Typically one wishes to determine S from $W(SX)/X$. A straightforward and rapidly converging application of

Newton's method enables us to find the value of the product SX for which $\psi = 0$ in the expression

$$\psi(SX) = W_M - W(SX), \tag{58}$$

where W_M is the measured equivalent width.

For purely Doppler broadening, the equivalent width takes the form

$$W(SX) = SX \sum_{n=0}^{\infty} \frac{[-\sqrt{(\ln 2)/\pi}\, SX/\Delta x_D]^n}{(n+1)!\,\sqrt{n+1}}. \tag{59}$$

The quantity Δx_D is the Doppler half-width encountered earlier. Newton's method requires knowledge of the first derivative, which is easily provided by multiplying each term in Eq. (59) by n/SX.

When pressure broadening dominates, the situation is more complicated because the resulting Lorentzian profile contributes significant area far from the line center. A further complication in this case is that the Lorentzian half-width cannot be accurately calculated and must be measured in other experiments. If both Doppler and pressure broadening are present, however, and if the Lorentzian to Doppler half-width ratio μ is small, the correction necessitated by pressure broadening is small. In this situation an accurate value of the Lorentzian half-width may not be needed. Line strength in the case of combined Doppler and pressure broadening may be obtained from the equivalent width by the use of tables (Jansson and Korb, 1968).

G. Determining the Spectrometer Response

In applying the technique of deconvolution, we take as known the spectrometer response function. It seems reasonable that the more accurately we know this function, the more accurate will be the deconvolved result. Although the nonlinear methods described in Chapter 4 are more tolerant of error, they too require a knowledge of the response function.

1. Difficulties in Use of Narrow Lines

Owing to aberrations, grating defects, and so on, it may not be adequate to approximate the response function by formulas based on idealized models. If a line source could be found having the spectrum that approximates a δ function, then perhaps the measurement of such a line would adequately determine the response function. We have learned, however, that the spatial coherence of the source plays an important part in the shape of the response function. This precludes the use of a laser line source to measure the response function applicable to absorption spectroscopy. Furthermore, we would have to be sure that both the pupil and the slit are filled with the same spatial distribution of flux as that used in the observation of the spectra to be deconvolved. The latter requirement seems to be a tall order, unless we use the same continuum source employed for observation of the absorption spectra themselves. This, in turn, implies that we might consider observing a very narrow absorption line. Let us consider such an observation.

An inherent difficulty in observing a narrow absorption line is that it does not absorb very much flux. The narrower the line, the less flux absorbed. In fact, the ideal δ-function line absorbs none at all. This statement may be verified by evaluating the integrated absorptance

$$\int_{\substack{\text{entire} \\ \text{line}}} [1 - e^{-XP(x)}] \, dx \tag{60}$$

of a narrow Gaussian absorption line $XP(x)$, and taking the limit as the $P(x)$ half-width Δx_G goes to zero, while holding the integrated absorption coefficient $X \int P(x) \, dx$ constant.

If we attempt to render a narrow absorption line observable by increasing the absorbing optical path, we succeed only in broadening the line as it is observed in the transmittance–absorptance regime. Although the shape and breadth of the absorption-coefficient curve $XP(x)$ have not been altered by this procedure, the spectrometer sees a broader spectral line because of the saturation of the absorption near the line center and increasing absorption in the wings. The term "half-width" as applied to absorption lines is almost universally used with respect to the absorption coefficient profile $P(x)$. Used alone it is *not meaningful* as a measure of the effective breadth of isolated lines employed to determine the response function. For this measure to be meaningful, the exponential relationship between absorption-coefficient and absorptance must be explicitly involved.

2. Narrow Absorption Lines

To illustrate this point, let us suppose that the half-width of a line in either the absorbance or absorptance regime is truly much narrower than the half-width of the instrument response function $r(x)$. The measured absorptance then cannot be more than a few percent, at the most.

Let us establish the required relationships more precisely. Consider a narrow idealized rectangular absorption line $A_T(x) = \text{rect}(x/2\,\Delta x_L)$ having half-width Δx_L and centered at $x = 0$. Its variance is easily found to be $\sigma_L^2 = (2\,\Delta x_L/3)^2$. Its area is $2\,\Delta x_L$. Now, let us assume that this line is being used to measure an instrument response function $\exp(-x^2/2\sigma_R^2)$ that has Gaussian shape and variance σ_R^2.

The resulting observed absorptance is given by the convolution of the gaussian and the narrow rectangle, and is also Gaussian to a high degree of approximation. The variance of this result is just the sum $\sigma_L^2 + \sigma_R^2$. The fractional increase in the square root of the variance is given by

$$\alpha = \frac{\sqrt{\sigma_R^2 + \sigma_L^2} - \sigma_R}{\sigma_R}. \tag{61}$$

For the observation to overestimate σ_R by a given fraction α, the linewidth is given by

$$\sigma_L = \sigma_R \sqrt{\alpha^2 + 2\alpha} \approx \sigma_R \sqrt{2\alpha}, \qquad \alpha \ll 1. \tag{62}$$

We may convert this relationship to half-width form by remembering that $\sigma_L = 2\,\Delta x_L/3$ and expressing σ_R in terms of the Gaussian instrument function half-width Δx_R:

$$\sigma_R = \Delta x_R / \sqrt{2 \ln 2}. \tag{63}$$

Doing this by making the appropriate substitutions and solving for Δx_R, we find

$$\Delta x_R = \tfrac{2}{3}\sqrt{(\ln 2)/\alpha}\,\Delta x_L. \tag{64}$$

Because the instrument response function must have unit area, the area under the narrow line is preserved by the measurement process. Recalling that this area is $2\,\Delta x_L$, we may write the absorptance amplitude A_L of the observation (which is Gaussian) in terms of its half-width and $2\,\Delta x_L$:

$$A_L = \sqrt{(\ln 2)/\pi}\,2\,\Delta x_L/\Delta x_R. \tag{65}$$

By making the appropriate substitution for Δx_R from Eq. (64), we obtain the amplitude of the observed line as a function of α, the fractional increase in breadth:

$$B_L = 3\sqrt{\alpha/\pi}. \tag{66}$$

To hold line broadening to less than 0.1%, we find that the maximum absorptance B_L of the line must be held to under 5.4%. For a maximum broadening of 1.0%, the absorptance cannot exceed 16.9%.

Usual noise levels produce considerable error in such measurements. Slit openings wide enough to yield the high signal-to-noise ratio needed would probably be unsuitable for data acquisition runs of the spectra to be deconvolved. Whether or not we can tolerate various errors in the response-function measurement depends on the way in which it is applied. Deconvolution with an artificially wide response function yields artificially narrow deconvolved lines and probably some artifacts as well. Experimentation may be the best guide in deciding this issue.

3. Successful Use of Absorption Lines

We may overcome the problems just described by measuring the possibly saturated absorption lines of finite breadth to which we are restricted and performing a deconvolution. In this case, we take the actual line shape as known and seek the spectrometer response function. The response function becomes the object $o(x)$, and the known line shape is identified with $s(x)$. Doppler-broadened lines with little or no Lorentzian component are best for this purpose owing to their localization of area near the line center. We may synthesize the line shape from Eqs. (9) and (16) if we know Boltzmann's constant k, the molecular mass m, the speed of light c, the absorption line frequency v_0, the temperature of the absorbing gas, T, and the product SX. The first four constants are known with more than adequate accuracy. The temperature T is easily determined, and the product SX is simply obtained from the area under the line by the equivalent-width method described in Section II.F. Only a rough estimate of pressure broadening is needed to determine its significance and account for its effect, if necessary. The optical path need not even be explicitly measured. Because only a modest narrowing is sought from the deconvolution process and because the true line shape does not impose the restriction of band limiting, a simple linear deconvolution method works adequately.

In conclusion, the observed spectrum of an isolated Doppler-broadened line, along with some well-known constants and the easily measured temperature, contains all the information needed to determine the response function. Application details for this method are available in the literature (Jansson, 1968, 1970).

III. RESOLUTION CRITERIA

When we say that objects are resolved, we mean that they are rendered so that their separate and distinguishable identities are apparent. The present section briefly formalizes this intuitive idea.

A. Two-Point Criteria

In the simplest possible case, we wish to distinguish only two objects. In optics the required specification is called a two-point resolution criterion. In spectroscopy a two-point criterion specifies the distance that must separate two spectral lines for them to be called resolved. Essentially similar criteria apply to other fields.

Previous sections have detailed phenomena that contribute to the degradation of resolution in optical spectra. Concepts useful in specifying resolution criteria have been established. Although transfer and point-spread functions of varying shape can yield identical numbers when a simple two-point criterion is applied, this many-to-one correspondence does not diminish the criterion's usefulness. More rigorous specification of the transfer function virtually requires graphical presentation for human interpretation. Its use therefore demands far more space in text and more time for study. Frequently, the functional form of the transfer function is well known anyway; systems being compared are often of similar type. In these cases, the two-point criterion is entirely adequate.

1. Optical Instruments and Rayleigh's Criterion

In optics and spectroscopy, resolution is often limited by diffraction. To a good approximation, the spread function may appear as a single-slit diffraction pattern (Section II). If equal-intensity objects (spectral lines) are placed close to one another so that the first zero of one sinc-squared diffraction pattern is superimposed on the peak of the adjacent pattern, they are said to be separated by the Rayleigh distance (Strong, 1958). This separation gives rise to a 19% dip between the peaks of the superimposed patterns.

It might seem that Lord Rayleigh's proposal is somewhat arbitrary; even if two lines are separated by a somewhat smaller distance, it will be apparent that two lines are present rather than one. Sparrow's criterion (Strong, 1958) does, in fact, say that two lines may be termed just resolved when spaced so that the dip first appears. The Rayleigh criterion, however, gives rise to particularly simple expressions for resolving power.

Resolving power at wavelength λ (wave number $\bar{v}$) is defined as $\lambda/\Delta\lambda = \bar{v}/\Delta\bar{v}$, where $\Delta\lambda$ is the Rayleigh distance in wavelength units ($\Delta\bar{v}$ is the Rayleigh distance in wave numbers). It turns out that the theoretical resolving power of a dispersive optical grating spectrometer is given simply by the product of the total number of lines on the grating, the number of times the flux is dispersed (number of passes), and the grating order in which the spectrum is observed. The Rayleigh resolution in wave numbers resulting from an optimally triangle-apodized Fourier interferogram is given by the reciprocal of the maximum optical path difference represented in the interferogram (Bell, 1972).

Let us cite an example to help us judge the equivalence between Fourier and dispersive instruments. A grating spectrometer employing a four-passed 8×10^4-line grating in the first order has a resolving power of $4 \times 8 \times 10^4 = 3.2 \times 10^5$. At 3200 cm^{-1} in the near infrared, this instrument has a Rayleigh resolution of 10^{-2} cm^{-1}. The same resolution can be achieved by a Fourier instrument having 10^2-cm maximum optical path difference. If the physical path in one leg of the interferometer is traversed twice, the total mechanical travel required between the central fringe position and the extreme is 50 cm.

Brief reflection on the sampling theorem (Chapter 1, Section IV.C) with the aid of the Fourier transform directory (Chapter 1, Fig. 2) leads to the conclusion that the Rayleigh distance is precisely two times the Nyquist interval. We may therefore easily specify the sample density required to recover all the information in a spectrum obtained from a band-limiting instrument with a sinc-squared spread function: evenly spaced samples must be selected so that four data points would cover the interval between the first zeros on either side of the spread function's central maximum. In practice, it is often advantageous to place samples somewhat closer together.

2. Other Two-Point Criteria

Though the Rayleigh criterion has particular meaning and simplicity when used with sinc or sinc-squared spread functions, the resulting 19% dip criterion may be applied to lines and objects having other shapes. Two such nonsinc equal-intensity lines, when separated so that a 19% dip appears, are said to be just resolved according to the Rayleigh criterion. Other two-point criteria may be more meaningful in these cases, however. For Gaussian and Lorentzian shapes and for unspecified peaklike objects, we often use criteria of full width at half maximum (FWHM—the breadth of the object at half of the peak height) or half-width at half maximum (HWHM—analogously defined). The variance or square root of the second moment sometimes also has value as a descriptor of breadth, and hence resolution. Fields other than spectroscopy have employed criteria such as optical line pairs, television lines, rise time, and limiting resolution.

3. Resolution "Limits" and Deconvolution Performance

What meaning do these two-point resolution criteria have in describing the deconvolution process, that is, resolution before and after deconvolution? Although width criteria may be applied to derive suitable before–after ratios, the Rayleigh criterion raises an interesting question. Because the diffraction pattern is an inherent property of the observing instrument, would it not be best to reserve this criterion to describe optical performance? The effective spread function after deconvolution is not sinc squared anyway.

An additional question arises concerning the way the Rayleigh criterion terminology has traditionally been used. There is a vague perception that the "Rayleigh limit" is a barrier to be penetrated only by supernatural means— that it is a limit of the most fundamental kind. And so it is, but only in a sense that requires some elaboration. The "Rayleigh limit" of both Fourier and dispersive instruments implies an upper limit Ω to the Fourier frequencies that may be measured. Beyond this limit, the instrument is incapable of passing any information. When the transfer function is triangular so that frequencies fall off linearly to zero at cutoff Ω (triangular apodization), the sinc-squared spread function results. The Rayleigh criterion then applies directly and meaningfully. After deconvolution, however, the effective transfer function is anything but triangular. It is thus possible to obtain the required 19% dip between lines that are spaced a good deal closer than the "Rayleigh limit" for a given instrument. Furthermore, in Chapter 4 we shall find that *even the instrumental Fourier frequency cutoff* Ω *does not pose a limit to the higher frequencies that may be restored.* This frequency extrapolation is achieved on the strength of additional information conveyed by simple constraints of physical realizability. Spectra having Fourier frequencies thus extended beyond cutoff Ω are said to be *superresolved.*

B. Conclusion

We recommend that the Rayleigh criterion and Fourier frequency cutoff Ω be used as fundamental specifications of optical performance, but that other criteria such as full width at half maximum be used in describing widths before and after deconvolution and in specifying spread functions having other than sinc-squared shape.

In this chapter we have detailed the processes that give rise to distortion of spectra. Although specifics have been limited to optical spectra, a number of the principles are easily generalized to other fields, including those for which data are acquired in more than one dimension. In the next chapter we introduce traditional methods of undoing the distortion, along with some of the concepts needed to develop the most successful modern methods described in Chapter 4.

REFERENCES

Armstrong, B. H. (1967). *J. Quant. Spectrosc. Radiat. Transfer.* **7,** 61–88.
Bell, R. J. (1972). "Introductory Fourier Transform Spectroscopy." Academic Press, New York.
Blass, W. E., and Halsey, G. W. (1981). "Deconvolution of Absorptions Spectra." Academic Press, New York.
Brodersen, S. (1953). *J. Opt. Soc. Am.* **43,** 1216–1220.
Dicke, R. H. (1953). *Phys. Rev.* **89,** 472–473.
Galatry, L. (1961). *Phys. Rev.* **122,** 1218–1223.

Herbert, F. (1974). *J. Quant. Spectrosc. Radiat. Transfer* **14**, 943–951.

Jansson, P. A. (1968). Ph.D. Dissertation, Florida State Univ., Tallahassee.

Jansson, P. A. (1970). *J. Opt. Soc. Am.* **60**, 184–191.

Jansson, P. A., and Korb, C. L. (1968). *J. Quant. Spectrosc. Radiat. Transfer* **8**, 1399–1409.

Kauppinen, J. K., Moffatt, D. J., Mantsch, H. H., and Cameron, D. G. (1981). *Appl. Spectrosc.* **35**, 271–276.

Kielkopf, J. F. (1973). *J. Opt. Soc. Am.* **63**, 987–995.

Korb, C. L., Hunt, R. H., and Plyler, E. K. (1968). *J. Chem. Phys.* **48**, 4252–4260.

Mielenz, K. D. (1967). *J. Opt. Soc. Am.* **57**, 66–74.

Penner, S. S. (1959). "Quantitative Molecular Spectroscopy and Gas Emissivities." Addison-Wesley, Reading, Massachusetts.

Pine, A. S. (1980). *J. Mol. Spectroscopy* **82**, 435–448.

Pliva, J., Pine, A. S., and Willson, P. D. (1980). *Appl. Opt.* **19**, 1833–1837.

Puerta, J., and Martin, P. (1981). *Appl. Opt.* **20**, 3923–3928.

Ramsay, D. A. (1952). *J. Am. Chem. Soc.* **74**, 72–80.

Rodgers, C. D. (1976). *Appl. Opt.* **15**, 714–716.

Stewart, J. E. (1967). *Infrared Phys.* **7**, 77–92.

Strong, J. (1958). "Concepts of Classical Optics." Freeman, San Francisco.

CHAPTER 3

Traditional Linear Deconvolution Methods

Peter A. Jansson

Engineering Physics Laboratory, E. I. du Pont de Nemours and Company (Inc.)
Wilmington, Delaware

I.	Direct Approach	69
II.	Van Cittert's Method	71
III.	Matrix Inversion	73
	A. Matrix Formulation	73
	B. Difficulties	74
	C. Iterative Methods	75
IV.	Inverse Filters	80
	A. Basic Method	80
	B. Wiener-Type Filter	81
	C. Other Minimum Mean-Square-Error Variations	82
	D. Relationship to Van Cittert's Method	83
	E. Stepwise Implementation	86
V.	Other Linear Methods	87
VI.	Overview of Linear Methods	89
	References	91

LIST OF SYMBOLS

a, b, g	functions $a(x), b(x), g(x)$, with no explicit dependences shown
$B(\omega)$	factor describing modification of inverse-filter transfer function by linear deconvolution method
C	relaxation factor in continuous formulation
$\mathbf{i}, \mathbf{o}, \hat{\mathbf{o}}$	column vectors containing elements $i_m, o_m, \hat{o}_m$
i_C, o_C	constant values of i_m, o_m
$i_m, s_m, o_m, \hat{o}_m^{(k)}, n_m$	discretely sampled versions of $i, s, o, \hat{o}^{(k)}, n$
$i(x), i$	"image" data that incorporate smearing by $s(x)$
$\hat{i}^{(k)}(x)$	kth estimate of $i(x)$
$I(\omega), \tau(\omega), O(\omega), N(\omega)$	Fourier transforms of $i(x), s(x), o(x), n(x)$
j	imaginary operator such that $j^2 = -1$
L	number of nonvanishing sampled spread-function values on either side of central maximum, spread function assumed symmetrical
N_s	number of components in composite inverse filter
$n(x)$	part of the image data $i(x)$ due to noise
$o(x), o$	"object" or function sought by deconvolution, usually the "true" spectrum, but also the instrument function when this is sought by deconvolution

67

DECONVOLUTION:
WITH APPLICATIONS IN SPECTROSCOPY

$\hat{o}(x)$, $\hat{o}^{(k)}(x)$, $\hat{O}(\omega)$, $\hat{O}^{(k)}(\omega)$	estimate and kth estimate of $o(x)$ and $O(\omega)$, respectively
rect(x)	rectangle function having half-width $\frac{1}{2}$
$\mathbf{s}$	matrix containing elements $[\mathbf{s}]_{nm} = s_{n-m}$
$[\mathbf{s}]_{nm}$	element of matrix $\mathbf{s}$
sinc(x)	$(\sin \pi x)/\pi x$
$s(x)$, s	spread function, usually the instrument function, but also spreading due to other causes
x, x'	generalized independent variables and arguments of various functions
$y(x)$	linear restoring filter such that $\hat{o}(x) = y(x) \otimes i(x)$; takes various forms depending on criteria of derivation
$y_s(x)$, $Y_s(\omega)$	component of inverse filter and its Fourier transform, respectively
$y(x, x')$	generalized shift-variant inverse-filtering kernel
$Y(\omega)$	Fourier transform of $y(x)$
β	parameter specifying influence of sharpness or smoothness criteria
$\otimes$	convolution operation
$\delta(x)$, δ	Dirac δ function or impulse
ζ	$1 - \tau(\omega)$;
$\theta(x)$	spurious part of solution
κ	relaxation factor in discrete formulation
$\phi_n(\omega)$, ϕ_n	power spectra of $n(x)$, n
$\phi_o(\omega)$, ϕ_o	power spectra of $o(x)$, o
ψ	Gaussian line-broadening function
ω	conjugate of x; Fourier frequency in radians per units of x
Ω	cutoff frequency such that $\tau(\omega) = 0$ for $|\omega| > \Omega$
Ω_p	cutoff frequency defining Frieden's optimum processing bandwidth

A linear deconvolution method is one whose output elements (the restoration) can be expressed as linear combinations of the input elements. Until recently, the only seriously considered methods of deconvolution were linear. These methods can be developed and analyzed in detail by use of long-standing mathematical tools. Analysis of linear methods tends to be simpler than that of nonlinear methods, and computations are shorter. This point is especially important, because deconvolution is inherently computation intensive. It is not surprising that linear methods have historically dominated deconvolution research and applications.

In spite of performance advantages in the use of nonlinear methods, it is instructive to start our deconvolution study by examining the linear methods; they will give us insight into the process. The ensuing development will also define the applicability domain of linear methods and reveal their limitations. We shall see that in some circumstances a linear method is the method of choice.

I. DIRECT APPROACH

Let us begin by looking at an approach that may be the most intuitive. Adhering to the notation that we introduced in Section V of Chapter 1, we write the equation of resolution distortion,

$$i = s \otimes o, \tag{1}$$

and its discrete counterpart

$$i_m = \sum_k s_{m-k} o_k. \tag{2}$$

In this equation, the image value i_m is the mth sample of the smeared version of o_k. We transform the subscripts in an obvious way that recognizes the commutative property of convolution and considers only spread functions that are nonvanishing over some finite domain having an odd number of points. We may then write

$$i_m = \sum_{l=-L}^{L} s_l o_{m-l}. \tag{3}$$

For convenience and simplicity in this section, we assume that the spread function s_l has its peak value at $l = 0$ and that it has L nonvanishing values on either side of s_0. The development based on these assumptions may easily be generalized.

Now let us assume that the true object is zero for all subscript values less than 1. When we do the convolution, then, i_m must have a value of zero for $m < 1 - L$. When we perform the numerical convolution starting with $m \ll 1 - L$, we compute zeros as we slide the spread function toward the region of finite o_k. Finally, when the end of the spread function encounters the first nonzero o_k, we obtain the first nonzero value of i_k. We illustrate this particular sum of products with an elementary example in which $L = 1$ and $m = 0$:

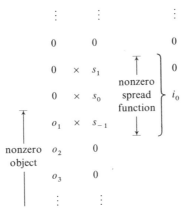

Here the first value of the image i_0 is given by

$$i_0 = o_1 s_{-1}, \tag{4}$$

and we may easily obtain an estimate of the first object value,

$$\hat{o}_1 = i_0/s_{-1}. \tag{5}$$

The next value of the image is given by sliding the spread function down and forming the appropriate sum of products:

$$i_1 = o_1 s_0 + o_2 s_{-1}. \tag{6}$$

Solving for o_2, we obtain the estimate

$$\hat{o}_2 = (i_1 - \hat{o}_1 s_0)/s_{-1}. \tag{7}$$

But we already know $\hat{o}_1$ and readily obtain by substitution

$$\hat{o}_2 = (i_1 - i_0 s_0/s_{-1})/s_{-1}. \tag{8}$$

Proceeding in similar fashion, we may obtain

$$\hat{o}_3 = (1/s_{-1})i_2 - (s_0/s_{-1}^2)i_1 + (s_0^2/s_{-1}^3 - s_1/s_{-1}^2)i_0, \tag{9}$$

and so forth.

We might generalize by noting that for this method to be considered, values of o_k (and hence of i_k) need not be vanishing for a stretch at one end but must merely be constant over an interval of breadth $2L + 1$. In this case, if i_C is the constant value of i_k, the constant value of o_k is given by

$$o_C = i_C \left/ \sum_{l=-L}^{L} s_l \right. . \tag{10}$$

When the spread function is normalized, we obtain $o_C = i_C$. The estimates of the object are thus easy to determine by starting at the ends.

We may obtain a general formula for the $(k + 1)$th object estimate in terms of the preceding $2L$ estimates and the image value i_{k-L+1} by constructing a more complete and general diagram analogous to our example for $L = 1$. By so doing we may derive

$$\hat{o}_{k+1} = \left(i_{k-L+1} - \sum_{l=-L+1}^{L} s_l \hat{o}_{k-L+1-l} \right) \left/ s_{-L} \right. . \tag{11}$$

Have we solved the problem? Is this all there is to deconvolution? Let us look a little closer at the three-point example by examining Eq. (5). Suppose that the first spread-function value s_{-1} is small relative to the other values of s, as is typical. The value i_0 would also then be small. We are dealing with data acquired from the real world, and no observation can be without error. Suppose that the observation of i_m is subject to the error n_m. We may then write our object estimate

$$\hat{o}_1 = (i_0 + n_0)/s_{-1}. \tag{12}$$

Even for good signal-to-noise ratios (small n_0), our estimate $\hat{o}_1$ could suffer considerably, being erroneous by an amount n_0/s_{-1}, where s_{-1} is small. We have calculated only the first element of the object estimate and already there is trouble.

We see that this trouble gets worse in the next step because the object estimate $\hat{o}_2$ is given by the magnified small difference between the uncertain quantity $\hat{o}_1 s_0$ and i_1, itself uncertain. The difficulties increase as we proceed, and it is clear that *uncertainty in the observations plays a fundamental role in our ability to restore the true object*. The naive approach with which we started exemplifies the way we may be misled when we do not provide explicitly for the effects of noise.

In spite of these difficulties, the method described is useful in certain special cases. An adaptation of it has been successfully applied to the removal of base-line distortion caused by inelastically scattered electrons in electron spectroscopy for chemical analysis (ESCA) (Chapter 5). As one might suspect, the method is most useful when the end value s_L is not small. In the ESCA example, we shall see that when s is written in its continuous formulation $s(x)$, it is actually terminated by a Dirac δ function.

II. VAN CITTERT'S METHOD

Van Cittert (1931) recognized that the image data $i(x)$ could be considered as a first approximation $\hat{o}^{(0)}(x)$ to the object $o(x)$. After all, in the absence of deconvolution, the spectroscopist often ignores (rightly or wrongly) instrumental spreading and uses the data as if they represent the true spectrum. This being the case, could we not blur the approximation $\hat{o}^{(0)}(x)$ to yield an "approximation" to the data, $\hat{i}^{(0)}(x)$? This is the form that the data would take if $\hat{o}^{(0)}(x)$ were the true object. Certainly we could, but we already have the data $i(x)$, and so what would be the purpose?

This blurring actually does serve a useful function. The difference $i(x) - \hat{i}^{(0)}(x)$, the easily computed error in the image estimate, is in some way related to the error in the object estimate, $o(x) - \hat{o}^{(0)}(x)$. Van Cittert recognized that this image-estimate error could be applied as a correction to the object estimate, thus producing a new object estimate:

$$\hat{o}^{(1)}(x) = \hat{o}^{(0)}(x) + [i(x) - s(x) \otimes \hat{o}^{(0)}(x)]. \tag{13}$$

This is illustrated in Fig. 1. Although $\hat{o}^{(1)}(x)$ is not the solution, the correction gives us some cause for optimism. The estimate $\hat{o}^{(1)}(x)$ is certainly narrower and taller than the data $i(x)$. One may continue the process, hoping for ever better estimates:

$$\hat{o}^{(k+1)}(x) = \hat{o}^{(k)}(x) + [i(x) - s(x) \otimes \hat{o}^{(k)}(x)]. \tag{14}$$

In this expression, $\hat{o}^{(k)}(x)$ and $\hat{o}^{(k+1)}(x)$ are the kth and $(k + 1)$th approximations to the object.

The fact that this method is somewhat intuitive may account for its rather wide use and apparently independent reinvention by workers following Van Cittert. The basic method has numerous variants (e.g., Herget *et al.*, 1962) and has given rise to constrained nonlinear methods that exhibit significantly better performance (Chapter 4, Sections IV.A and V.A). It has been firmly identified with both inverse filtering (see Section IV.D) and the solution of linear equations by relaxation (DiCola *et al.*, 1967; Jansson 1968, 1970; see Section III.C). Its convergence has been analyzed by Hill and Ioup (1976) and others (Section IV.D). Some of its variants are concerned with damping the noise that grows with each iteration when the method is applied to real data. These spurious solution components invariably have high frequency and often give rise to solutions that are not physically possible, such as transmittance less than zero or in excess of 100%.

The method has utility where only modest correction is required, such as in the determination of the spectrometer response function when a narrow line has been measured (Chapter 2, Section II.G).

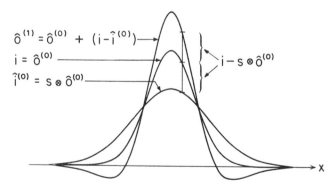

Fig. 1 Corrections applied in Van Cittert's method.

III. MATRIX INVERSION

We have presented two deconvolution methods from an intuitive point of view. The approach that suits the reader's intuition best depends, of course, on the reader's background. For those versed in linear algebra, methods that stem from a basic matrix formulation of the problem may lend particular insight. In this section we demonstrate a matrix approach that can be related to Van Cittert's method. In Section IV.D, both approaches will be shown to be equivalent to Fourier inverse filtering. Similar connections can be made for *all* linear methods, and many limitations of a given linear method are common to all.

A. Matrix Formulation

As we have seen earlier, we may write the convolution integral [Chapter 1, Eq. (86)] in discrete form:

$$i_n = \sum_m s_{n-m} o_m, \qquad (15)$$

where i_n are the image values and o_m the object values; the response function is represented by the values s_{n-m}. We have omitted the limits of summation. They are understood to represent the finite range of m that is useful in describing the object over its range of interest. The values of n are usually those for which data are available. As indicated in Section I, sometimes end effects need to be considered.

Equation (15) can be expressed in a matrix formulation:

$$\mathbf{i} = \mathbf{so}. \qquad (16)$$

In this equation, $\mathbf{i}$ and $\mathbf{o}$ are column vectors such that each element $[\mathbf{i}]_n = i_n$ and $[\mathbf{o}]_m = o_m$. The elements of $\mathbf{s}$ are given by $[\mathbf{s}]_{nm} = s_{n-m}$.

It is now possible to see that the matrix formulation has the potential for describing the more-general Fredholm integral equation. This equation corresponds in spectroscopy to the situation where the functional form of $s(x)$ varies across the spectral range of interest. In these circumstances, s is expressed as a function of *two* independent variables. Although we proceed with the present treatment formulated in terms of convolutions, the reader should bear the generalization in mind.

Equation (15) reveals that, in the case of convolution, the matrix $\mathbf{s}$ has a certain property of symmetry. Each row is the same as the row above except that it is shifted one element to the right. Usually we choose the indices so that the largest element is on the diagonal. This shifting property may be expressed as

$$[s]_{jk} = [s]_{mn} \quad \text{if} \quad j - k = m - n. \tag{17}$$

Under these conditions, the spread function is termed shift invariant and **s** is called a Toeplitz matrix. When Eq. (17) does not hold, the spread function is called shift variant, and the spreading can no longer be described by simple convolution.

If the ends of the rows wrap around so that the leftmost element of each row is the rightmost element of the row above, and so on, the matrix is called circulant. The foregoing properties have significance when noniterative solutions to the matrix-posed problem are sought by discrete Fourier transform. Andrews and Hunt (1977) provide numerous details and references, especially for the case in which these linear methods are applied to the restoration of two-dimensional images.

B. Difficulties

For the present we turn our attention to a naive approach that proposes a direct solution to Eq. (16), which, after all, is only a set of linear equations in unknowns o_m. Solving this set, we should obtain the object estimate

$$\hat{\mathbf{o}} = [s]^{-1}\mathbf{i}. \tag{18}$$

Apparently, all we need to do is invert the matrix **s**.

One computational problem here is that for long runs of spectra the matrix **s** is very large, and we might feel the limitation of computer time and storage. However, this problem pales in comparison with the fundamental difficulty usually encountered with **s**. Typically each row in the matrix is similar to the one above. Indeed, these rows do operate upon neighboring points. Each equation, therefore, brings only a small amount of added information to the system. We do not have a set of robustly independent equations. If **s** has no inverse at all, it is said to be *singular*. At best, **s** is *ill conditioned*. In this case even the smallest noise perturbations in **i** cause large and possibly oscillatory fluctuations in the estimate $\hat{\mathbf{o}}$. When solution is attempted by certain noniterative methods such as Gaussian elimination, small differences are computed between similar large numbers. Errors, introduced by roundoff, for example, can become a serious problem. When the inverse can be computed, the solution $\hat{\mathbf{o}}$ typically exhibits large spurious fluctuations that originate in the noise that must always be present in any physical measurement.

Any linear deconvolution method that is to achieve even modest success must deal with these difficulties. Numerous aspects of the matrix formulation as applied to image restoration have been reviewed by Andrews and Hunt (1977), who devote spectial attention to the difficulties mentioned here.

C. Iterative Methods

An iterative approach has the advantage of allowing control of spurious fluctuations by interaction with the solution as it evolves. This may be done either automatically in the algorithm or by the exercise of human judgment. Although we reserve discussion of the most-powerful methods of interaction for the next chapter, it is appropriate here to set the stage by developing a family of iterative methods that have seen much use.

1. *Point Simultaneous*

It is well known that large systems of linear equations can be solved iteratively provided that the diagonal element is larger than any of its neighbors in the same row. Let us rewrite Eq. (15) with the diagonal element terms pulled outside:

$$i_n = [\mathbf{s}]_{nn}o_n + \sum_{n \neq m} [\mathbf{s}]_{nm}o_m. \tag{19}$$

We have returned, for the time being, to the general case of the shift-variant spread function. Solving for o_n, we obtain

$$o_n = \frac{1}{[\mathbf{s}]_{nn}} \{i_n - \sum_{n \neq m} [\mathbf{s}]_{nm}o_m\}, \tag{20}$$

so that we could exactly determine o_n if we knew all of its neighbors. Now suppose that by some means we had obtained estimates $\hat{o}_m^{(k)}$ of all its neighbors. We could then obtain a refined estimate of o_n,

$$\hat{o}_n^{(k+1)} = \frac{1}{[\mathbf{s}]_{nn}} \{i_n - \sum_{n \neq m} [\mathbf{s}]_{nm}\hat{o}_m^{(k)}\}, \tag{21}$$

and cycle through all the values of n until we have a complete set of the $\hat{o}_n^{(k+1)}$. It was appropriate that we chose originally to solve for o_n and not one of its neighbors, because o_n is the single object element that is weighted most heavily by the spreading process that yields i_n; that is, the diagonal element in each row of $\mathbf{s}$ is the largest.

Because it is also presumably available, why not include the old estimate of o_n as well? This eliminates the awkwardness of a summation with a missing element and allows us to express our new estimate in terms of the old estimate and a correction term:

$$\hat{o}_n^{(k+1)} = \hat{o}_n^{(k)} + \frac{1}{[s]_{nn}} \{ i_n - \sum_m [s]_{nm}\hat{o}_m^{(k)} \}. \tag{22}$$

Repeated application of this equation constitutes the point-simultaneous relaxation method, also known as the point Jacobi method or the method of simultaneous displacements.

There is no guarantee that the correction term is scaled correctly to ensure the most-rapid convergence. Furthermore, if it is scaled too large, the method might not converge at all. Accordingly, we introduce a relaxation factor κ, so that

$$\hat{o}_n^{(k+1)} = \hat{o}_n^{(k)} + \frac{\kappa}{[s]_{nn}} \{ i_n - \sum_m [s]_{nm}\hat{o}_m^{(k)} \}. \tag{23}$$

This is called point-simultaneous overrelaxation. If we set $\kappa = [s]_{nn}$, we have obtained the discrete formulation of Van Cittert's method. This connection between Van Cittert's method and the classic iterative methods of solving simultaneous equations was demonstrated in an earlier work (Jansson, 1968, 1970).

2. Point Successive

To compute a single point $\hat{o}_n^{(k+1)}$ in the estimate, we need the old estimate $\hat{o}_n^{(k)}$ at that point and a number of points to either side as well. Usually the $\hat{o}_n^{(k)}$ are computed in the order of increasing n. Why must we use old values for $\hat{o}_n$ in the summation for values of $m < n$ when more-recent estimates are available? The answer is that we need not. Accordingly, we break the summation into two parts:

$$\hat{o}_n^{(k+1)} = \hat{o}_n^{(k)} + \frac{\kappa}{[s]_{nn}} \{ i_n - \sum_{m<n} [s]_{nm}\hat{o}_m^{(k+1)} - \sum_{m \geq n} [s]_{nm}\hat{o}_m^{(k)} \}. \tag{24}$$

As we might expect, this iteration typically converges faster because of the continual use of the most-"recent" information. It is called point-successive overrelaxation. For $\kappa = 1$, it is known as the method of successive displacements or the point-successive relaxation or Gauss–Seidel method. Note, however, that whereas Eq. (23) could be applied with either ascending or descending n, yielding identical results either way, Eq. (24) can be applied only in the order of ascending n. It is a fundamentally asymmetrical method and may produce slightly asymmetrical spectral-line artifacts when employed in the deconvolution of spectra.

Would it be possible to make use of the most-"recent" information and still preserve the symmetry of the method? This author believes that it could be done by computing the new estimates $\hat{o}_n^{(k+1)}$, not in sequence of ascending n, but in a binary interleaved fashion. Such an interleaved sampling can be obtained by a bit-reversed scheme. In this scheme, the sampling positions are generated in the sequence that results when bits are reversed in each number of a simple ascending binary count. For example, the sequence 0, 1, 2, 3, 4, 5, 6, 7 generates 0, 4, 2, 6, 1, 5, 3, 7. Allebach (1981) describes the bit-reversed scheme developed by Deutsch (1965) and other schemes that may be useful in the present application. Other alternatives might be to compute new values in order of ascending n for n odd and to compute new values in order of descending n for n even. One could also use a Monte Carlo method of selecting n. These modifications might even result in more-rapid convergence.

With all of the point-successive methods, only one array in the computer is needed to store both $\hat{o}^{(k)}$ and $\hat{o}^{(k+1)}$ because the element $\hat{o}_n^{(k)}$ is never again needed once $\hat{o}_n^{(k+1)}$ is computed. This contrasts with the point-simultaneous methods, where, for each n, the elements of $\hat{o}^{(k)}$ must be available corresponding to the full range over which $[s]_{nm}$ is finite.

3. Memory-Saving Point Simultaneous

In a discrete convolution such as $a = b \otimes g$, where a and g are stored in similar-sized arrays, the ends of the a array are often not used. This occurs if a elements are computed only for those positions at which a full set of N nonvanishing g values are available, where N is the number of nonvanishing b values being used. It is thus possible to use only one array to store both a and g. One replaces g_1 with the first value of a obtained, continuing in an ascending sequence in the indices. When the operation is complete, the a result may be shifted back to the desired position. The values of g are lost by this method. The loss of g is of no consequence in some cases.

This technique is applicable in the case of Eqs. (22) and (23), where $\hat{o}^{(k)}$ values are not needed once they are outside the range of corresponding finite $[s]_{nm}$. Therefore, we see that, like the point-successive method, *point-simultaneous relaxation does not require more than one principal computer array for storing the solution estimate.* This fact is frequently not appreciated in the programming of these methods. Furthermore, when spectra are smoothed by polynomial convolution (see Section III.C.5), the same considerations apply, and only one array is needed to store the data before and after smoothing. If shifting the convolution product values back into position is judged undesirable, the array containing a and g, or $\hat{o}^{(k)}$ and $\hat{o}^{(k+1)}$, may be treated as a circular buffer (Knuth, 1973). Appropriate pointers are then adjusted when convolution is completed.

4. Convergence and Optimum Relaxation Factors

For all relaxation methods, theorems can be proved (Wendroff, 1966) that convergence is guaranteed provided that the diagonal of **s** is dominant:

$$[s]_{nn} > \sum_{n \neq m} |s_{nm}|. \tag{25}$$

In deconvolution problems, such dominance is rarely the case. It is a very modest resolution correction indeed when the response function is so narrow that its central element dominates all the others. Fortunately, the necessary conditions are far less severe, and convergence is usually not a serious problem. Equation (25) seems to tell us, however, that each new row in the matrix brings more independent information when $[s]_{nn}/\sum_{n \neq m} |[s]_{nm}|$ is as large as possible.

Kawata et al. (1979) and Kawata and Ichioka (1980a, 1980b) discuss the convergence of these methods from both linear algebra and Fourier transform points of view. They also introduce a reblurring method that guarantees convergence. We pursue this idea in Section IV.D.3.

The optimum value of the relaxation factor κ can be obtained from the largest eigenvalue of **s**. This eigenvalue may, in turn, be obtained approximately from the relative magnitude of vectors $\hat{o}^{(k)}$ and $\hat{o}^{(k+1)}$. Improvements discussed in the next section, and amplified in the next chapter, significantly alter the method. Trial-and-error choice of κ is therefore preferable and probably necessary.

5. Band Limits and Polynomial Filters

At this point, we note that there is no mechanism presently built into the relaxation methods to prevent undesirable high-frequency noise from growing with each iteration. Any spurious solution $\theta(x)$ satisfies Eq. (1) (see also Chapter 1, Sections V.A and V.B) for ω beyond the band limit. If we know that the object $\hat{o}$ is truly band limited, with frequency cutoff $\omega = \Omega$, we can band-limit both data i and first object estimate $\hat{o}^{(1)}$. The relaxation methods cannot then propagate noise having frequencies greater than Ω into an estimate $\hat{o}^{(k)}$. (One possible exception involves computer roundoff error. Sufficient precision is usually available to avoid this problem.)

If we know that the instrument response is band limiting with frequency cutoff Ω, we may likewise process the estimate. The resulting solution will then also be band limited. *The high-frequency spectral structure beyond cutoff Ω that we would wish to restore is forever lost to these linear methods. The data contain no information about the high-frequency content.* We must wait until Chapter 4 to see how straightforward and seemingly unimportant constraints carry sufficient information to allow restoration of these "lost" frequencies.

Band limiting of the data can be accomplished either in Fourier space or, to a useful approximation, directly by convolution with least-squares polynomial filters. Although these filters had been employed in previous numerical work (Whittaker and Robinson, 1967), Savitzky and Golay (1964) are responsible for their current widespread use in spectroscopy and have developed valuable differentiating methods based on them. Numerical errors in the Savitzky–Golay tables are corrected by Steinier *et al.* (1972). Willson and Edwards (1976) and Willson and Polo (1981) give excellent spectroscopy-oriented reviews of their properties, including frequency behavior. Porchet and Günthard (1970), Yamashita and Minami (1969), Bromba and Ziegler (1981), and others also discuss their properties. Jansson (1972) has extended their application to the two-dimensional continuous regime and suggested analogous two-dimensional discrete filters.

6. Relaxation Summary

We reemphasize that the foregoing relaxation equations containing the general shift-variant response-function element denoted by $[s]_{nm}$ are equally valid for the special case of convolution, whether discrete or continuous. Cast in the continuous notation for convolution, the relaxation methods are epitomized by the repeated application of

$$\hat{o}^{(k+1)}(x) = \hat{o}^{(k)}(x) + C[i(x) - s(x) \otimes \hat{o}^{(k)}(x)] \tag{26}$$

to obtain successive estimates of the object, given the observed data $i(x)$ and a spread function $s(x)$. The object is usually considered to be the spectrum as it would be observed by a hypothetical, perfectly resolving spectrometer. In that case $s(x)$ is the spectrometer response function. As an alternative, $o(x)$ could be the spectrometer response function, which we seek to determine from data $i(x)$ representing observations of a spectral-line of known shape. In the latter case, $s(x)$ represents the known true spectral line shape.

For successive overrelaxation, we understand Eq. (26) to incorporate the use of $\hat{o}^{(k+1)}$ values in place of $\hat{o}^{(k)}$ values in the convolution product as soon as they are formed for preceding x values. This adaptation can be explicitly displayed by the appropriate use of the Heaviside step function in a modified version of Eq. (26). The method of Van Cittert is a special case of simultaneous relaxation in which $C = 1$.

We have also learned that the point-successive methods need not demand more computer memory than point-simultaneous methods, and that the seemingly inherent asymmetry of the point-successive methods can be overcome. Furthermore, the linear methods described in this section are incapable of restoring object Fourier frequencies beyond the cutoff of the observing spectrometer.

We defer additional analysis of the relaxation method until we have properly introduced the concept of Fourier inverse filtering.

IV. INVERSE FILTERS

A. Basic Method

We might ask whether it is possible to create a linear filter function $y(x)$ that could undo, by simple convolution, the smearing in the data caused by the spectrometer. We express the behavior of such a filter by the equation

$$o(x) = y(x) \otimes i(x). \tag{27}$$

In the previous section, we used a matrix formulation to express such a filter. It is also possible to take a Fourier transform approach.

The equation that describes instrumental spreading,

$$i(x) = s(x) \otimes o(x), \tag{28}$$

may be transformed to the Fourier domain with the aid of the convolution theorem. Doing so, we obtain

$$I(\omega) = \tau(\omega)O(\omega). \tag{29}$$

At first glance, it would seem trivial to obtain the object transform

$$O(\omega) = \frac{1}{\tau(\omega)} I(\omega). \tag{30}$$

The transfer function corresponding to the filter in Eq. (27) would then be given simply by the inverse Fourier transform of

$$Y(\omega) = \frac{1}{\tau(\omega)}. \tag{31}$$

Typically, $\tau(\omega)$ is small for ω large. A spectrometer suppresses high frequencies. If the data $i(x)$ have appreciable noise content at those frequencies, it is certain that the restored object will show the noise in a more-pronounced way. It is clearly not possible to restore frequencies beyond the band limit Ω by this method when such a limit exists. (Optical spectrometers having sinc or sinc-squared response-function components do indeed band-limit the data.) Furthermore, where the frequencies are strongly suppressed, the signal-to-noise ratio is poor, and $Y(\omega)$ will amplify mainly the noise, thus producing a noisy and unusable object estimate.

We should, however, be able to carry out the processing within some band limit $|\omega| < \Omega_p$. Frieden (1975) has shown that the processing bandwidth

$$\Omega_p = \Omega[1 - (\phi_n/\phi_o)^{1/2}] \tag{32}$$

is optimum in the sense of minimum mean-square error

$$\left\langle \int |o(x) - \hat{o}(x)|^2 \, dx \right\rangle \tag{33}$$

when $\tau(\omega)$ is triangular (sinc-squared response function) with cutoff Ω and the ratio ϕ_n/ϕ_o is constant. In expression (33), $\hat{o}(x)$ is the object estimate produced by such filtering, and the angle brackets denote the ensemble average. The quantities ϕ_o and ϕ_n are the power† spectra of object and noise, respectively, given by the ensemble averages

$$\phi_o(\omega) = \langle |O(\omega)|^2 \rangle, \tag{34}$$

$$\phi_n(\omega) = \langle |N(\omega)|^2 \rangle. \tag{35}$$

Here we have introduced $N(\omega)$ as the Fourier transform of the noise $n(x)$. Frieden assumed a simple triangular form for $\tau(\omega)$ and constant noise-to-signal ratio for all frequencies. Goldman (1953) provides an appropriate discussion of ensemble averages and the like.

B. Weiner-Type Filter

It is suitable, at this point, to repair our formulation of the problem in a realistic way that accounts explicitly for the nature of the additive noise $n(x)$:

$$i(x) = s(x) \otimes o(x) + n(x). \tag{36}$$

When we solve numerically for $o(x)$, we obtain only an estimate $\hat{o}(x)$.

Faced with the problem of noise corruption alone $[s(x) = \delta(x)]$, Norbert Wiener (Goldman, 1953) devised his well-known *smoothing* filter

$$Y(\omega) = \frac{\phi_o(\omega)}{\phi_o(\omega) + \phi_n(\omega)}. \tag{37}$$

In deriving the Wiener smoothing filter, one asks what filter $y(x)$ results in the smallest mean-square error between the object estimate

$$\hat{o}(x) = y(x) \otimes [o(x) + n(x)] \tag{38}$$

---

† This term may be a misnomer when it is given dimensions of $|O(\omega)|^2$. The quantity $O(\omega)$ itself has dimensions of spectral power if, for example, it is considered to be photon flux. On the other hand, if we let $O(\omega)$ be the transform of the detector voltage, then the power designation is appropriate.

and the true object. Thus, we may obtain Eq. (37) by minimizing the ensemble average of the mean-square error,

$$\left\langle \int |y \otimes (o + n) - o|^2 \, dx \right\rangle. \tag{39}$$

Here the limits of integration are understood to include the entire spectrum,
and we use the angle brackets to denote the average over many spectra
acquired under similar conditions. For brevity, we have refrained from
explicitly showing the x dependence of o, n, and y.

This filter is not an inverse filter of the type that we seek, being intended
only for noise reduction. It does not undo any spreading introduced by
$s(x)$. It is, however, an optimum filter in the sense that no better linear filter
can be found for noise reduction alone, provided that we are restricted to
the knowledge that the noise is additive and Gaussian distributed.

How can this approach be adapted to deconvolution? The problem is
similar, but now we ask that $y(x)$ also incorporate the inverse of $s(x)$. Both
Bracewell (1958) and Helstrom (1967) have derived this variant of the
Wiener filter. Accordingly, we may minimize

$$\left\langle \int |y \otimes (s \otimes o + n) - o|^2 \, dx \right\rangle \tag{40}$$

and obtain the linear *inverse* filter

$$Y(\omega) = \frac{\tau^* \phi_o}{|\tau|^2 \phi_o + \phi_n}. \tag{41}$$

Here again, and frequently henceforth, we suppress from the notation the
dependence of the variables. Provided that we know only the noise and
object power spectra and that the noise is additive and Gaussian distributed,
no better linear filter can be found than this one.

It is true that the power spectra are not always easy to obtain and that
sometimes other types of information are available. Marmolin *et al.* (1978)
have optimized a restoring filter of this form for optical images, where
human observers were called upon to judge the visual quality of the
restorations.

C. Other Minimum Mean-Square-Error Variations

Frieden (1975) has considered a similar problem wherein he has added a
sharpness criterion

$$\left\langle \int_{-\infty}^{\infty} \left| \frac{d\hat{o}}{dx} \right|^2 \, dx \right\rangle = \left\langle \int_{-\Omega}^{\Omega} \omega^2 |\hat{O}(\omega)|^2 \, d\omega \right\rangle. \tag{42}$$

In this expression Ω is the cutoff frequency above which the data contain no information about $o(x)$; that is, $I(\omega) = 0$ for $|\omega| > \Omega$. We see that the sharpness criterion is a measure of the "steepness" of the solution $o(x)$. The previous criterion, expression (40), is replaced with a sum of two terms. It includes both the mean-square-error criterion and sharpness. The filter is then sought that minimizes

$$\left\langle \int |y \otimes (s \otimes o + n) - o|^2 \, dx \right\rangle + \beta \left\langle \left|\frac{d\hat{o}}{dx}\right|^2 \, dx \right\rangle \tag{43}$$

within the allowed bandpass. This results in Frieden's sharpness-constrained filter

$$Y(\omega) = \left(\frac{\tau^* \phi_o}{|\tau|^2 \phi_o + \phi_n}\right)\left(\frac{1}{1 + \beta \omega^2}\right). \tag{44}$$

It is similar to the previous filter but contains a sharpness-controlling factor that either enhances or suppresses the high frequencies, depending on whether $\beta < 0$ or $\beta > 0$.

Other modifications are possible to the same basic approach of seeking a filter that is optimum in the sense of least mean-square error. Backus and Gilbert (1970), for example, derive a linear filter by minimizing a sum of terms in which noise and resolution criteria are separately weighted. Frieden (1975) discusses variations of this technique.

A chief advantage of all the filters described in Section IV is convenience. We have written the Fourier transfer function of each one. Certainly it is possible to perform the filtering in the Fourier space. It is also possible, however, and often even more convenient, to convolve the data with the filter function itself. This is especially true if the filter can be adequately approximated by a convolution kernel that vanishes except over a relatively small domain.

D. Relationship to Van Cittert's Method

The entire discussion of relaxation methods was conducted without examining Fourier space consequences. Van Cittert's method is easy to study this way and has been treated by Burger and Van Cittert (1933), Bracewell and Roberts (1954), Sakai (1962), and Frieden (1975). By applying the convolution theorem to Eq. (14), we may write

$$\hat{O}^{(k+1)}(\omega) = \hat{O}^{(k)}(\omega)[1 - \tau(\omega)] + I(\omega). \tag{45}$$

Repeated application of this formula, starting with $\hat{O}^{(0)}(\omega) = I(\omega)$, results in the series

$$\hat{O}^{(k)} = I \sum_{n=0}^{k} (1 - \tau)^n. \tag{46}$$

We may use the identity

$$(1 - \zeta) \sum_{n=0}^{k} \zeta^n = 1 - \zeta^{k+1} \tag{47}$$

with $\zeta = 1 - \tau$ to obtain

$$\hat{O}^{(k)}(\omega) = \frac{I(\omega)}{\tau(\omega)} B(\omega), \tag{48}$$

where $B(\omega) = 1 - [1 - \tau(\omega)]^{k+1}$.

For $k \to \infty$, $B(\omega)$ approaches unity, provided that $|1 - \tau(\omega)| < 1$. In this case, $\hat{O}^{(k)}$ is just the object estimated by inverse filtering. For k finite, the inverse-filter estimate is modified by a factor that suppresses frequencies for which $\tau(\omega)$ is small. The larger k is, the less is this suppression. For typical transfer functions $\tau(\omega)$ that suppress high frequencies, the factor $B(\omega)$ controls the high-frequency content of $\hat{o}^{(k)}$. In the spectrum domain, it is also possible to derive simple expressions for filters $y(x)$ that are fully equivalent to an arbitrary number of relaxation iterations. Blass and Halsey (1981) have done so, but the highly useful nonlinear modifications of these methods cannot be incorporated.

1. Noise and Lost Information

With the aid of Eq. (48), we can show that $\hat{O}^{(k)}(\omega) = (k + 1)N(\omega)$ for $\tau(\omega) = 0$. The object estimate consists of noise at frequencies that τ does not pass. The noise grows with each iteration. This problem can be alleviated if we bandpass-filter the data to the known extent of τ to reject frequencies that τ is incapable of transmitting. Practical applications of relaxation methods typically employ such filtering. Least-squares polynomial filters, applied by finite discrete convolution, approximate the desired characteristics (Section III.C.5). For k finite and $\tau \neq 0$, but nevertheless small,

$$\hat{O}^{(k)}(\omega) = (k + 1)I(\omega). \tag{49}$$

We thus boost the strongly suppressed frequencies in a way that grows linearly with iteration.

Let us not forget, at this point, that well-designed bandpass filtering can only prevent the appearance of solution frequencies totally rejected by $\tau(\omega)$. Noise in the frequency range that we wish to restore cannot be thus rejected, however. In the limit of simple inverse filtering we find that

$$\hat{O}(\omega) = [1/\tau(\omega)][I(\omega) - N(\omega)] = O(\omega) - N(\omega)/\tau(\omega). \tag{50}$$

Noise is a serious problem where $\tau(\omega)$ is small. We see that the estimate $\hat{O}(\omega)$ carries with it none of the sensible noise treatment that modified filters like those of the Wiener type provide. We may, however, supply the required noise suppression between iterations by smoothing. The polynomial filters described in Section III.C.5, for example, may be used. Very little has been done, however, to determine how suppression might be accomplished optimally. Usually it is treated in an ad hoc way.

Where $\tau(\omega) = 0$—beyond the band limit Ω, for example—$I(\omega)$ contains no information about $O(\omega)$. Clearly it is impossible to restore these lost frequencies based on $\tau(\omega)$ and the information in $I(\omega)$ alone. We see in the next chapter how a simple modification to the various relaxation methods brings about a dramatic improvement.

2. Relaxation Factors and Convergence

We saw earlier that Van Cittert's method converges only for frequencies at which $|1 - \tau(\omega)| < 1$. This restriction is alleviated in the case of point-simultaneous overrelaxation. Referring to Eqs. (14) and (26), we see that we may deal with the case where $C \neq 1$ by simply replacing $I(\omega)$ and $\tau(\omega)$ with $CI(\omega)$ and $C\tau(\omega)$ in the analysis of the Van Cittert method. Doing this, we obtain the new requirement that $|1 - C\tau(\omega)| < 1$ or $0 < C\tau(\omega) < 2$, for $\tau(\omega)$ real.

We may also rederive $B(\omega)$ for the case of overrelaxation, with a choice of $\hat{O}^{(0)}(\omega) = I(\omega)$, as Kawata *et al.* (1979) and Kawata and Ichioka (1980a) have done:

$$B(\omega) = 1 - [1 - \tau(\omega)][1 - C\tau(\omega)]^k. \tag{51}$$

Alternatively, we point out that the choice of $\hat{O}^{(0)}(\omega) = CI(\omega)$ is possible, and we may obtain the somewhat simpler

$$B(\omega) = 1 - [1 - C\tau(\omega)]^{(k+1)}. \tag{52}$$

When $\tau(\omega)$ is real, it is easy to adjust C so that we satisfy the required $C\tau(\omega) < 2$ for the largest $\tau(\omega)$, but what can we do if $\tau(\omega)$ has negative values? Being applied in the data space, the constant C is not a function of ω. We must find a single C that works. Yet, certain spreading phenomena *do* give rise to negative $\tau(\omega)$. The simple moving average is an example. It is equivalent to convolution with rect(x), giving $\tau(\omega) = (1/\sqrt{2\pi})$ sinc($\omega/2\pi$). Both coherent and incoherent optical response functions for spectrometers having *unequal* entrance- and exit-slit spectral widths present a similar situation. These cases may be analyzed by the methods of Section II of Chapter 2. In the next section we introduce a method of accommodating phenomena that give rise to $\tau(\omega)$ having negative values.

3. Reblurring and Smoothing

Kawata *et al.* (1979) introduced a concept of reblurring to guarantee that the transfer function does not have negative values. They recognized that the convolution of the spread function with a reversed version of itself yields, in Fourier space, a function that cannot have negative values:

$$s(x) \otimes s(-x) \leftrightarrow |\tau(\omega)|^2. \tag{53}$$

This may be easily proved by applying the definition of the Fourier transform. It is a version of the autocorrelation theorem. The authors elaborate in a later paper (Kawata and Ichioka, 1980b).

It is thus possible to convolve both spread function and data $i(x)$ with $s(-x)$. We may then use the relaxation methods as before. This time, however, we replace $i(x)$ with $s(-x) \otimes i(x)$ and $s(x)$ with $s(-x) \otimes s(x)$. Not only are we assured convergence, but we have also succeeded in band-limiting the data $i(x)$ in such a way as to guarantee that all noise is removed from $i(x)$ at frequencies where $i(x)$ contains no information about $o(x)$. Furthermore, Ichioka and Nakajima (1981) have shown that reblurring reduces noise in the sense of minimum mean-square error.

The analysis in this chapter is not valid when constraints and/or smoothing between iterations are employed as described in Chapter 4. These valuable techniques complicate the situation considerably. We may, however, employ the analysis already given as a guide. In spite of attendant complications, the pragmatic truth is that these techniques, especially constraints, are well worth the effort. They produce better resolution and show reduced sensitivity to both noise in $i(x)$ and errors in $s(x)$. Results achieved are sometimes little short of astonishing when judged by the standards of the traditional linear methods described in this chapter.

E. Stepwise Implementation

We have noted the noise-sensitivity problem of the simple inverse filter and introduced modifications to alleviate these difficulties. Modifications yielded different functional forms for $Y(\omega)$. The convenient single-step property of the basic method was nevertheless retained. This property contrasts with the need for possibly arbitrary stopping criteria when we use iterative methods, which are computationally more expensive. The iterative methods do, however, allow the user to control the signal-to-noise versus resolution tradeoff by stopping the process when the growth of spurious components begins to negate the benefit of further resolution improvement. Unfortunately, the result (Section IV.D) is not usually equivalent to optimum tailoring of $Y(\omega)$. Would it be possible to achieve the linear inverse-filtered result in a preset fixed number of stages? If so, the users could obtain a true inverse-filtered result when justified by high data quality, but otherwise they could stop short of that objective, achieving smoother results with somewhat compromised resolution.

Rendina and Larson (1975)† satisfied these requirements by employing the N_sth root of $Y(\omega)$ as the transfer function of a component filter $Y_s(\omega)$. Application of the component filter N_s times yields the result obtained by a single application of $Y(\omega)$. Specifically, the component filter is given by

$$Y_s(\omega) = [Y(\omega)]^{1/N_s}. \tag{54}$$

By converting to polar coordinates we obtain

$$Y_s = |Y|^{1/N_s} \exp[(j/N_s)\arctan(\text{Im } Y/\text{Re } Y)], \tag{55}$$

which is suitable for computation. In Eq. (55), Re and Im denote the real and imaginary parts of their arguments, respectively. The component filter may be applied either as written in Eq. (55) by multiplication in the Fourier domain or by convolution with the use of $y_s(x)$, the Fourier transform of $Y_s(\omega)$. In both cases, the x-domain result may be inspected after each pass.

The composite filter $Y(\omega)$ may either be the true inverse filter, truncated for ω large if necessary, or any of the variations described in Section IV. In their original work, Rendina and Larson chose $Y(\omega) = \psi(\omega)/\tau(\omega)$, where $\psi(\omega)$ is a Gaussian line-broadening function that limits the ultimate resolution obtainable but yields a manageable $Y(\omega)$. For their studies Rendina and Larson used $N_s = 4$.

V. OTHER LINEAR METHODS

A large number of linear methods have been developed with particular characteristics that tend to suit them to specific deconvolution problems. None of these adaptations shows beneficial results nearly so profound as those resulting from the imposition of the physical-realizability constraints discussed in the next chapter. Furthermore, the present work is not intended

---

† Abstract only; complete text obtained from authors in 1975.

as an exhaustive review, so we only mention only a few additional linear methods, because of their popularity, the insight conveyed, or their adaptability to constraints.

The demand that the solution $\hat{o}$ be consistent with the data i results in the improved resolution that we expect from a deconvolution method. As we have explained, however, it also results in the amplification of high-frequency noise. The smoothing of this noise to some extent defeats the purpose of deconvolution. The tradeoff between smoothness and consistency is explicit in the formulation of a method first described by Phillips (1962) and further developed by Twomey (1965). In this method, we minimize the quantity

$$\beta \sum_{n=1}^{N} (\hat{o}_{n+1} - 2\hat{o}_n + \hat{o}_{n-1})^2 + \sum_{m=1}^{M} \left\{ i_m - \sum_{n=1}^{N} [\mathbf{s}]_{mn}\hat{o}_n \right\}^2. \tag{56}$$

The first term governs smoothness through the second differences of $\hat{o}_n$. The second term imposes the consistency between the solution values $\hat{o}_n$ and the data values i_m. The tradeoff is controlled by varying parameter β. Frieden (1975) explores the method briefly. Hunt (1973) applies it to images in a computationally efficient way that uses the fast Fourier transform.

Confronted with a problem in which two data sets were available, Breedlove et al. (1977) chose a solution that minimizes a sum of terms not unlike expression (56). Available were two images: one a blurred representation of the object, the other a superposition of sharp renderings. In this sum, the right-hand term accommodates the blurred image as in expression (56). The other term incorporates the multiple exposure via the Lagrange multiplier technique. Solutions obtained by this method illustrated the desirability of using all the available data.

Huang et al. (1975) and Huang (1975) described a method based on iteratively correcting estimate $\hat{o}$ by adding a term that involves a hyperplane projection operation. Like the relaxation methods discussed in Sections II and III, it has the potential to be upgraded by the implementation of constraints.

The choice of sampling rate is sometimes governed by a tradeoff between frequency aliasing (low rate) and noise sensitivity (high rate). Quick and Bolgiano (1976) have employed the Poisson transformation to avert this complication. Consult Piovoso and Bolgiano (1970) for additional background on the Poisson transform.

Matrix methods such as singular value decomposition and others based on pseudoinverses have been described in the digital image processing literature. A book chapter by Andrews (1975) and texts by Pratt (1978), Andrews and Hunt (1977), and Hall (1979) treat this subject and give further references. These works also contain abundant references and considerable detail regarding other linear deconvolution methods and their variations as applied to the computer restoration of images.

VI. OVERVIEW OF LINEAR METHODS

All of the methods described in this chapter are linear methods because each element of $\hat{o}(x)$ can be obtained by a linear combination of the elements of $i(x)$. In the continuous regime, a linear estimate $\hat{o}(x)$ can always be expressed by the integral

$$\hat{o}(x) = \int y(x, x')i(x') \, dx', \tag{57}$$

where $y(x, x')$ is a filtering kernel, shift variant in general. Sometimes this formulation is not immediately apparent; many linear methods are well disguised by mathematical foliage (Wells, 1980). However well obscured, a linear method is a linear method, and it is severely limited in its capability relative to modern constrained nonlinear methods. Sometimes, however, the researcher does not have the a priori information required for application of the nonlinear methods, or this information may not be applicable. Under these conditions linear methods are useful. In addition, there are other occasions where linear methods are preferable. Included are situations where computational speed is of paramount importance, as in real-time deconvolution, and where linear filtering hardware is available, whether analog or digital. Performance can be satisfactory if the spectrometer transfer function allows passage of very high frequencies, even though they may be somewhat suppressed. The Wiener-type filter may be a good choice if one *must* use a linear restoring method because it deals explicitly with the problem of noise and with achieving a solution that is, at least in a limited sense, optimum.

In spectroscopy, however, it is often the case that solutions having negative values are not physically possible. On occasion, an upper bound may also be imposed. These highly valuable adaptations take the discussion into the realm of nonlinearity and are deferred until the next chapter.

Nevertheless, certain types of prior knowledge can be introduced within the context of a linear method. Probabilities, signal and noise statistics, power spectra, and the like may be incorporated. Often this type of prior knowledge is difficult to obtain. In any case, it rarely exerts an influence nearly so profound as that of simple bounds on the amplitude of the solution. If the observing spread function obliterates all frequencies beyond the cutoff Ω, they are forever lost to the linear restoration methods. No linear filter's amplification factor can ever make anything but a zero out of the zero that the observing instrument has left us. All it can do is amplify noise at these frequencies.

It is a good rule to use all the prior knowledge in which one has high confidence. If, for example, it is known that an unresolved group of spectral lines has exactly four components and that one of the components is narrower than Δx, this knowledge may be employed for curve fitting in the manner covered in Section V.C of Chapter 1. Once these assumptions are made, however, an unexpected fifth spectral line may be lost. As stated earlier, the curve-fitting procedures are not true deconvolution methods, although they can be valuable under certain circumstances.

All deconvolution methods involve an implicit tradeoff of signal-to-noise ratio for resolution. Even the optical spectrometer itself exhibits such a tradeoff when slit width is varied. When slits are made narrow to boost resolution, much of the dispersed photon flux (and its information content) is lost to the detection system. A different situation is found in the use of multiplex spectrometry. In some spectral regions, photographic plates or multiple detectors may be employed. Multiple-slit spectrometers with various slit encoding schemes and single detectors can also be used (Stewart, 1970). Furthermore, it is possible to omit the disperser entirely and use a Fourier transform spectrometer. Would it be possible to defeat the conventional spectrometer tradeoff by running with wider slits and then deconvolving? This is indeed possible; orders-of-magnitude improvement in the productivity of a conventional spectrometer can thus be achieved. Blass and Halsey (1981) address this question.

If at the outset the data are very noisy and if the noise predominates in the Fourier frequency range needed to effect a restoration, constraints provide the only hope for improvement. The reason is that many of the noise values in the data would "restore" to physically unrealizable values by linear deconvolution. The constrained methods are inherently more robust because they must find a solution that is consistent with both data *and* physical reality.

As a final note, we wish to caution the reader concerning evaluation of the literature. Many authors display the results that their methods produce when tested with noise-free, computer-simulated "data." Almost all linear methods will produce outstanding results in these circumstances. In particular, virtually perfect restorations are easy to obtain when these "data" are not band limited. The only uncertainty troubling such restorations is the round-off noise in the computer. If the computer utilizes eight significant figures, the signal-to-noise ratio is on the order of 10^8. What integration time must the spectroscopist employ to achieve this phenomenal data quality?

REFERENCES

Allebach, J. P. (1981). *J. Opt. Soc. Am.* **71,** 99-105.
Andrews, H. C. (1975). *In* "Top. Appl. Phys.: Picture Processing and Digital Filtering," Vol. 6 (T. S. Huang, ed.), pp. 21-68. Springer-Verlag, New York.
Andrews, H. C., and Hunt, B. R. (1977). "Digital Image Restoration." Prentice-Hall, Englewood Cliffs, New Jersey.
Backus, G., and Gilbert, F. (1970). *Philos. Trans. R. Soc. London,* **A266,** 123-192.
Blass, W. E., and Halsey, G. W. (1981). "Deconvolution of Absorption Spectra." Academic Press, New York.
Bracewell, R. N. (1958). *Proc. Inst. Radio Eng.* **46,** 106-111.
Bracewell, R. N., and Roberts, J. A. (1954). *Aust. J. Phys.* **7,** 615-640.
Breedlove, J. R., Kruger, R. P., Trussell, H. J., and Hunt, B. R. (1977). *SPIE* **119,** 258-263.
Bromba, M. U. A., and Ziegler, H. (1981). *Anal. Chem.* **53,** 1583-1586.
Burger, H. C., and van Cittert, P. H. (1933). *Z. Phys.* **81,** 428-434.
Deutsch, S. (1965). *IEEE Trans. Broadcast.* **11,** 11-21.
DiCola, G., Rota, A., and Bertolini, G. (1967). *IEEE Trans. Nucl. Sci.* **14,** 640-653.
Frieden, B. R. (1975). *In* "Top. Appl. Phys.: Picture Processing and Digital Filtering," Vol. 6 (T. S. Huang, ed.), pp. 177-248. Springer-Verlag, New York.
Goldman, S. (1953). "Information Theory." Dover, New York.
Hall, E. L. (1979). "Computer Image Processing and Recognition." Academic Press, New York.
Helstrom, C. W. (1967). *J. Opt. Soc. Am.* **57,** 297-303.
Herget, W. F., Deeds, W. E., Gailar, N. M., Lovell, R. J., and Nielsen, A. H. (1962). *J. Opt. Soc. Am.* **52,** 1113-1119.
Hill, N. R., and Ioup, G. E. (1976). *J. Opt. Soc. Am.* **66,** 487-489.
Huang, T. S., ed. (1975). "Top. Appl. Phys.: Picture Processing and Digital Filtering," Vol. 6. Springer-Verlag, New York.
Huang, T. S., Barker, D. A., and Berger, S. P. (1975). *Appl. Opt.* **14,** 1165-1168.
Hunt, B. R. (1973). *IEEE Trans. Comput.* **C-22,** 805-812.
Ichioka, Y., and Nakajima, N. (1981). *J. Opt. Soc. Am.* **71,** 983-988.
Jansson, P. A. (1968). Ph.D. Dissertation, Florida State Univ., Tallahassee.
Jansson, P. A. (1970). *J. Opt. Soc. Am.* **60,** 184-191.
Jansson, P. A. (1972). *J. Opt. Soc. Am.* **62,** 195-198.
Kawata, S., and Ichioka, Y. (1980a). *J. Opt. Soc. Am.* **70,** 762-768.
Kawata, S., and Ichioka, Y. (1980b). *J. Opt. Soc. Am.* **70,** 768-772.
Kawata, S., Ichioka, Y., and Suzuki, T. (1979). *Proc. of the 4th Int. Joint Con. on Pattern Recognition,* 525-529.
Knuth, D. E. (1973). "The Art of Computer Programming," Vol. 1, Fundamental Algorithms, 2nd Ed. Addison-Wesley, Reading, Massachusetts.
Marmolin, H., Nyberg, S., and Berggrund, U. (1978). *Photogr. Sci. Eng.* **22,** 142-147.
Phillips, D. L. (1962). *J. Assoc. Comput. Mach.* **9,** 84-97.
Piovoso, M. J., and Bolgiano, L. P. Jr. (1970). *In* Proc. Symposium Computer Processing in Communications, New York, April 8-10, 1969 (J. Fox, ed.). Polytechnic Press, Brooklyn, N.Y.
Porchet, J. P., and Günthard, H. H. (1970). *J. Phys.* **E3,** 261-264.
Pratt, W. K. (1978). "Digital Image Processing." Wiley, New York.
Quick, L. T., and Bolgiano, L. P., Jr. (1976). Proc. Int. Conf. Acoustics, Speech and Signal Processing, Philadelphia, Pa., IEEE. pp. 350-353.
Rendina, J., and Larson, P. (1975). Paper Number 66, Pittsburgh Conference on Analytical

Chemistry and Applied Spectroscopy, Cleveland Convention Center, Cleveland, Ohio, March 3–7 (abstract only).

Sakai, H. (1962). "A Slit Function Correction and an Application to the Study of the Absorption Lines in the H_2O Pure Rotation Spectrum." U. S. Armed Services Technical Information Agency Report AD287897.

Savitzky, A., and Golay, M. J. E. (1964). *Anal. Chem.* **36,** 1627–1639.

Steinier, J., Termonia, Y., and Deltour, J. (1972). *Anal. Chem.* **44,** 1906–1909.

Stewart, J. E. (1970). "Infrared Spectroscopy Experimental Methods and Techniques." Dekker, New York.

Twomey, S. (1965). *J. Franklin Inst.* **279,** 95–109.

Van Cittert, P. H. (1931). *Z. Phys.* **69,** 298–308.

Wells, D. C. (1980). *SPIE* **264,** 148–154.

Wendroff, B. (1966). "Theoretical Numerical Analysis." Academic Press, New York.

Whittaker, E., and Robinson, G. (1967). "The Calculus of Observations: An Introduction to Numerical Analysis," 4th ed. Dover, New York.

Willson, P. D., and Edwards, T. H. (1976). *Appl. Spectrosc. Rev.* **12,** 1–81.

Willson, P. D., and Polo, S. R. (1981). *J. Opt. Soc. Am.* **71,** 599–603.

Yamashita, K., and Minami, S. (1969). *Jpn. J. Appl. Phys.* **8,** 1505–1512.

CHAPTER 4

Modern Constrained Nonlinear Methods

Peter A. Jansson

Engineering Physics Laboratory, E. I. du Pont de Nemours and Company (Inc.)
Wilmington, Delaware

I.	Introduction	96
II.	Meaning of Constraints	97
III.	The Promise of Analytic Continuation	97
IV.	Early Constrained Methods	99
	A. Gold's Ratio Method	99
	B. Schell's Method	101
V.	Greater Realization of Constraint Benefits	102
	A. Jansson's Method	102
	B. Biraud's Method	111
	C. Frieden's Method of Maximum Likelihood	115
	D. Alternating Projection Methods	121
	E. Howard's Methods	123
	F. Monte Carlo Methods	125
VI.	Other Methods	128
VII.	Concluding Remarks	130
	A. Cautions	130
	B. The Future	131
	References	132

LIST OF SYMBOLS

$a(x)$	real Fourier series such that $\hat{o}^+(x) = [a(x)]^2$
A, B	lower and upper bounds to $\hat{o}$, respectively
$A(\omega)$	Fourier transform of $a(x)$
b	constraint operator
$B_W(\omega)$	$\tau(\omega)Y_W(\omega)$, factor by which simple inverse filter must be modified to give Wiener version
C	constant
E	mean-square error within specified frequency limits
$\hat{\varepsilon}$	maximum-likelihood solution for noise
E_{res}	residual squared error due to noise alone
$g(x), g$	component of $\hat{o}^+(x)$ such that $\hat{o}^+(x) = g^*(x)g(x)$
$G(\omega), G$	Fourier transform of $g(x)$
h	Planck's constant
H	Shannon-type entropy $-\sum n_m \ln n_m$
H_1	Burg-type entropy $\sum \ln n_m$

93

DECONVOLUTION:
WITH APPLICATIONS IN SPECTROSCOPY

$H(\ )$	Heaviside step function
$\mathbf{i}, \mathbf{o}, \hat{\mathbf{o}}$	column vectors containing elements $i_m, o_m, \hat{o}_m$
$\otimes$	denotes convolution operation
$i_m, o_m, \hat{o}_m, \hat{o}_m^{(k)}$	discretely sampled versions of $i, o, \hat{o}, \hat{o}^{(k)}$
$\hat{\imath}^{(t)}$	$s \otimes \hat{o}^{(t)}$, "estimate" of the data
$i(x), i$	"image" data that incorporate smearing by $s(x)$
$\hat{\imath}(x)^{(k)}, \hat{\imath}^{(k)}$	kth "estimate" of $i(x)$ based on $\hat{o}^{(k)}$
$I(\omega), \tau(\omega), O(\omega), \hat{O}(\omega)$	Fourier transforms of $i(x), s(x), o(x), \hat{o}(x)$
k	Boltzmann constant
L	number of nonlinear filtering stages; index of limit of sought $A(\omega_l)$
M	index of maximum s_m; measure of number of samples of $\hat{O}^+(\omega)$ over range $\pm\Omega_c$ such that $2M + 1$ samples cover the range; number of cells or elements constituting object o_m
n	integer, usually an exponent or index
n_m	o_m/hv_m, number of photons counted in each object element
$\hat{n}_m$	estimate of n_m
N	$\sum_{m=1}^{M} \mathrm{n}_m$, total number of photons in object
$N_o(\omega)$	Fourier transform of noise
$o(x), o$	"object" or function sought by deconvolution, usually the true spectrum, but also the instrument function when this is sought by deconvolution
$\hat{o}(x)$	estimate of $o(x)$
$\hat{o}^{(k)}(x), \hat{o}^{(k)}$	kth estimate of $o(x)$
$\hat{o}^+(x), \hat{o}^+$	nonnegative estimate of $o(x)$
$\hat{o}_e(x), \hat{O}(\omega)$	bandwidth-extrapolating part of $\hat{o}_p(x)$ and its Fourier transform
$\hat{o}_p(x)$	physically acceptable estimate of $o(x)$
$p(\mathrm{q}_1, \ldots, \mathrm{q}_M)$	probability of given set of q_m
$p_c(q_m)$	one of M identical components of $p(q_1, \ldots, q_m)$
P	measure of number of samples required to specify $A(\omega)$ for $\hat{O}^+(\omega)$ over range $\pm\Omega_c$ $(P = M/2)$
$P(n_1, \ldots, n_M)$	probability of given number-count set $\{n_m\}$
q'	particular value that $q(x)$ may assume
q_0	peak height of Doppler-broadened emission line
q_1	height of Doppler emission line at distance Δx from center
q_m	relative probability that a photon occupies cell m of the object
$q(x)$	continuous form of prior spectrum
Q	constant value of all elements of flat prior spectrum
Q_m	values of prior spectrum in which user has highest possible conviction
r_0	constant in relaxation function
$r[\hat{o}], r$	relaxation function that weights correction terms—a function of $\hat{o}$
$r_x^{(k)}(x)$	$r[\hat{o}^{(k)}(x)]$, relaxation function as function of x
$R^{(k)}(\omega)$	Fourier transform of $r^{(k)}(x)$

$\mathbf{s}$	matrix containing elements $[\mathbf{s}]_{nm} = s_{nm}$
s_m	sampled value of shift-invariant spread function
s_{nm}	general spread-function matrix element; for convolution, sampled version of function $s(x_n - x_m)$, where x_n and x_m are sampled values of x
$s(x), s$	spread function, usually the instrument function, but also spreading due to other causes
t	sequential number of grain
T	absolute temperature
W_m	Bose–Einstein degeneracy factor
x	independent variable, typically wave number (cm^{-1}), wavelength, or spatial coordinate
x_n, x_m	discrete sampled value of x
x_p	location at which trial grain is added
X	limit of object extent such that $|x| \leq X$
$y(x)$	linear restoring filter such that $\hat{o}(x) = y(x) \otimes i(x)$
$y_W(x), Y_W(\omega)$	linear restoring filter of Wiener type and its Fourier transform
$Y(\omega)$	Fourier transform of $y(x)$
z_m	normal modes or degrees of freedom available for occupation by photons of frequency v_m
δ_{mp}	Kronecker delta
$\Delta\hat{o}$	object estimate increment due to added grain
Δx	distance from center of model line, defining region over which probability will be computed
$\Delta\omega$	sample spacing required to specify object completely
κ_0	constant in discrete formulation of relaxation function
$\kappa[\hat{i}^{(k)}]$	relaxation function in discrete formulation
μ, λ_m	Lagrange multipliers
v_m	photon frequency in hertz
ρ	factor affecting grain allocation
σ^2	variance of additive noise
Φ	objective function to be minimized
Φ_T	equilibrium value of function Φ at temperature T
$\Phi[\hat{o}(x_1), \ldots, \hat{o}(x_M)]$	objective function with object estimate elements as parameters
ω	conjugate of x; Fourier frequency in radians per units of x
Ω	cutoff frequency such that $\tau(\omega) = 0$ for $|\omega| > \Omega$
Ω_c	limit of ω such that restoration $\hat{O}^+(\omega)$ is required to be consistent with $I(\omega)$ within $\pm\Omega_c$
Ω_H	upper limit of frequencies $|\omega|$ used for input in Howard's method
Ω_p	limit of extended Fourier bandwidth required of $\hat{o}^+$
$\star$	correlation operation

I. INTRODUCTION

In preceding chapters we laid a foundation for the study of deconvolution. We presented several linear methods that exemplify the groundwork available before recent developments revolutionized the deconvolution field. Why, in their simplicity and elegance, did the linear methods fail to stimulate the wide adoption of deconvolution methods in spectroscopy?† After all, available instrumental resolution is limiting in many applications, and the simplicity of the microcomputer makes numerical processing attractive.

The answer lies in the relatively poor performance of linear methods, especially with band-limited data. Frequently a linear restoration reveals little true structure that could not have been seen in the original data. Even worse, noise-based artifacts often call the result into question. One might even say that the linear methods have helped to give deconvolution a bad reputation in spectroscopy.

To be sure, linear methods have value where fast computation is necessary. They perform reasonably well when the experimental data are not band limited, and in trials with computer-generated "data" devoid of noise. Spectroscopic data are often band limited, however, and computation time is becoming less of a problem with advances in computer hardware. The quantity of data required in spectroscopy is far less than that in image processing, for example, another field that has given much attention to deconvolution. Image processing problems are two and sometimes three-dimensional, whereas spectral problems are usually one dimensional.

Knowing that the better nonlinear constrained methods are now available, why have researchers generally been reluctant to accept them? Perhaps the linear approach has an attraction that is not related to performance. Early in a technical career the scientist–engineer is indoctrinated with the principles of linear superposition and analysis. Indeed, a rather large body of knowledge is based on linear methods. The trap that the linear methods lay for us is the existence of a beautiful and complete formalism developed over the years. Why complicate it by requiring the solution to be physically possible?

Perhaps the benefits of physical-realizability constraints, particularly ordinate bounds such as positivity, have not been sufficiently recognized. Surely everyone agrees in principle that such constraints are desirable. Even the early literature on this subject frequently mentions their potential advantages. For one reason or another, however, the earliest nonlinear constrained methods did not fully reveal the inherent power of constraints.

---

† This is not to say that all linear methods are simple. On the contrary, the linearity of some methods is remarkably obscure.

In the following pages, we trace the development of the nonlinear methods and two concurrent themes that underlie this work: first the physical-realizability theme already mentioned, and then the concept that one should be able to restore the Fourier frequencies obliterated by the finite Fourier bandpass of the spectrometer, that is, frequencies not present in the data. The extent to which these two themes are closely coupled was not fully appreciated in early work.

II. MEANING OF CONSTRAINTS

Throughout this chapter, and indeed the entire volume, we often use the word "constraint" to indicate bounds placed on the solution. We frequently mean positivity; that is, the solution $\hat{o}(x)$ cannot have negative values. This constraint is the must powerful. It is usually the easiest of the amplitude bounds to implement and is sometimes inherent in the way the solution is represented. We also consider spatial bounds; the solution is known to vanish over certain values of the independent variable x, for example, outside a single region within well-defined limits. In practice, this type of constraint exerts a far weaker influence on the solution than the amplitude bound.

Some authors consider nonlinearities in the processing of data values, that is, photometric nonlinearities. They term the resulting restoration nonlinear. Here we assume that such effects are either not present or have been corrected. We reserve the term "nonlinear" to describe the situation in which the solution $\hat{o}(x)$ cannot be expressed as a linear function of the irradiance data $i(x)$. For situations where it is proper to introduce photometric nonlinearity, we refer the reader to Hunt (1975, 1977) and Andrews and Hunt (1977).

III. THE PROMISE OF ANALYTIC CONTINUATION

Sometimes the spectrometer completely obliterates the information at all Fourier frequencies ω beyond some finite cutoff Ω. This is specifically true of dispersive optical spectrometers, where the aperture determines Ω. The cutoff Ω may be extended to high Fourier frequencies by multipassing the dispersive element or employing the high orders from a diffraction grating.

Whatever the method, the cutoff is finite and absolute. Similar considerations prevail in the Fourier interferometer, where the maximum path difference determines Ω. It would seem foolish to suggest that the information at these frequencies could be restored when it is not even present in the data.

Recall that inverse filtering (Chapter 3) produces the estimate $\hat{O}(\omega) = I(\omega)/\tau(\omega)$, which is undefined for $\tau(\omega) \geq \Omega$. Indeed, we cannot ever completely know $O(\omega)$ if we are restricted to using the information in $I(\omega)$ alone. If we add the modest bit of knowledge that the spectrum $o(x)$ is finite in extent, however, the lost frequency components no longer appear irretrievable (Wolter, 1961). The reasoning goes as follows. It is known (Guillemin, 1949) that the Fourier transform $O(\omega)$ of a function $o(x)$ having finite extent is analytic for all ω. If $O(\omega)$ is known only over the interval $-\Omega$ to Ω, but is known with absolute accuracy, then it is known for all ω. This may be clarified if we note that an analytic function has a unique Taylor-series representation. Once its coefficients are specified for $\omega < |\Omega|$, they are valid for all ω. The reconstruction of the lost information may be done by the Taylor-series route (Frieden, 1975) or other means (Harris, 1964; Goodman, 1968).

Attempts to implement the technique have not met with much success. The difficulty lies in the knowledge available about $O(\omega)$ for $|\omega| < \Omega$. Like all quantities derived by measurement, the estimate $\hat{O}(\omega)$ for $|\omega| < \Omega$ contains error. In most practical cases, $\tau(\omega)$ is small for ω close to Ω. Consequently, $\hat{O}(\omega)$ is most uncertain where its influence over the region to be restored is greatest. The method is essentially one of extrapolation and is subject to the difficulties that beset all extrapolation procedures when they are applied to uncertain data.

If, on the other hand, $O(\omega)$ were known to high precision over some neighborhood near the cutoff Ω, then its derivatives would also be known. Because an analytic function is uniquely specified for all ω by knowledge of all its derivatives at a single ω, a modest extrapolation should be possible for $|\omega|$ not too much larger than Ω. Unfortunately, dispersive optical spectrometers severely degrade the signal-to-noise ratio near the cutoff Ω. Unapodized Fourier interferograms, however, do not show this fault.

The constraint of finite extent has in practice proved to be of only limited value for improving the quality of restorations, whether the process is one of analytic continuation or otherwise. This is borne out, for example, by the work of Howard (Chapter 9). We have now presented the idea of recovering frequencies beyond the cutoff Ω, however. We should not find it too surprising if other types of information bring with them the ability to restore these presumably lost frequencies.

IV. EARLY CONSTRAINED METHODS

A. Gold's Ratio Method

It has been noted that deconvolution methods, most of which were linear, had a propensity to produce solutions that did not make good physical sense. Prominent examples were found when negative values were obtained for light intensity or particle flux. As noted previously, the need to eliminate these negative components was generally accepted. Accordingly, Gold (1964) developed a method of iteration similar to Van Cittert's but used multiplicative corrections instead of additive ones.

Given the equation describing spectrometer distortion,

$$i = s \otimes o, \tag{1}$$

and a spectrum estimate $\hat{o}^{(k)}$, this method proposes to estimate the observed data by use of

$$\hat{i}^{(k)} = s \otimes \hat{o}^{(k)}. \tag{2}$$

A ratio containing this quantity and the data may be used to construct a new estimate of the true spectrum,

$$\hat{o}^{(k+1)} = \hat{o}^{(k)} [i/\hat{i}^{(k)}]. \tag{3}$$

The iteration may be given in a single equation:

$$\hat{o}^{(k+1)} = \hat{o}^{(k)} \frac{i}{s \otimes \hat{o}^{(k)}}. \tag{4}$$

As with Van Cittert's method, we may take $\hat{o}^{(0)} = i$. We see that the successive estimates $\hat{o}^{(k)}$ cannot be negative, provided that s and i are everywhere positive. The reader should note that this assumption may be violated in a base-line region. Here the data i probably contain negative values arising from noise. A means of dealing with this problem is needed.

Gold's method has been used by a number of workers, including Siska (1973), who applied it to molecular-beam scattering data, MacNeil and Dixon (1977), who applied it to photoelectron spectra, and Jones et al. (1967), who restored infrared spectra of condensed-phase samples. The author is unaware of any experimental results with this method, however, that illustrate the full potential achievable by constrained methods to be described later in this chapter. In the work of Jones et al., the resulting resolution is probably limited by the inherent breadth of spectral lines observed with condensed-phase samples.

In the previous chapter we found that when Van Cittert's method converges, it converges to the linear inverse filter. We are therefore led to inquire about the convergence properties of the ratio method. Gold's original consideration of the problem was based on the matrix formulation given in the previous chapters. In that formulation, the spectrometer distortion is given by

$$\mathbf{i} = \mathbf{so}. \tag{5}$$

Here $\mathbf{o}$ and $\mathbf{i}$ are the object and image vectors, respectively, and $\mathbf{s}$ is the spread-function matrix as defined in Chapter 3. Gold was able to show that proper convergence is assured if the following conditions hold:

(1) all principal minors of $\mathbf{s}$ are positive,
(2) all eigenvalues of $\mathbf{s}$ are real, and
(3) all eigenvalues of $\mathbf{s}$ are positive.

These conditions prevail with all positive definite matrices, that is, all matrices having positive (only) elements and determinants greater than zero. Gold observed that even if $\mathbf{s}$ does not obey these conditions, $\mathbf{s}^T\mathbf{s}$ does, where $\mathbf{s}^T$ is the transpose of $\mathbf{s}$. We may therefore multiply Eq. (5) by $\mathbf{s}^T$ and obtain

$$\mathbf{s}^T\mathbf{i} = \mathbf{s}^T\mathbf{so}. \tag{6}$$

If $\mathbf{i}$ is positive, then $\mathbf{s}^T\mathbf{i}$ is also positive and therefore meets the requirement for a positive solution. The added smoothing of $\mathbf{i}$ imposed by $\mathbf{s}^T$ also helps to reduce or eliminate possibly troublesome negative components. Confident of convergence, we may seek the solution, using $\mathbf{s}^T\mathbf{i}$ in place of $\mathbf{i}$ and $\mathbf{s}^T\mathbf{s}$ in place of $\mathbf{s}$. The rate of convergence depends on the conditioning of $\mathbf{s}$ and the choice of $\hat{\mathbf{o}}^{(0)}$.

The matrix notation serves to stress that the technique is applicable to shift-variant spread functions, that is, where $s_{jk} \neq s_{lm}$ for $j - k = l - m$. Many of the deconvolution methods described here with the convolution notation may thus be generalized. In the convolution notation, the present method may be expressed by the equation

$$\hat{o}^{(k+1)}(x) = \hat{o}^{(k)}(x)\frac{s(-x) \otimes i(x)}{s(-x) \otimes s(x) \otimes \hat{o}^{(k)}(x)}. \tag{7}$$

We note that transposing $\mathbf{s}$ is equivalent to reversing the abscissa of $s(x)$. The reader will be struck by the similarity of this treatment to the reblurring method of Kawata et al. (Chapter 3, Section IV.D.3). Their method is, in fact, an adaptation of the present use of $\mathbf{s}^T$ to Van Cittert's method.

Because the solution $\hat{o}^{(k)}$, $k \to \infty$, is independent of choice $\hat{o}^{(0)}$, it is unique. But how close is it to the ideal solution $o(x)$? The fact that $\hat{o}^{(k)}$ is nonnegative means, at least, that the method does not converge to the inverse-filter estimate.

Gold's method does not appear to be widely known or used at present. Perhaps the future literature will answer some of the questions that may be asked about its ultimate performance, both theoretically and with respect to practical applications.

B. Schell's Method

Focusing our attention once again on Fourier space, recall that the Wiener inverse filter $Y_W(\omega)$ is obtained by finding the function $Y_W(\omega)$ that minimizes the mean-square error

$$\int \langle |y_W(x) \otimes i(x) - o(x)|^2 \rangle \, dx = \int_{-\Omega}^{\Omega} \langle |Y_W(\omega)I(\omega) - O(\omega)|^2 \rangle \, d\omega. \quad (8)$$

Schell (1965) recognized that the major deficiency of the Wiener inverse filter is the nonphysical nature of the partially negative solutions that it is prone to generate. He sought to extrapolate the band-limited transform $O(\omega)$ by seeking a nonnegative physical solution $\hat{o}^+(x)$ through minimization of

$$\int_0^{\Omega} B_W^2(\omega) |Y(\omega)I(\omega) - \hat{O}^+(\omega)|^2 \, d\omega + \int_{\Omega}^{\Omega_p} |\hat{O}^+|^2 \, d\omega. \quad (9)$$

In this expression, $Y(\omega)$ is the simple inverse filter $1/\tau(\omega)$ and $B_W(\omega)$ the factor by which the simple filter must be modified to give the Wiener inverse filter, so that $Y_W(\omega) = B_W(\omega)/\tau(\omega)$. The quantity $\hat{O}^+(\omega)$ is the Fourier transform of $\hat{o}^+(x)$. The quantity Ω_p specifies the limit of the extended bandwidth required of the solution $\hat{o}^+(x)$.

We are permitted to specify the integrals for positive ω only, because of the even property of the integrand. This simplication, in turn, stems from the real nature of all the x-space components of the integrand. Minimizing expression (9) is equivalent to asking that the physical solution conform to the Wiener inverse-filter estimate in the sense of minimum mean-square error *after* suitable weighting of the positive solution to ensure best conformance at frequencies of greatest certainty.

A positive, physically realizable spectrum can be represented as the inverse Fourier transform of an autocorrelation function. This may be verified by noting that such a spectrum $\hat{o}^+(x)$ can be represented as the modulus of a complex function $g(x)$:

$$\hat{o}^+(x) = |g(x)|^2 = g^*(x)g(x). \quad (10)$$

By the convolution theorem, its Fourier transform is

$$\hat{O}^+(\omega) = G^*(-\omega) \otimes G(\omega) = \int_{-\infty}^{\infty} G^*(\omega')G(\omega' + \omega)\,d\omega' = G \star G, \quad (11)$$

where G is the Fourier transform of g and $G \star G$ the autocorrelation of G as defined by Eq. (11). Schell (1965) found that an analytic function $G(\omega)$, and hence $\hat{o}^+(x)$, minimizing expression (9) may be obtained by the iterative procedure

$$G^{(k+1)}(\omega) = \text{const} \times \int_{-\Omega p/2-\omega}^{\Omega p/2-\omega} B_W^2(\omega')[Y(\omega')I(\omega') - G^{(k)}(\omega') \star G^{(k)}(\omega')]$$

$$\times G^{(k)}(\omega + \omega')\,d\omega', \quad (12)$$

where $G^{(k)}$ and $G^{(k+1)}$ are the kth and $(k+1)$th approximations to G, respectively.

The solutions illustrated in Schell's original publication were indeed entirely positive and showed some resolution improvement over the inverse-filter estimates. The improvement in these examples was not, however, as great as we have come to expect from the best of the newer methods and may not in fact demonstrate the method's real potential. The method does bring with it in a very explicit way, however, the idea that the Fourier spectrum may be extended, on the basis of a knowledge of positivity. Previous studies had focused on the finite extent constraint to achieve this objective.

V. GREATER REALIZATION OF CONSTRAINT BENEFITS

A. Jansson's Method

Just as others who have used linear methods, this author was disappointed to note the appearance of spurious nonphysical components when he applied the linear relaxation methods (Chapter 3) to the deconvolution of infrared spectra. Infrared absorption spectra, and other types of spectra as well, must lie in the transmittance range of zero to one. Spurious peaks appeared to nucleate on specific noise fluctuations in the data and grow with successive iterations, even though the mean-square error

$$\Phi = \int [i(x) - s(x) \otimes \hat{o}^{(k)}(x)]^2 \, dx$$

became smaller and smaller. Indeed, negative-area contributions from nonphysical parts of the solution served to counterbalance the positive area erroneously allocated in the physical region.

Thus, the author considered forcing the solution to lie within the physical bounds. Besides eliminating the objectionable nonphysical result, this approach, it was hoped, would also improve the accuracy of the solution within the physical bounds. Because of the limited performance evidenced in previous literature available on deconvolution, the author was unprepared for the magnitude of the improvement that resulted.

1. Initial Attempt

The nonlinear iterative methods described here are based on the linear relaxation methods developed in Sections III. C.1 and III.C.2 of Chapter 3. Initially, the correction term was set equal to zero in regions where $\hat{o}^{(k)}$ was nonphysical. To illustrate this, we may rewrite the point-simultaneous equation [Chapter 3, Eq. (23)] with a relaxation parameter that depends on the estimate $\hat{o}^{(k)}$:

$$\hat{o}^{(k+1)} = \hat{o}^{(k)} + r[\hat{o}^{(k)}][i - s \otimes \hat{o}^{(k)}].\tag{13}$$

For clipping or simple truncation of the nonphysical part, we define the relaxation function

$$r[\hat{o}^{(k)}] = \begin{cases} r_0 & \text{for } \hat{o}^{(k)} \text{ physical (in this case } 0 \leq \hat{o}^{(k)} \leq 1) \\ 0 & \text{elsewhere.} \end{cases}$$

It is generally desirable to choose r_0 large so that large corrections will be made to $\hat{o}^{(k)}(x)$, allowing it to approach $o(x)$ with the smallest possible amount of computation. Let us assume that r_0 is chosen fairly large, so that some initial corrections will be overcorrections. This procedure is normal in the linear relaxation method. The overcorrections typically "damp out" with continued iteration and a solution results. If r_0 is chosen too large, however, the method fails to converge. Either trial and error, or theorems concerning the eigenvalues of s serve to establish the optimum r_0.

With the present definition of r, however, an overcorrection that would normally disappear gradually through ensuing iterations results in a value of $\hat{o}^{(k)}(x)$ that vanishes for all subsequent iterations. This behavior occurs because further corrections to that value are prohibited. To use the method, the investigator is compelled to take small values for r_0. Even in this case, erroneously nonphysical values of $\hat{o}^{(k)}$ that have been forced to zero are never allowed to return to the finite range that might better represent the true spectrum $o(x)$. This form of the method therefore demands excessive computation and yields a solution that, although physically realizable, is not the best achievable estimate.

2. The Method and Its First Application

Instead of clipping the correction term, we introduced the concept of a relaxation function. This function modulates the correction term so that it assumes full value in the center of the physical region (transmittance = 0.5) and falls to zero at the amplitude bounds. Seeing the method as one of gently "molding" the solution to lie in the physical regime, we sought a way to force the nonphysical part back toward the bound. Previous experience had shown that noise nucleated nonphysical peaks that grew with successive iteration. Would it not be possible just to reverse the sign of the correction for these nonphysical components? To refine the idea further, we considered that larger nonphysical elements require larger "reverse" corrections. Accordingly, we adopted a functional form for r that varied linearly from a maximum value at transmittance = 0.5 to zero at the bounds, and negative beyond. Specifically,

$$r[\hat{o}^{(k)}(x)] = r_0[1 - 2|\hat{o}^{(k)}(x) - \tfrac{1}{2}|]. \tag{14}$$

This relaxation function is shown in Fig. 1. It yielded what was perhaps the first deconvolution result that demonstrated the real power of the physical-realizability constraint, and possibly the first to make use of both upper and lower bounds.

In its first application, this relaxation function gave an excellent estimate *after only 10 iterations.* Figure 2 illustrates the result obtained from processing the Q branch of N_2O found at about 2798 cm^{-1} in the infrared. This spectrum

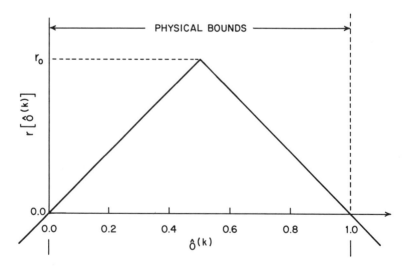

Fig. 1   Relaxation function $r[\hat{o}^{(k)}]$.

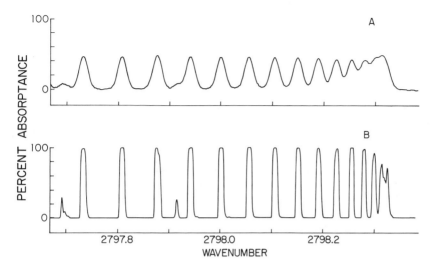

Fig. 2 Deconvolution of the N_2O infrared Q branch at 2798 cm^{-1} by Jansson's method: (a) raw data, (b) restoration. Flat-topped lines in the restoration faithfully represent the saturated absorption present (Jansson et al., 1970).

is a natural resolution target when the sample gas pressure is low; the ever more-closely spaced lines in the band head pose the challenge. Two small "hot-band" lines are clearly resolved, as are the most closely spaced lines in the band head. Some of the spectral-line shapes are flat topped in absorptance or transmittance, as one would expect for these strongly absorbed, Doppler-broadened lines. (Remember that inherently Gaussian shape prevails only in the absorption coefficient regime; see Chapter 2, Section I.E.)

For this work, the spectrometer function $s(x)$ was determined by the method outlined in Section II.G.3 of Chapter 2. In digitizing the data, a sample density was chosen to accommodate about 70 samples taken across the full width at half maximum of $s(x)$. A 25-point cubic polynomial smoothing filter was used in the deconvolution procedure to control high-frequency noise. Instead of the convolution in Eq. (13), the point-successive modification described in Section III.C.2 of Chapter 3 was employed. In Eq. (24) of Chapter 3, we replaced κ with the expression

$$\kappa[\hat{o}_n^{(k)}] = \kappa_0(1 - 2|\hat{o}_n^{(k)} - \tfrac{1}{2}|). \tag{15}$$

This resulted in the iterative procedure described by

$$\hat{o}_n^{(k+1)} = \hat{o}_n^{(k)} + \frac{\kappa[\hat{o}_n^{(k)}]}{[\mathbf{s}]_{nn}} \{i_n - \sum_{m<n} [\mathbf{s}]_{nm}\hat{o}_m^{(k+1)} - \sum_{m\geq n} [\mathbf{s}]_{nm}\hat{o}_m^{(k)}\}. \tag{16}$$

In this equation i_n is the nth element of the sampled image or observed

spectrum, $\hat{o}_n^{(k)}$ and $\hat{o}_n^{(k+1)}$ are the nth elements of the kth and $(k+1)$th estimates of the object or true spectrum, and s is the matrix characteristic of
instrumental spreading. The quantity κ_0 is a constant. Like any other method
that drives the error $i - s \otimes \hat{o}$ toward a minimum, this procedure tends to
conserve the area under impulsive objects. That is, deconvolved spectral
lines contain the same area as their counterparts in the raw data.

In the first application of this method we adopted the following procedure.
The data were smoothed once and used as $i(x)$. The initial estimate $\hat{o}^{(0)}(x)$
was taken to be $i(x)$ with one additional pass of smoothing. Estimates $\hat{o}^{(1)}$
through $\hat{o}^{(5)}$ resulted from using the relaxation function of Eq. (15) with the
iteration of Eq. (16), taking $\kappa_0 = 0.1$ and smoothing $\hat{o}$ between iterations.
Estimates $\hat{o}^{(6)}$ through $\hat{o}^{(10)}$ (the final estimate) were obtained with $\kappa_0 = 0.2$
and no smoothing. The program, which was coded in FORTRAN, ran for
approximately 2 min on a CDC-6400 computer for each 3000-point spectral
segment. The original publications (Jansson, 1968, 1970; Jansson, Hunt, and
Plyler, 1968, 1970) may be consulted for additional experimental detail.†

Shortly after the original work, Barnes et al. (1972) applied the method to
the measurement and analysis of the v_3 band of methane. Their analysis pre-
dicted virtually every detail of the observed spectrum and verified many details of the deconvolved spectrum that were not apparent in the observed data.

3. Sampling, Bandwidth Extrapolation,
   and the Relaxation Function

The Fourier frequency bandpass of the spectrometer is determined by the
diffraction limit. In view of this fact and the Nyquist criterion, the data in
the aforementioned application were oversampled. Although the Nyquist
sampling rate is sufficient to represent all information in the data, it is not
sufficient to represent the estimates $\hat{o}^{(k)}$ because of the bandwidth extension
that results from information implicit in the physical-realizability constraints. Although it was not shown in the original publication, it is clear
from the quality of the restoration, and by analogy with other similarly
bounded methods, that Fourier bandwidth extrapolation does indeed occur.
This is sometimes called superresolution. The source of the extrapolation
should be apparent from the Fourier transform of Eq. (13) with $r(x)$ specified
by Eq. (14).

---

† This program, essentially as used in the original work, is available as program 212 from
the Quantum Chemistry Program Exchange, Chemistry Department, Indiana University,
Bloomington, Indiana 47401. It does not, however, incorporate the refinements described in
sections that follow.

It is seen that the dependence of r on x is determined by $\hat{o}^{(k)}(x)$, which changes as k increases. Consider $r_x^{(k)}(x) = r[\hat{o}^{(k)}(x)]$ and define $R^{(k)}(\omega)$ as its Fourier transform. The Fourier transform equivalent of Eq. (13) is then

$$\hat{O}^{(k+1)}(\omega) = \hat{O}^{(k)}(\omega) + R^{(k)}(\omega) \otimes [I(\omega) - \tau(\omega)\hat{O}^{(k)}(\omega)]. \tag{17}$$

Although the bracketed term contains no frequencies beyond the bandlimit Ω, convolution with $R(\omega)$ ensures the presence of such components. To carry the analysis further, we need to specify the form of r so that the explicit dependence on x can be determined through its connection with $\hat{o}^{(k)}(x)$.

In the case of a lower bound only, we may select the simple relaxation function $r[\hat{o}^{(k)}] = r_0\hat{o}^{(k)}$ so that

$$\hat{o}^{(k+1)} = \hat{o}^{(k)} + r_0\hat{o}^{(k)}[i - s \otimes \hat{o}^{(k)}], \tag{18}$$

which has the Fourier transform

$$\hat{O}^{(k+1)} = \hat{O}^{(k)} + r_0\hat{O}^{(k)} \otimes [I - \tau\hat{O}^{(k)}]. \tag{19}$$

This expression tells us that successive estimates $\hat{O}^{(k)}$ will have nonzero values for $|\omega| \geq \Omega$.

Although we have not proved that $\hat{O}(\omega)$ approaches $O(\omega)$ for $\omega > \Omega$, experimentation tends to bear this out. This is entirely consistent with current knowledge of the general properties of constraints applied to various deconvolution methods (Biraud, 1969; Frieden, 1972; Howard, 1981a, 1981b).

Other choices are possible for the relaxation function r. For these choices, expressions similar to Eq. (19) may be derived. Although it is awkward to deal with the absolute value in Eq. (14), a parabolic form meeting the principal requirements is more readily expressed in Fourier space. A simple quadratic (Willson, 1973) and its generalization to other powers have been used by Blass and Halsey (1981):

$$r[\hat{o}] = 4r_0[\hat{o}(1 - \hat{o})]^n. \tag{20}$$

Here n is an integral exponent. Frieden (1975), on the other hand, generalizes the original relaxation function [Eq. (14)] to treat the case of arbitrary upper bound B and lower bound A:

$$r[\hat{o}] = r_0[1 - 2(B - A)^{-1}|\hat{o} - (A + B)/2|]. \tag{21}$$

Maitre (1981) has experimented with trapezoidal forms and powers of expressions involving the cosine. Jansson and Davies (1974) have used both exponential- and straight-line-based functions in two versions adapted to emission spectra (Chapter 5, Section III.B).

Considerable imagination can be exercised in devising the relaxation function. As noted previously, however, inferior results will be obtained unless care is taken to ensure that $r[\hat{o}]$ does not abruptly set the correction term equal to zero when $\hat{o}^{(k)}$ is nonphysical, but rather allows $\hat{o}^{(k)}$ to be corrected gradually as iteration proceeds. This error is frequently observed in applications of this method reported in the literature.

4. Smoothing, Reblurring, and Other Adaptions

In the author's original work, presmoothing of the data was found to be essential for acceptable restoration. Smoothing, however, potentially limits the resolution achievable. When applied to the data (only) before deconvolution, the smoothing function enters as additional and uncompensated spreading. This undesirable effect may be alleviated by including the additional spreading in $s(x)$ too. This is accomplished by smoothing $s(x)$ before it is used for deconvolution (Richards *et al.*, 1979). Kawata *et al.* (1978) have carried the concept a step further in their development of a reblurring method to guarantee convergence with linear iterative methods. (See also Chapter 3, Section IV.D.3, and Kawata and Ichioka 1980b.) Future applications of the present method could possibly benefit from consistent application of prefiltering both $i(x)$ and $s(x)$ with the abscissa-reversed spread function $s(-x)$. We note, however, that when an idealized optical spectrometer has equal entrance- and exit-slit spectral widths, the optical part of the transfer function is guaranteed real and positive. This condition, by analysis similar to that in Section IV.D.2 of Chapter 3, assures convergence.

Thomas (1981) and Schafer *et al.* (1981) have also discussed the reblurring method. In addition, Schafer *et al.* have studied a generalized class of iterative deconvolution algorithms. They examined the convergence properties of the iteration

$$\hat{o}^{(k+1)} = b\hat{o}^{(k)} + r\{i - s \otimes [b\hat{o}^{(k)}]\}, \tag{22}$$

where b is a constraint operator that, in a simple form, may supply bounds to the solution by clipping. Here, the quantity r may be either a constant parameter or a function of x or $\hat{o}^{(k)}$. The development is based on the contraction mapping theorem of functional analysis and is actually more general than that specified in Eq. (22). The results are therefore not limited to distortions of the convolution type. In addition, constraints of finite extent in x may be applied. It is shown that, under the proper circumstances, convergence to a solution consistent with both data and constraints is guaranteed.

Because the present nonlinear method is an outgrowth of the linear ones, it is not surprising that such useful techniques as the reblurring discussed in Chapter 3 may be adapted. Other elements of Chapter 3 may be borrowed as well, including memory-conserving methods of programming (Section

III.C.3) and the use of special sampling methods to gain the rapid-convergence advantage of point-successive iteration without inducing asymmetry (Section III.C.2).

The line breadth remaining after deconvolution may be influenced primarily by broadening for which compensation was not attempted. Nevertheless, the maximum breadth-reduction ratio attainable when the method is really challenged indicates its potential. The original application (Section V.A.2) showed reduction on the order of 2.5 times, limited as described earlier. Subsequent applications have yielded substantially greater improvements (Chapter 5, Chapter 7).

Later chapters detail application of the present method to electron spectroscopy for chemical analysis (Chapter 5), high-resolution dispersive infrared spectroscopy (Chapter 6), and tunable-diode-laser spectroscopy (Chapter 7). Because the heart of the method is the repeated application of simple convolution, the method has been adapted to the processing of images (Kawata *et al.*, 1978; Kawata and Ichioka, 1980a; Saghri and Tescher, 1980; Maitre, 1981; Gindi, 1981).

In one application (Maitre, 1981), the parallelism of noncoherent optical convolution methods is combined with video techniques in a totally analog application to images. Matsuoka *et al.* (1982) have developed a hybrid optical–video–digital system based on similar principles. Pipeline processors that perform image convolution at high speed are ideal for digital applications. The inherent simplicity of the method adapts it to high-speed implementations by use of other hybrid techniques. It is even possible to devise a transversal filter that has all the desirable nonlinear properties of the method. We describe this filter in the next section.

5. *High-Speed Filter Hardware*

For simplicity of computer implementation, and in almost all practical cases, $s(x)$ can be taken as zero outside some limited range of x. Using filter terminology, we may say that it has a finite impulse response. Let us consider the discrete version. For discretely sampled data, we write the sampled response function as s_m. As in Sections V. A. 1–V. A.4, we take its output at the center of the filter. That is, the output corresponds to the Mth finite value, where M is the index at which s_m is maximum. Because data are almost always sampled sequentially, we may take the index m as being directly proportional to time. Visualizing the convolution as in Section II.A of Chapter 1, we readily see that the filter's output lags its input by precisely M samples.

Note now that the present iterative method relies on sequential application of convolution by s. All these convolutions could be linearly combined into a a single transversal filter (Blass and Halsey, 1981) were it not for the non-linearity introduced between stages, which is essential for implementing bounds and achieving superresolution. We now propose that the present method could be implemented in a cascaded design, the required function $r[\hat{o}^{(k)}]$ being applied between stages. The approach is illustrated schematically in Fig. 3. Here the $\times$'s represent the sampled values of the functions indicated on the left side. The bracket denotes application of Eq. (16). The span of the bracket indicates the span of the finite values of s, that is, the elements of $\hat{o}^{(k)}$ needed for each summation in the convolution. From the figure, it is apparent that the total lag of this nonlinear superresolving transversal filter is just LM samples, where L is the number of stages corresponding to the number of iterations in the conventional software implementation. Smoothing between stages is easily accomplished by analogy with this design. Bearing in mind that only 10 iterations were needed to produce the restoration in Fig. 2, we observe that the hardware requirements are indeed rather modest.

Specifically what hardware is required? Both digital shift registers and analog delay lines could be utilized in the convolution. The latter are based on either sample-and-hold circuits or charge-coupled "bucket brigades." Analog voltages derived from the tapped delay stages are easily weighted by appropriate coefficient resistors and summed into an operational amplifier. The products and sums required are thus performed in parallel and, with

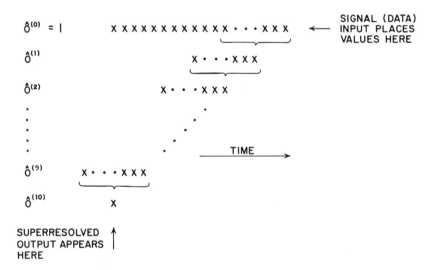

Fig. 3 Schematic representation of a superresolving filter. Brackets show convolution by s and its use in nonlinear corrections resulting in successively improved estimates.

proper choice of wide-band amplifier, are essentially instantaneous. Many hybrid analog–digital variations on this theme are possible.

The filter version of the method is a truly continuous process, as opposed to the "batch" process that the iterative representation implies. Furthermore, no approximation is necessary in deriving the filter version. The output of the filter is precisely the same as that of the basic iterative constrained software method. Bearing in mind that only a modest lag results, one may think of it as a real-time implementation. In an alternative approach, a laboratory computer might apply a purely software version of this filter to spectra continuously as data are acquired. The filter may also be packaged in firmware as part of a microprocessor-based instrument. Other applications also suggest themselves.

With suitable scan-line delay circuits, the filter may be applied in real time to live video images. In data communications, bandwidth extrapolation offers the opportunity to make far better use of channel capacity than is now possible. The bounded methods produce their most impressive restorations when $o(x)$ spends a lot of time at or near the bounds. The present method was designed for use with both upper and lower bounds, which makes it ideal for the bilevel signals used in digital transmission.

Substantial bandwidth extrapolation may be anticipated on the basis of the powerful a priori knowledge that the signal can assume only one of two values. Although the nonlinear hardware filter described here is readily feasible, and perhaps practical with today's standard commercial electronic technology, experimental work along these lines is unknown to the author as of this writing.

B. Biraud's Method

Like Schell, Biraud forced the solution to be positive by using a representation in which negative values are not possible. He chose to represent the solution as the square of a real Fourier series. In the continuous regime, we have

$$\hat{o}^{+}(x) = [a(x)]^2, \tag{23}$$

where $a(x)$ is real. In Fourier space, we have

$$\hat{O}^{+}(\omega) = A(\omega) \otimes A(\omega), \tag{24}$$

where $A(\omega)$ is the Fourier transform of $a(x)$. Biraud formulated his method for application when the predominant noise affects the measurement before blurring. This is unlike the situation considered in Eqs. (12) and (36) of

Chapter 3. Consistency of solution $\hat{o}^+(x)$ with the observed data in the presence of additive noise of this type requires minimization of

$$E = \int_{-\Omega_c}^{\Omega_c} |Y(\omega)I(\omega) - \hat{O}^+(\omega)|^2 \, d\omega \tag{25}$$

by picking the best function $A(\omega)$. In this equation $\pm\Omega_c$ defines the frequency limits within which consistency is desired. The quantity Ω_c is usually chosen to be slightly smaller than Ω to ensure that the inverse-filter estimate $Y(\omega)I(\omega)$ is not excessively affected by the noise. The discrete representation is central to Biraud's method of solution. Accordingly, we let the finite frequency band of the data be represented by $2J + 1$ samples so that we have $J \Delta\omega = \Omega$, where $\Delta\omega$ is the sampling interval required to specify the object completely. This, in turn, is established by knowledge of the object's finite extent with limits $\pm X$, so that $|x| \leq X$. We set $\Delta\omega = \pi/X$, consistent with the sampling theorem. Let us define $\omega_n = n \Delta\omega$ so that the $2J + 1$ samples of the solution over the finite frequency band are given by $\hat{O}^+(\omega_n)$. Now, a function that is nonvanishing only over a finite interval has a self-convolution that is contained in an interval twice as large. In this case we have only $2P + 1$ samples of $A(\omega_n)$, where $P = J/2$.

2. Procedure

Biraud's method starts by considering the centermost value of $\hat{O}^+(0)$ as known, because it is at $\omega = 0$ that the inverse filter gives its most reliable estimate. Furthermore, this assumption guarantees that the total flux in the restoration is the same as that measured. Thus we find

$$\hat{O}^+(0) = Y(0)I(0) = A(\omega) \otimes A(\omega)\big|_{\omega=0}. \tag{26}$$

The centermost values $A(\omega_{-1})$, $A(\omega_0)$, and $A(\omega_1)$ are considered to be unknown, and all other $A(\omega_n)$ are taken as zero. Three cubic equations for determining these unknowns may be derived by using Eq. (26) as a constraint term and applying Lagrange's method of undetermined multipliers to Eq. (25) with limits $\Omega_c = \Omega$ (Frieden, 1975). The three partial derivatives $\partial/\partial A(\omega_{-1})$, $\partial/\partial A(\omega_0)$, and $\partial/\partial A(\omega_1)$ are set equal to zero, and the resulting equations are then solved. In the next step, $A(\omega_{-1})$ and $A(\omega_1)$ are taken as known, and $A(\omega_{-2})$, $A(\omega_0)$, and $A(\omega_2)$ are sought, all other A's being taken as zero. The process is continued until $A(\omega_{-P})$, $A(\omega_0)$, and $A(\omega_P)$ are obtained.

The entire procedure is now repeated, starting again with the centermost trio, this time using the previous iteration's $A(\omega_l)$ values instead of zeros for all the "outer" A's. For a time, each such cycle through the $A(\omega_l)$ produces ever smaller minima. Eventually, however, saturation is reached and little reduction is noted with continuing iteration. Even in the ideal case of a perfect solution, noise restricts the minimum achievable to

$$E_{\text{res}} = \int_{-\Omega_c}^{\Omega_c} |N_o(\omega)Y(\omega)|^2 \, d\omega, \tag{27}$$

where $N_o(\omega)$ is the Fourier transform of the noise, here considered to affect measurement of the true object or spectrum before blurring. The expected value $\langle E_{\text{res}} \rangle$ of this quantity is known if estimates of the statistics of $N_o(\omega)$ can be obtained. If this expected minimum is reached before saturation, the process is terminated and the solution is taken as the self-convolution of the prevailing $A(\omega_l)$. If saturation occurs before $\langle E_{\text{res}} \rangle$ is attained, however, additional degrees of freedom are needed to satisfy the concurrent conditions of positivity, finite extent, and consistency with the data.

The required degrees of freedom result from allowing the A's *beyond the data band limit* Ω to vary. First $A(\omega_0)$ and $A(\omega_{\pm(P+1)})$ are sought in a full repetition of everything up to this point. Because the error specified by Eq. (25) is evaluated for frequencies less than the band limits $\pm\Omega$, the added degrees of freedom are required within this band. One might well ask how degrees of freedom can be increased within the error computation band $\pm\Omega_c$ by adding coefficients to A for $|\omega| > \omega_P$, when ω_P was chosen to provide $\hat{O}(\omega)^+$ values just up to the band limit Ω, but no further. The answer lies in the mixing of the band-extending A coefficients by the self-convolution operation. Values of $\hat{O}^+(\omega)$ within the band $\pm\Omega$ are in fact formed by products containing the new coefficients of A, thus allowing E to be further reduced.

After coefficients $A(\omega_{\pm(P+1)})$ are determined, the process is continued by adding values of A until the expected residual $\langle E_{\text{res}} \rangle$ is achieved. The procedure is summarized in the accompanying structured flowchart.

Set limit of sought $A(\omega_l)$ to data bandlimit ($L = P$).

DO until error E is less than expected residual $\langle E_{\text{res}} \rangle$.

    DO until saturation or until $E \leq \langle E_{\text{res}} \rangle$.

        DO for $l = 1$ to sought limit L.

            Fix $A(\omega_m) = 0$ for $m > l$.

            Solve cubic equations for $A(\omega_0)$ and $A(\omega_{\pm l})$.

    Increase solution frequency band ($L = L + 1$).

3. Performance

The method employs a gradual increase in frequency beyond the data band limit. High-frequency components are not sought until the best values of low-frequency components are found. Because frequencies are not sought above the lowest needed to satisfy the data, the method is inherently smooth. Furthermore, Biraud's method appears to be the first to have simultaneously utilized both the constraint of positivity and that of finite extent with specific limits, the latter being inherent in the sampling. These facts are probably responsible for the impressiveness of the restoration in the original publication (Biraud, 1969), which is reproduced in Fig. 4.

The original publication, in contrast to the treatment given here which follows Frieden (1975), described the use of an assumed trial function $A(\omega)$ together with a perturbation method needed to adjust it for consistency with the data. Biraud noted empirically that, for simulated test data, the method gave an $o(x)$ estimate that was quite accurate regardless of the chosen trial $a(x)$. He also observed that the need for a high signal-to-noise ratio was greatly reduced as compared with the requirements of a linear method. These observations are consistent with those that one obtains in working with all of the better deconvolution methods constrained to physically realizable solutions.

A study and evaluation of Biraud's method was the subject of a thesis by Wong (1971). A method developed by Robaux and Roizen-Dossier (1970) is similar to those of Schell and Biraud in that it seeks to extrapolate $\hat{O}(\omega)$ for

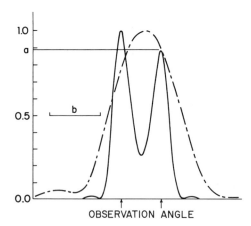

Fig. 4 Restoration of Cygnus A astronomical radio source at 11 cm by Biraud's method. Dash-dotted curve, recorded measurement; solid curve, deconvolution, a = 0.89, b = beam width at half power. The ordinate is intensity given in arbitrary units. The peak separation is 104 arcsec (Biraud, 1969).

$|\omega| > \Omega$ by fitting over $|\omega| < \Omega$ a function the Fourier transform of which must be everywhere positive.

C. Frieden's Method of Maximum Likelihood

So far, we have asked only that solution $\hat{o}(x)$ satisfy the data plus bounds on amplitude and/or bounds on spatial extent. Although bounds help considerably in reducing the number of possible solutions, we are still left with the possibility that more than one solution may satisfy the data. By satisfying the data, we mean that the expression

$$\Phi = \int (i - s \otimes \hat{o})^2 \, dx$$

represents a local minimum in the space having coordinates that are the elements of $\hat{o}$. In discrete form, these coordinates are the $\hat{o}(x_i)$. Criteria other than minimum mean-square error may be employed, but the difficulty remains: which of the numerous potential solutions should we pick?

1. The Most-Likely Estimate

Frieden (1971, 1972) attacked this problem by seeking the solution that has the greatest probability of occurrence. The lower-bound condition [$\hat{o}(x)$ positive] was a natural consequence of the physical random-grain model that he used. This work was unique in that separate solutions were found for $o(x)$ and the noise. In its simplest form, the original method required the data to be all positive. Even though $o(x)$ satisfies this requirement, $i(x)$ often does not, because of noise excursions that, in spectra, may be found in the base-line region. A bias term was added to the data to ensure positivity.

Subsequent research by Herschel (1971), Kikuchi and Soffer (1977), and Frieden (Chapter 8) has refined the concept in a way that provides an explicit and sensible accounting for noise contributions. This work also provides solutions that incorporate a type of prior knowledge not used before. In particular, the users may express their bias by proposing a prior spectrum or guess as to what the true spectrum $o(x)$ might look like. Furthermore, they may express their relative confidence in the guess by specifying a probability of occurrence for each value that may be assumed by an element of the estimate $\hat{o}(x)$. Both the prior spectrum and its associated user-conviction probability function may be obtained from past experience by statistical analysis. In Chapter 8, Frieden examines the possibilities of maximum and minimum conviction in connection with the types of prior knowledge available. He even considers the case of user bias, with high user conviction in that bias, but no empirical data. (Can we discard our spectrometers?) The survey of the various limiting cases serves to reveal important properties of the estimates obtained under practical circumstances. We now summarize Frieden's approach and consider one case of interest to the spectroscopist.

In his derivation, Frieden uses the concept of a number-count set $\{n_m\}$, each member representing the number of photons counted in a spectral interval. The total number of photons $\sum_{m=1}^{M} n_m$ is taken as known to be N. In terms of frequencies v_m, the values of the object spectrum are given by $o_m = n_m h v_m$, where h is Planck's constant. The number of normal modes or degrees of freedom available for occupation by photons of frequency v_m is labeled z_m. The Bose–Einstein degeneracy factor

$$W_m = (n_m + z_m - 1)!/n_m!(z_m - 1)! \tag{28}$$

is introduced to express the number of ways that n_m photons may be distributed among z_m degrees of freedom. For the case of electrons, the reader may wish to consider the corresponding Fermi–Dirac case. For either case, in the limit of sparsely populated degrees of freedom, the Boltzmann factor results.

Frieden finds the probability of a given number-count set $\{n_i\}$ in the absence of empirical data to be given by an Mfold integral:

$$P(n_1, \ldots, n_M)$$

$$= \left(\prod_{m=1}^{M} W_m \right) \int \cdots \int_{\sum q_m = 1} q_1^{n_1} \cdots q_M^{n_M} p(q_1, \ldots, q_M)\, dq_1 \cdots dq_M. \tag{29}$$

Here the values q_m represent the prior spectrum and $p(q_1, \ldots, q_M)$ is the known probability of a given spectrum (set of q_m). The number-count-set estimate $\{\hat{n}_m\}$, and hence the object estimate $\{\hat{o}_m\}$, is obtained by maximizing P by the usual methods of finding extrema.

The influence of data i_m and total photon count $N = \sum n_m$ is exerted by adding the appropriate constraints to $\ln P$ in proportions controlled by Lagrange multipliers λ_m and μ. The most likely solution $\{\hat{n}_m\}$ is then found by maximizing

$$\ln P(n_1, \ldots, n_M) + \mu \left(\sum_{m=1}^{M} n_m - N \right) + \sum_{m=1}^{M} \lambda_m h \sum_n (n_n v_n s_{mn} - i_m). \tag{30}$$

Suppose that the user has the highest-possible conviction that his or her prior spectrum is flat. Specifically, let the user have the greatest certainty that each probability q_m has constant value Q. Furthermore, suppose that the degrees of freedom are sparsely populated and that the noise is additive, independent Gaussian, and strict-sense stationary of order 1 having variance σ^2 (Frieden, 1983). Then, maximizing expression (30), with added terms resulting from noise, results in solutions

$$\hat{n}_m = Q \exp(-1 - \mu - h\nu_m \sum_n \lambda_n s_{nm}), \tag{31}$$

$$\hat{\varepsilon}_m = -\sigma^2 \lambda_m \tag{32}$$

for signal and noise, respectively, expressed in terms of the Lagrange multipliers μ and λ_m. To determine these quantities the solutions are substituted into the data and total-photon-count constraint equations. Accordingly, we have

$$i_m = hQ \sum_n \nu_n s_{mn} \exp(-1 - \mu - h\nu_n \sum_k \lambda_k s_{kn}) - \sigma^2 \lambda_m, \quad m = 1, \ldots, M, \tag{33}$$

and

$$N = Q \sum_n \exp(-1 - \mu - h\nu_n \sum_k \lambda_k s_{kn}), \tag{34}$$

where s_{kn} is the spread function.

These $M + 1$ equations in $M + 1$ unknowns μ and $\{\lambda_m\}$ may be solved by the Newton–Raphson method, in which the unknowns are iteratively adjusted until the right and left sides of the equations agree. The object spectrum number-count set $\{\hat{n}_m\}$ and noise $\{\hat{\varepsilon}_m\}$ are then computed by substitution of μ and the $\{\lambda_m\}$ into Eqs. (31) and (32).

2. Entropy

In the derivation of the foregoing formulas by the methods used by Frieden in Chapter 8, the first term of expression (30) becomes

$$\sum_m n_m \ln(Q/n_m) = \ln Q \sum_m n_m - \sum_m n_m \ln n_m. \tag{35}$$

The quantity $-\sum n_m \ln n_m$, one term of the expression to be maximized, is the Shannon entropy H as used by Jaynes and familiar to those who have studied thermal physics. For a given set $\{n_m\}$ obeying $\sum_{m=1}^M n_m = N$, maximum H is attained when all the n_m have the same value $n_m = N/M$. The requirement that H be maximum, even when other constraints are attached, tends to force the n_m toward this constant value and hence inhibits large excursions. This property of maximum-entropy restorations is certainly desirable.

Otherwise, note that $i = s \otimes \hat{o}$ can be satisfied by *large spurious* fluctuations of neighboring elements n_m provided that they are of opposite polarity, equal in area, and contain only frequencies not passed by s. That is, the smoothing influence of s will average the fluctuations of the n_m, allowing them to develop without affecting the quantity $i - s \otimes \hat{o}$. Consistency with the data is, however, effective in preventing the development of spurious fluctuations having low-frequency content. It is the maximum-entropy requirement that prevents such large excursions generally, including high-frequency behavior.

Thus, although maximum entropy inhibits large excursions, it does not by itself exert a smoothing influence in the sense of suppressing high- *versus* low-frequency components of the n_m. This fact is readily noted in that neither $\sum n_m$ nor $\sum_m n_m \ln n_m$ depends on the order in which the n_m are placed; they may be freely interchanged without altering these quantities. This property is also desirable, because retrieval of the suppressed, even obliterated, high frequencies is our objective.

When the degree-of-freedom sites are highly occupied ($n_m \gg z_m$), the entropy term becomes $\sum (z_m - 1) \ln n_m$. This is a weighted form of the entropy $H_1 = + \sum \ln n_m$ that was maximized by Burg (1967, 1975) in his restoration scheme. Burg's entropy also discriminates against large excursions in n_m as described previously, although to a lesser degree (Lacoss, 1971). To guarantee a positive solution, Burg regarded the n_m, and hence $o(x)$, as the square of another function in a manner reminiscent of the approaches of Schell (1965) and Biraud (1969). He also assumed the first two moments of that function to be constant. As in the approaches of Schell and Biraud, inverse-filter estimates were used as data inputs.

A closed-form result involving only matrix solution and substitution operations was obtained. Under certain circumstances, particularly in the low-noise case, application of the Burg method results in spurious resolution or "spontaneous line splitting" (Fougere *et al.*, 1976). One solution to this problem (Fougere, 1977) introduces iteration.

3. *Choosing the Prior Spectrum*

A strength of the maximum-likelihood approach that was not fully exploited in the foregoing analysis is the possibility of specifying the user's bias toward a nontrivial preferred prior spectrum $\{Q_m\}$ *and* the user's conviction about that choice through the probability $p(q_1, \ldots, q_M)$. Through choice of p it is possible to input the predisposition of the spectrum ordinates toward values in the wing and flat peak-top regions of the spectral lines. There is always a suitable p function, because it describes the user's state of knowledge in a given experiment. Examples are given in Chapter 8.

One way that probability p may be estimated is by statistical analysis of previous experimental data. Various spectral models could also be used to specify p. Both regular models such as that of Elsasser and random models are described by Goody (1964). They have long been used in predicting overall band contours resulting from the superposition of the large numbers of lines in atmospheric IR absorption spectra. We choose to illustrate the estimation of p by using a very simple model composed of a single emission line.

We ask what is the probability of a given object spectrum value when the functional form of the spectrum $o(x)$ is known, but its x displacement is unknown. Because of the unknown displacement, the probability distribution at each location is the same. Call this component probability p_c such that

$$p(q_1, \ldots, q_M) = \prod_{m=1}^{M} p_c(q_m).$$

The component p_c may be evaluated by examining a typical spectrum. The desired probability $p_c(q')$ at a given ordinate value q' is just proportional to the "amount" of x for a given "amount" of q in the typical spectrum, or the absolute value of the increment in x corresponding to the increment in q where q has the value q'. Because the value q' may generally occur more than once in the spectrum, a summation is in order. The desired probability is thus $\sum |dx/dq|$, where the summation indicates contributions from all locations at which $q(x)$ crosses q'.

For simplicity, suppose that we have an emission spectrum consisting of a single, predominantly Doppler-broadened line. This spectrum may be taken as an approximation to the case of widely separated and nonoverlapping lines of equal intensity. Again for simplicity, now consider the typical emission spectrum to be continuous, not sampled. Thus $q(x)$ is given by $q(x) = q_0 \exp(-x^2)$ for a typical line, where q_0 is the peak height. We have chosen the abscissa to be measured in either wavelength or wave number relative to a typical line center. For illustrative purposes only, the x interval for the observation is taken to be $2 \, \Delta x$. If the line is centered in this interval at $x = 0$, we can never have $q(x) < \exp(-\Delta x^2)$.

The derivative dq/dx is given by

$$dq/dx = -2xq(x) = -2q\sqrt{\ln(q_0/q)}, \tag{36}$$

and the probability of q is given by

$$p_c(q) = C \, dx/dq \tag{37}$$

$$= C[\ln(q_0/q)]^{-1/2}/2q, \tag{38}$$

where C is a normalization constant chosen so that

$$\int_{q_0}^{q_1} p_c(q) \, dq = 1, \tag{39}$$

where $q_1 = q_0 \exp(-\Delta x^2)$. For simplicity we have selected half of the line as representative.

A plot of $p_c(q)$ is shown in Fig. 5. It is clear that consideration of the integral only over $x < \Delta x$ is critical to having a finite contribution to the normalization of $p(q)$ in the base-line region. At the peak(s) of the spectral line(s) there is a large contribution due to the flatness of $q(x)$ there. This causes no problem in the evaluation of the integral in Eq. (39) provided that Eq. (37) is used. The value of C displayed implicitly in Fig. 5 is obtained as follows:

$$\frac{1}{C} \int_{q_0}^{q_1} p_c(q) \, dq = x \Big|_0^{\Delta x} = \Delta x, \tag{40}$$

finally resulting in

$$p_c(q) = [\ln(q_0/q)]^{-1/2}/2q \, \Delta x. \tag{41}$$

The analysis may be extended to the use of absorption lines, other line shapes, and, most important, spectra comprising many lines of varying intensity whose intensity statistics are available.

4. Additional Remarks

Like the other nonlinear constrained methods, the maximum-entropy method has proved its capacity to restore the frequency content of $\hat{o}$ that has not survived convolution by s and is entirely absent from the data (Frieden, 1972; Frieden and Burke, 1972). Its importance to the development of deconvolution arises from the statistical concept that it introduced. It was the first of the nonlinear methods explicitly to address the problem of selecting a preferred solution from the multiplicity of possible solutions on the basis of sound statistical arguments.

Both varieties of maximum entropy have been discussed by Frieden (1975). Frieden's original work was stimulated by an important earlier paper by Jaynes (1968). Frieden's original report (Frieden, 1971) and journal publication (Frieden, 1972) are well worth consulting. Hershel (1971) provided an excellent early treatment of the method. Maximum entropy has been applied to restoration of images of planetary objects (Frieden and Swindell, 1976) and nebulae in the presence of Poisson noise (Frieden and Wells, 1978), as well as data from numerous other fields. Gull and Daniell (1978) have used a computationally efficient version to restore radio maps of galaxies. Kikuchi and Soffer (1977) have contributed significantly to rigorous development of the method from quantum-statistical principles. Their work is summarized, further developed, and adapted to the spectroscopic case in Chapter 8.

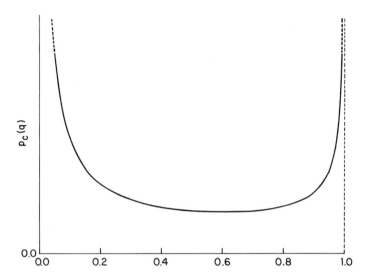

Fig. 5   Component probability $p_c(q)$ for a single Doppler-broadened emission line. The ordinate scale depends on the q domain chosen for normalization.

D. Alternating Projection Methods

The principal thrust of the present chapter is to describe methods that facilitate superior restoration through the use of bounds. Sometimes bounds are implemented in the object domain, for example, positivity or finite extent, and sometimes in the Fourier domain, for example, band limitedness. The effectiveness of this type of prior knowledge was apparent from the early work with constraints. A number of researchers therefore focused their efforts on making use of all possible combinations of constraints and partial data. What could be more suggestive than direct application of bounds to trial solutions by truncation, first in the object domain and then in the Fourier domain?

The author first learned of this approach when he heard a paper given by J. L. Harris (1971), who was concerned with the reconstruction of a three-dimensional microscopic object (the complex refractive index) from optical interference microscope image data. Subsequent to Harris's talk, Gerchberg and Saxton (1972) independently published a similar algorithm designed to solve the "phase problem" of electron microscopy. This was the very same problem that stimulated Gabor's discovery of holography, which, ironically, has never seen much use in its original field of intended application. The problem of reducing speckle in holograms generated by computer provided the motivation for a version of the method disclosed a year earlier in a patent filed by Hirsch *et al.* (1971). Thus it is apparent that a number of workers were led independently to the same general approach.

In the Gerchberg–Saxton application, random numbers constitute an initial estimate of the phase of the emerging wave front. They are combined with the observed image amplitudes and then Fourier transformed by fast Fourier transform (FFT). The amplitudes resulting from the transform are discarded. The phases of the transformed function are combined with the observed diffraction plane amplitudes and then inverse transformed. The amplitude of the transformed function is then discarded and its phases are combined with the observed image amplitudes. The cycle of transforming and combining of the most-recent phase estimate with observed amplitude is repeated until convergence is achieved.

Subsequently, other workers developed numerous variations and generalizations of the basic method. Most methods can be summarized by the accompanying generalized error-reduction algorithm.

Assume a trial solution (possibly by inverse filtering).

DO until error is acceptably small.

> Make smallest possible change needed to satisfy constraints and/or data in present domain.
>
> Fourier transform.
>
> Make smallest possible change needed to satisfy constraints and/or data in present domain.
>
> Inverse Fourier transform.

This algorithm has been called an error-reduction algorithm because it has been shown for a particular case (and it may be generally true) that the error can only decrease with succeeding iterations (Liu and Gallagher, 1974). Note, however, that the error does not necessarily decrease to zero or even to the desired minimum residual value that is based on noise.

For a basic deconvolution problem involving band-limited data, the trial solution $\hat{o}^{(0)}$ may be the inverse- or Wiener-filtered estimate $y(x) \otimes i(x)$. Application of a typical constraint may involve chopping off the non-physical parts. Transforming then reveals frequency components beyond the cutoff, which are retained. The new values within the bandpass are discarded and replaced by the previously obtained filtered estimate. The resulting function, comprising the filtered estimate and the new superresolving frequencies, is then inverse transformed, and so forth.

The method has been thoroughly analyzed in the electron microscopy application by Ferwerda and colleagues [see references cited by Van Toorn and Ferwerda (1977)]. Convergence properties, uniqueness of solutions, and generation of the algorithm have all been investigated with prolate spheroidal wave functions, the properties of Hilbert spaces, and Von Neumann's alternating projection theorem. These developments appear in papers by Gerchberg (1974), DeSantis and Gori (1975), Papoulis (1975), Youla (1978), Sato *et al.* (1981), Lent and Tuy (1981), Fiddy and Hall (1981), Montgomery (1982), Marks (1981, 1982), Stark *et al.* (1981a,b, 1982), Cahana and Stark (1981), and Marks and Smith (1981). Rushforth and Frost (1980) analyzed and compared these algorithms by using prolate spheroidal wave functions. They also introduced a version that extrapolates the bandwidth by a small explicit increment for each successive iteration. In this sense, their version is similar to Biraud's method.

These algorithms, though mostly guaranteed to converge, often converge with excruciating slowness. Fienup (1979, 1980) has a similar approach, which he calls the input–output method, that converges more quickly. He has recently (Fienup, 1982) compared his method with the earlier algorithms and gradient search methods as well. His conclusion favors the input–output method. Fienup has suggested to Frieden (private communication) that switching strategies between iterations is often effective in improving the rate of convergence because it avoids a saturation or plateau effect that has been noted (Fienup, 1982). Schafer *et al.* (1981) review and analyze constrained iterative algorithms and include the alternating projection methods as a special case.

Interest primarily in alternating projection methods has been sufficient to schedule a special "Topical Meeting on Signal Recovery and Synthesis with Incomplete Information and Partial Constraints" (Lake Tahoe, Nevada, January 12–14, 1983).

E. Howard's Methods

In previous sections we have seen how constraints of boundedness have enabled recovery of frequencies beyond the band limit of the observing instrument. Starting with the inverse-filter estimate

$$\hat{o}(x) = y(x) \otimes i(x), \tag{42}$$

Howard (1981a, 1981b) asked precisely what additional frequency components were required to ensure a physically acceptable result $\hat{o}_p(x)$. Specifically, he represented this result in the following way:

$$\hat{o}_p(x) = \hat{o}(x) + \hat{o}_e(x), \tag{43}$$

where $\hat{o}_e(x)$ is the bandwidth-extension function that supplies the needed but missing components. In Howard's method, only the most-certain components of the inverse-filter estimate $\hat{o}$ are used, say, for frequencies ω such that $|\omega| < \Omega_H$. Cutoff Ω_H may be Frieden's optimum processing bandwidth (Frieden, 1975) or it may be chosen by other means. It could be as large as band limit Ω, but is usually less because of relative suppression by $\tau(\omega)$ near Ω.

Solutions for objects of finite extent are obtained through the familiar minimum mean-square-error criterion by selecting the Fourier components that minimize

$$\Phi = \int [\hat{o}(x) + \hat{o}_e(x)]^2 \, dx, \tag{44}$$

where the integral covers the region outside the finite object extent.

In the practical numerical case $\hat{o}(x)$ and its transform $\hat{O}(\omega)$ are both sampled. Thus, setting equal to zero the partial derivatives of Φ with respect to the Fourier coefficients results in a linear set of normal equations that can easily be solved in closed form. The FFT is then used to obtain the sampled $\hat{o}_e(x)$ from its Fourier coefficients. Estimate $\hat{o}_p(x)$ results from combining $\hat{o}_e(x)$ and $\hat{o}(x)$ via Eq. (43). Only those coefficients are sought that are needed to produce an acceptable solution. By ignoring components of very high frequency, the ill-posed problem of deconvolution is regularized (Rushforth *et al.*, 1982) through smoothing. Note that owing to the retention of only a finite number of Fourier coefficients, the method does not produce $\hat{o}_p(x) = 0$ for x outside the known finite object extent. The mean-square value of $\hat{o}_p(x)$ is, however, the smallest that it can be with a finite number of Fourier coefficients.

Although restorations by this method are superior to those obtained by inverse filtering alone, it should be no surprise that they do not show the dramatic improvements obtainable through the positivity constraint. When, in fact, the integration limits of Eq. (44) are taken well outside the true extent of the object, very little improvement is noted. The applied finite-extent bound must literally butt against the true object for greatest effect. How, then, can the more powerful and useful constraint of positivity be incorporated?

Howard (1981b) attacked this problem by considering the minimization of

$$\Phi = \int_{-\infty}^{\infty} [\hat{o}_p H(-\hat{o}_p)]^2 \, dx. \tag{45}$$

Here H is the Heaviside step function. Physical realizability now implies positivity for $\hat{o}_p$. With this modification, only negative values of $\hat{o}_p$ contribute to the integral. Unfortunately, Φ is no longer differentiable. The restriction on Φ expressed in Eq. (45) inhibits derivation of the least-squares normal equations. Instead, Howard considered the Heaviside step as a limit:

$$H(-\hat{o}_p) = \lim_{K \to \infty} [1 - \exp(K\hat{o}_p)]^{-1}. \tag{46}$$

By making the substitution for H and after some algebra taking the limit, he obtained a set of simultaneous nonlinear equations for the Fourier coefficients. Howard solved these equations by an efficient iterative method. His solutions are comparable to the best that we have come to expect from the positive-constrained restoring methods.

Note that other constraints may be added. Finite extent is particularly easy. Combining the two ideas in this section is done merely by using the finite-extent limits with the integral in Eq. (45).

Howard has applied his method to high-resolution infrared spectroscopic data obtained by dispersive techniques and to both experimental and simulated Fourier interferograms. The method in the latter application explicitly renders the data as they would be observed by an interferometer having a path difference exceeding the mechanical limits of the instrument used for the observation. Details on both the method and its application constitute Chapters 9 and 10.

Zhou and Rushforth (1982) and Rushforth *et al.* (1982) have explored Howard's finite-extent-bounded method via matrix techniques. Zhou and Rushforth demonstrated the effectiveness of utilizing the maximum amount of prior knowledge.

F. Monte Carlo Methods

We first mentioned the applicability of optimization (minimization) methods in Section V.C of Chapter 1. Constraints pose no particular problem to many of these methods. It would seem that the deconvolution problem with object amplitude bounds should be a straightforward application. The most general case, however, deals with each sampled element $\hat{o}_m$ of the estimate as a parameter of the objective function and hence the solution. Excessive computation is then required. The likelihood is great that only local minima of the objective function Φ will be found. Nevertheless, the optimization idea may be teamed with a Monte Carlo technique and a decision rule to yield a method having some promise.

1. *A Decision Rule*

The decision-rule approach tried by Frieden (1974, 1975) will serve to introduce the concept of an object built up from grains. In this approach, *both* $\hat{o}$ and x are taken to be discrete. That is, at a particular value of independent variable x_m, we permit $\hat{o}$ to be only an integral multiple of $\Delta\hat{o}$, the grain size. We may consider the use of random numbers to select locations for grain placement. A decision rule provides the basis for acceptance or rejection of a given grain. Frieden's decision rule places the tth grain of intensity increment $\Delta\hat{o}$ at location x_p if it results in a spectrum estimate $\hat{i}^{(t)}$ such that

$$\hat{i}^{(t)}(x_m) \leq \rho i(x_m) \qquad \text{for all } m. \tag{47}$$

The estimate is appropriately incremented by the relation

$$\hat{i}^{(t)}(x_m) = \hat{i}^{(t-1)}(x_m) + \Delta\hat{o}s(x_m - x_p). \tag{48}$$

The factor ρ is understood to have the smallest possible value permitting grain allocation to *any* solution element. The object is updated at location x_p (only) via

$$\hat{o}^{(t)}(x_p) = \hat{o}^{(t-1)}(x_p) + \Delta\hat{o}. \tag{49}$$

Because ρ is made to be minimum, the object tends to be built up in its brightest and broadest regions first. The grains that form the tails are placed last. Because the requirement on ρ assures allocation of the grain where it is *most* needed, chance really does *not* play a role in the use of the decision rule.

Other decision rules are, of course, possible. As noted earlier, local minima in the objective function surface tend to trap optimization algorithms into globally nonoptimum solutions. When an algorithm is trapped, small perturbations in the estimate elements $\hat{o}(x_n)$ only raise the value of the objective function $\Phi[\hat{o}(x_1), \ldots, \hat{o}(x_M)]$. In conventional optimization algorithms, large perturbations are sometimes used in an attempt to jump from the trap. A perturbation is accepted only if diminished Φ results. A particular Monte Carlo approach having strong parallels in statistical physics goes against this basic philosophy. The principles are detailed in the following section.

2. Annealing

If we increment solution $\hat{o}$ grain by grain and if the algorithm is trapped in a local minimum, it is impossible for a single grain to produce the jump needed to yield a reduced Φ. This barrier accrues from the definition of a grain as a small increment, and the concept of a local minimum having finite breadth. A means of provisionally accepting grains that *raise* Φ is needed.

When machine computation was still young, Metropolis *et al.* (1953) developed an algorithm that overcame the local-minimum trapping problem encountered in the search for the equilibrium configuration of a collection of atoms at a given temperature. This algorithm accepted steps that both decreased and *increased* the hamiltonian. Upward steps were accepted with a probability that was given by the Boltzmann factor: the higher the temperature, the more likely the acceptance. Downward steps were always accepted. After many computer cycles at a given temperature, the system evolved toward thermal equilibrium. Its macroscopic parameters had assumed the Boltzmann distribution appropriate to the specified temperature.

Annealing is a physical process by which one gradually lowers the temperature of a system until its parameters "freeze" into their final values. Kirkpatrick et al. (1983) have simulated this process by starting with a "melted" system and stepwise-lowering the temperature in the algorithm of Metropolis et al. (See also Physics Today, May 1982, pp. 17–19.) They applied this method to finding the mimima of "hamiltonians" describing problems of optimum integrated circuit design and computer wiring, and to the classic traveling salesman problem. Smith et al. (1983) have used the method to attack the problem of reconstructing radio-labeled objects imaged in two projections by aperture arrays.

We propose here to apply the technique to the bounded deconvolution problem. Trial grain positions may be selected at random or by some other process, such as bit-reversed sampling (Allebach, 1981; Deutsch, 1965). The trial solution after t grains have been placed is $\hat{o}_m^{(t)}$. The corresponding objective function may be, but is not limited to, a sum of squares:

$$\Phi^{(t)} = \sum_n \left[i(x_n) - \sum_m s(x_n - x_m)\hat{o}_m^{(t)} \right]^2. \tag{50}$$

The influence of trial grain t at location x_p is assessed by evaluating the change in Φ. We may express the object change at an arbitrary location x_m by the relation

$$\hat{o}^{(t)}(x_m) = \hat{o}^{(t-1)}(x_m) + \Delta\hat{o}\,\delta_{mp}, \tag{51}$$

where δ_{mp} is the Kronecker delta. This relation may be employed with the definition of Φ given in Eq. (50) to evaluate the change

$$\Delta\Phi = \Phi^{(t)} - \Phi^{(t-1)}$$
$$= \Delta\hat{o}\left\{ \sum_n \left[s(x_n - x_p) \right]^2 \right\} - 2\,\Delta\hat{o}\left[\sum_n i(x_n)s(x_n - x_p) \right]$$
$$+ 2\,\Delta\hat{o}\left[\sum_n s(x_n - x_p)i^{(t-1)}(x_n) \right]. \tag{52}$$

In the last term we identify

$$i^{(t-1)}(x_n) = \sum_m s(x_n - x_m)\hat{o}_m^{(t-1)}. \tag{53}$$

In optimizing the computation we note that only this term has a quantity that changes with iteration. Furthermore, i needs to be updated only over the range of finite s via

$$i^{(t)}(x_n) = i^{(t-1)}(x_n) + s(x_n - x_p)\,\Delta\hat{o}. \tag{54}$$

Having now a value for $\Delta\Phi$, we may express the "Boltzmann factor" as $\exp(-\Delta\Phi/kT)$. The "Boltzmann constant" k is given the units of $T/\Delta\Phi$. In the manner of Kirkpatrick et al., we start with T large, a "melted" system. If a trial grain produces a reduction in Φ and if it does not give rise to a constraint violation, it is unequivocally accepted. If Φ grows, the grain's probability of acceptance is $\exp(-\Delta\Phi/kT)$, contingent on the value of a random number.

In one possible implementation, the temperature may be held fixed at temperature T until little further overall reduction in Φ is observed. The objective function will fluctuate about an equilibrium value Φ_T. (Note that, for this process to work, grains must have both positive *and negative* values.) The temperature may then be repeatedly dropped in fixed increments, this process being repeated according to an annealing schedule. Alternatively, very small decrements in T may be made with each trial grain. At some point, lowering T will provide no further reduction of Φ. This occurrence indicates that $\hat{o}$ is in the "ground state"—that we have found the $\hat{o}$ best accommodating both data and constraints. How are constraints implemented? Very simply, a grain is rejected whenever it would give rise to a violation.

The inherent simplicity and flexibility of the method are appealing. It would seem that there is much room for development. Many interesting questions can be posed. Would varying the grain size be useful? Perhaps large grain sizes would be useful in the beginning, at high temperatures, to approximate the solution while economizing on computation. What annealing schedules are best? How can the rich parallel with statistical mechanics be best exploited? Some of these questions were addressed in the original work by Kirkpatrick et al. (1983). It is likely that all aspects of these and many other questions will shortly be explored.

VI. OTHER METHODS

The linear methods have no provisions for bounds on $\hat{o}(x)$. In principle, solutions are free to assume any value, as long as they satisfy the data $i(x)$. We have shown that a problem of uniqueness results. Many functions are thus possible candidates—they all satisfy the data, or nearly so. Noise, which is always present in measurements, unfortunately provides the basis by which a linear method prefers one solution over another.

It has been demonstrated both theoretically and computationally that a simple bound of positivity produces superior solutions, by eliminating many of the undesired solutions from the selection of possible solutions. Another way of reducing the selection is to employ a norm or solution criterion, typically a quantity to be extremized. In the foregoing we used norms of minimum mean-square error, maximum likelihood, and maximum entropy. All these norms force a unique answer to the problem. Frieden (1979, 1980, 1981) has recently developed a new norm, that of maximum information. This norm selects the $ô(x)$ that conveys the most information about itself into the data $i(x)$. Frieden found empirically that solutions based on maximum information closely resemble the maximum-entropy solution for impulsive objects (such as spectral lines), but show differences for objects having edges and plateaus.

Since the value of bounds has come to be widely accepted, numerous other effective bounded methods have appeared. Linear programming has provided the basis for a method presented by Mammone and Eichmann (1982a, 1982b). In a method loosely related to linear programming, MacAdam (1970) exploited the relationship between polynomial multiplication and convolution. His method is particularly suited to human interactive adjustment of constraints.

Constraints were used by Ichioka et al. (1981) in an application of the method of steepest descent. The rapid-convergence properties of the conjugate gradient method have been applied to deconvolution by Angel and Jain (1978). Maitre (1981) achieved superresolution by adapting constraints to the conjugate gradient method. Lawson and Hanson (1974) developed the algorithm NNLS to solve the least-squares problem in the presence of the nonnegativity constraint. Chambless and Broadway (1981) have successfully applied it to nuclear radiation spectra.

In general, problems having solutions that vary radically or discontinuously for small input changes are said to be ill-posed. Deconvolution is an example of such a problem. Tikhonov was one of the earliest workers to deal with ill-posed problems in a mathematically precise way. He developed the approach of regularization (Tikhonov, 1963; Tikhonov and Arsenin, 1977) that has been applied to deconvolution by a number of workers. See, for example, papers by Abbiss et al. (1983), Chambless and Broadway (1981), Nashed (1981), and Bertero et al. (1978). Some of the methods that we have previously described fall within the context of regularization (e.g., the method of Phillips and Twomey, discussed in Section V of Chapter 3). Amplitude bounds, such as positivity, are frequently used as key elements of regularization methods.

In this chapter, we have detailed Frieden's maximum-likelihood approach. Other probability-based methods based on Bayes' theorem have been explored by Richardson (1972), Lucy (1974), Hunt (1975, 1977), Mendes and DePolignac (1975), Mendes et al. (1975), Trussell (1976), Kennett et al. (1978a, 1978b, 1978c), Hunt and Trussell (1976), and Trussell and Hunt (1979).

VII. CONCLUDING REMARKS

This chapter has attempted to give some flavor of the historical development of nonlinear methods. Early investigators of these methods expended great effort in overcoming the popular notion that bandwidth extrapolation was not possible or practical. It was, for example, believed that the Rayleigh limit of resolution was a limit of the most fundamental kind—unassailable by mathematical means. To be sure, the popular notion was reinforced by a long history of misfortune with linear techniques and hypersensitivity to noise. Anyone who still needs to be convinced of the virtues of the nonlinear methods would benefit from reading the paper by Wells (1980); the nonlinear point of view is nowhere else more clearly stated.

Now that the efficacy and potential of the nonlinear methods have gained substantial acceptance, we need not fear that a few cautionary notes will spoil the sensible trend in their direction.

A. Cautions

First let us deal with deconvolution in general. We have a few admonitions to the reader of a literature report on a new method. They should ask, does the writer deal fairly with noise? Even the most volatile of the linear methods can produce a reasonable restoration when noise is limited to roundoff error in the seventh significant figure of the "data." A method's capability of yielding acceptable restorations in the presence of realistic noise is critical to its practicality.

There is also the question of convergence. Proof of convergence does not constitute proof that a given method is practical. A method that converges quickly and consistently in trials with real data wins hands down over a proved converging method that requires hundreds or even thousands of iterations, even though the former method may lack the required proof. All this is contingent, of course, on convergence to the *correct* solution. Employment of the proper norm and/or bound goes a long way toward assuring validity.

An amplitude bound has the added virtue of producing solutions having reduced noise sensitivity, fewer artifacts, superior resolution, and possible bandwidth extrapolation. In contrast, methods having an output that is linear in the irradiance data $i(x)$ either produce artifacts or trade off resolution to suppress artifacts. If a bound makes physical sense and can be computationally afforded, use it. Simple clipping of unphysical parts does not always work well, however. Subtle techniques may be more desirable.

In spite of authors' usual claims, the newest algorithm is not necessarily the best. Nevertheless, a new algorithm, even if it performs poorly, may contain the seed of an idea that can be adapted and improved. Such an algorithm, however, is not for the reader who needs a proved effective method. It is material for the deconvolution researcher.

Next, we ask whether anything is lost by use of the bounded methods. Although the answer is yes, the loss is almost always vastly outweighed by the benefits. In using bounded methods, we do, in fact, lose the ability to observe a certain kind of unexpected result. Consider, for example, a fluorescing sample placed between a source and an optical absorption spectrometer. A deconvolution method imposing an upper transmittance bound of 100% will be confused by the data, to say the least. The moral? Be sure of the validity of constraints before you apply them.

Finally, we suggest that a backlog of data on objects (spectra) of known properties be analyzed until the characteristics of the deconvolution method used are completely familiar. Only then should results yielded by unknown objects be judged. As in any other experimental work, the experiment should be repeated to verify reproducibility and develop confidence in the result. In this sense, the deconvolution process may be treated just like any piece of laboratory apparatus. Indeed, it takes on that identity when packaged in a laboratory microcomputer.

B. The Future

Linear deconvolution methods have served to educate us as to the pitfalls of the deconvolution problem. Their occasional successful applications both tantalized and discouraged us. Now, there are fewer and fewer circumstances in which use of linear methods is justified. The more-generally useful nonlinear methods have teamed with the powerful hardware that they demand to enhance future prospects for wide application of deconvolution methods.

There is no doubt that the next few years will continue to see further improvements through the incorporation of very general bounds and the fullest utilization of statistical prior knowledge. Computing time and cost will also be reduced through improvements in both algorithms and hardware.

We anticipate a proper answer to the question, how far can we go?— theoretically and experimentally. We might expect that instruments will be specifically designed to produce data for subsequent deconvolution. Parameters of interest in such a design might be high signal-to-noise ratio and precise control of the independent variable x. There will be interest in optimum tailoring of the instrument response $s(x)$. Perhaps integrated hardware–software systems will define the state of the art.

Future experimentalists will come to regard deconvolution as an essential tool in the standard repertoire. Through deconvolution they will make otherwise-expensive observations with inexpensive instruments. The limits of state-of-the-art experimental apparatus will be extended. This progress, in turn, will produce data that cannot fail to reveal phenomena that would otherwise lie hidden. The new experimental results will stimulate advances in physical theory, thereby providing incentives for better and better apparatus. Because of its applicability across disciplinary lines, deconvolution will play a key role in the advancement of all branches of science that yield to the tools of measurement.

REFERENCES

Abbiss, J. B., DeMol, C., and Dhadwal, H. S. (1983). *Opt. Acta* **30**, 107–124.

Allebach, J. P. (1981). *J. Opt. Soc. Am.* **71**, 99–105.

Andrews, H. C., and Hunt, B. R. (1977). "Digital Image Restoration." Prentice-Hall, Englewood Cliffs, New Jersey.

Angel, E. S., and Jain, A. K. (1978). *Appl. Opt.* **17**, 2186–2190.

Barnes, W. L., Susskind, J., Hunt, R. H., and Plyler, E. K. (1972). *J. Chem. Phys.* **56**, 5160–5172.

Bertero, M., DeMol, C., and Viano, G. A. (1978). *Opt. Lett.* **3**, 51–53.

Biraud, Y. (1969). *Astron. Astrophys.* **1**, 124–127.

Blass, W. E., and Halsey, G. W. (1981). "Deconvolution of Absorption Spectra." Academic Press, New York.

Burg, J. P. (1967). *In* Proc. 37th Meeting Society Exploration Geophysicists, Oklahoma City, Oklahoma.

Burg, J. P. (1975). Ph.D. Thesis, Stanford Univ.

Cahana, D., and Stark, H. (1981). *Appl. Opt.* **20**, 2780–2786.

Chambless, D. A., and Broadway, J. A. (1981). *Nucl. Instrum. Methods* **179**, 563–571.

DeSantis, P., and Gori, F. (1975). *Opt. Acta* **22**, 691–695.

Deutsch, S. (1965). *IEEE Trans. Broadcast.* **11**, 11–21.

Fiddy, M. A., and Hall, T. J. (1981). *J. Opt. Soc. Am.* **71**, 1406–1407.

Fienup, J. R. (1979). *Opt. Eng.* **18**, 529–534.

Fienup, J. R. (1980). *Opt. Eng.* **19**, 297–305.

Fienup, J. R. (1982). *Appl. Opt.* **21**, 2758–2769.

Fougere, P. F. (1977). *J. Geophys. Res.* **82**, 1051–1054.

Fougere, P. F., Zawalick, and Radoski, H. R. (1976). *Phys. Earth Planet. Inter.* **12**, 201–207.

Frieden, B. R. (1971). "Restoring with Maximum Likelihood." Technical Report Number 67, Optical Sciences Center, University of Arizona, Tucson.

Frieden, B. R. (1972). *J. Opt. Soc. Am.* **62**, 511–518.

Frieden, B. R. (1974). *J. Opt. Soc. Am.* **64**, 561 (abstract).

Frieden, B. R. (1975). *In* "Topics in Applied Physics: Picture Processing and Digital Filtering," Vol. 6 (T. S. Huang, ed.), pp. 177–248. Springer-Verlag, New York.

Frieden, B. R. (1979). SPIE Vol. 207, Applications of Digital Image Processing III, pp. 14–25.

Frieden, B. R. (1980). *Opt. Eng.* **19**, 290–296.

Frieden, B. R. (1981). *J. Opt. Soc. Am.* **71**, 294–303.

Frieden, B. R. (1983). "Probability, Statistical Optics, and Data Testing: A Problem Solving Approach." Springer-Verlag, New York.

Frieden, B. R., and Burke, J. J. (1972). *J. Opt. Soc. Am.* **62**, 1202–1210.

Frieden, B. R., and Swindell, W. (1976). *Science* **191**, 1237–1241.

Frieden, B. R., and Wells, D. C. (1978). *J. Opt. Soc. Am.* **68**, 93–103.

Gerchberg, R. W. (1974). *Opt. Acta* **21**, 709–720.

Gerchberg, R. W., and Saxton, W. O. (1972). *Optik* **35**, 237–246.

Gindi, G. (1981). Ph.D. Dissertation, Univ. Arizona, Tucson.

Gold, R. (1964). "An Iterative Unfolding Method for Response Matrices." AEC Research and Development Report ANL–6984, Argonne National Laboratory, Argonne, Illinois.

Goodman, J. W. (1968). "Introduction to Fourier Optics." McGraw-Hill, New York.

Goody, R. M. (1964). "Atmospheric Radiation." Oxford University Press, London.

Guillemin, E. A. (1949). "The Mathematics of Circuit Analysis," p. 288. Wiley, New York.

Gull, S. F., and Daniell, G. J. (1978). *Nature* **272**, 686–690.

Harris, J. L. (1964). *J. Opt. Soc. Am.* **54**, 931–936.

Harris, J. L., Sr. (1971). *J. Opt. Soc. Am.* **61**, 1563 (abstract).

Herschel, R. S. (1971). Ph.D. Dissertation, University of Arizona, Tucson. Available as Technical Report Number 72, Optical Sciences Center, University of Arizona, Tucson.

Hirsch, P. M., Jordan, J. A., and Lesem, L. B. (1971). U.S. Patent No. 3,619,022.

Howard, S. J. (1981a). *J. Opt. Soc. Am.* **71**, 95–98.

Howard, S. J. (1981b). *J. Opt. Soc. Am.* **71**, 819–824.

Hunt, B. R. (1975). International Optical Computing Conference, pp. 11–13.

Hunt, B. R. (1977). *IEEE Trans. Comput.* **C-26**, 219–229.

Hunt, B. R., and Trussell, H. J. (1976). Proc. Int'l. Conf. Acoustics, Speech, and Signal Processing, Philadelphia, April, IEEE. pp. 354–356.

Ichioka, Y., Takubo, Y., Matsuoka, K., and Suzuki, T. (1981). *J. Opt. (Paris)* **12**, 35–41.

Jansson, P. A. (1968). Ph.D. Dissertation, Florida State Univ., Tallahassee.

Jansson, P. A. (1970). *J. Opt. Soc. Am.* **60**, 184–191.

Jansson, P. A., and Davies, R. D. (1974). *J. Opt. Soc. Am.* **64**, 1372.

Jansson, P. A., Hunt, R. H., and Plyler, E. K. (1968). *J. Opt. Soc. Am.* **58**, 1665–1666.

Jansson, P. A., Hunt, R. H., and Plyler, E. K. (1970). *J. Opt. Soc. Am.* **60**, 596–599.

Jaynes, E. T. (1968). *IEEE Trans. Syst. Sci. Cybernetics* **SSC-4**, 227–241.

Jones, R. N., Venkataraghavan, R., and Hopkins, J. W. (1967). *Spectrochim. Acta* **23A**, 925–939.

Kawata, S., and Ichioka, Y. (1980a). *J. Opt. Soc. Am.* **70**, 762–768.

Kawata, S., and Ichioka, Y. (1980b). *J. Opt. Soc. Am.* **70**, 768–772.

Kawata, S., Ichioka, Y., and Suzuki, T. (1979). *Proc. of the 4th Int. Joint. Conf. on Pattern Recognition*, pp. 525–529.

Kennett, T. J., Prestwich, W. V., and Robertson, A. (1978a). *Nucl. Instrum. Methods* **151**, 285–292.

Kennett, T. J., Prestwich, W. V., and Robertson, A. (1978b). *Nucl. Instrum. Methods* **151**, 293–301.

Kennett, T. J., Brewster, P. M., Prestwich, W. V., and Robertson, A. (1978c). *Nucl. Instrum. Methods* **153**, 125–135.

Kikuchi, R., and Soffer, B. H. (1977). *J. Opt. Soc. Am.* **67**, 1656–1665.

Kirkpatrick, S., Gelatt, C. D., and Vecchi, M. P. (1983). *Science* **220**, 671–680.

Lacoss, R. T. (1971). *Geophysics* **36**, 661–675.

Lawson, C. L., and Hanson, R. J. (1974). "Solving Least Squares Problems." Prentice-Hall, Englewood Cliffs, New Jersey.

Lent, A., and Tuy, H. (1981). *J. Math. Anal. Appl.* **83**, 554–565.

Liu, B., and Gallagher, N. C. (1974). *Appl. Opt.* **13**, 2470–2471.

134

Lucy, L. B. (1974). *Astron. J.* **79**, 745–754.

MacAdam, D. P. (1970). *J. Opt. Soc. Am.* **12**, 1617–1627.

MacNeil, K. A. G., and Dixon, R. N. (1977). *J. Electron Spectrosc. Relat. Phenom.* **11**, 315–331.

Maitre, H. (1981). *Comput. Graphics Image Proces.* **16**, 95–115.

Mammone, R., and Eichmann, G. (1982a). *Appl. Opt.* **21**, 496–501.

Mammone, R., and Eichmann, G. (1982b). *J. Opt. Soc. Am.* **72**, 987–991.

Marks, R. J., II (1981). *Appl. Opt.* **20**, 1815–1820.

Marks, R. J., II (1982). *Opt. Lett.* **7**, 376–377.

Marks, R. J., II, and Smith, M. J. (1981). *Opt. Lett.* **6**, 522–524.

Matsuoka, K., Shigematsu, Y., Ichioka, Y., and Suzuki, T. (1982). *Appl. Opt.* **21**, 4493–4499.

Mendes, M., and DePolignac, C. (1975). *Nucl. Instrum. Methods* **127**, 413–419.

Mendes, M., DePolignac, C., and Delestre, C. (1975). *Nucl. Instrum. Methods* **127**, 405–411.

Metropolis, N., Rosenbluth, A. W., Rosenbluth, M. N., Teller, A. H., and Teller, E. (1953). *J. Chem. Phys.* **21**, 1087–1092.

Montgomery, W. D. (1982). *Opt. Lett.* **7**, 1–3.

Nashed, M. Z. (1981). *IEEE Trans. Antennas Propag.* **AP-29**, 220–231.

Papoulis, A. (1975). *IEEE Trans. Circuits Syst.* **9**, 735–742.

Richards, M. A., Schafer, R. W., and Mersereau, R. M. (1979). IEEE International Conference on Acoustics, Speech, and Signal Processing, pp. 401–404.

Richardson, W. H. (1972). *J. Opt. Soc. Am.* **62**, 55–59.

Robaux, O., and Roizen-Dossier, B. (1970). *Opt. Acta* **17**, 733–746.

Rushforth, C. K., and Frost, R. L. (1980). *J. Opt. Soc. Am.* **70**, 1539–1544.

Rushforth, C. K., Crawford, A. E., and Zhou, Y. (1982). *J. Opt. Soc. Am.* **72**, 204–211.

Saghri, J. A., and Tescher, A. G. (1980). Proc. SPIE Conf. Advances in Image Transmission II, Vol. 249, pp. 71–77.

Sato, T., Norton, S. J., Linzer, M., Ikeda, O., and Hirama, M. (1981). *Appl. Opt.* **20**, 395–399.

Schafer, R. W., Mersereau, R. M., and Richards, M. A. (1981). *Proc. IEEE* **69**, 432–450.

Schell, A. C. (1965). *Radio Electron. Eng.* **29**, 21–26.

Siska, P. E. (1973). *J. Chem. Phys.* **59**, 6052–6060.

Smith, W. E., Barrett, H. H., and Paxman, R. G. (1983). *Opt. Lett.* **8**, 199–201.

Stark, H., Cahana, D., and Webb, H. (1981a). *J. Opt. Soc. Am.* **71**, 635–642.

Stark, H., Cahana, D., and Habetler, G. J. (1981b). *Opt. Lett.* **6**, 259–260.

Stark, H., Cruze, S., and Habetler, G. (1982). *J. Opt. Soc. Am.* **72**, 993–1000.

Thomas, G. (1981). Proc. IEEE Conf. on Acoustics, Speech and Signal Processing, Vol. 1, pp. 47–49.

Tikhonov, A. N. (1963). *Sov. Math.* **4**, 1624–1627.

Tikhonov, A. N., and Arsenin, V. Y. (1977). "Solutions of Ill-Posed Problems." Winston and Sons, Washington, D.C.

Trussell, H. J. (1976). Ph.D. Thesis, University of New Mexico, Albuquerque.

Trussell, H. J., and Hunt, B. R. (1979). *IEEE Trans. Comput.* **C-27**, 57–62.

Van Toorn, P., and Ferwerda, H. A. (1977). *Optik* **47**, 123–134.

Wells, D. C. (1980). *SPIE* **264**, 148–154.

Willson, P. D. (1973). Ph.D. Thesis, Michigan State Univ., East Lansing.

Wolter, H. (1961). *In* "Progress in Optics," Vol. 1 (E. Wolf, ed.).

Wong, H. C. F. (1971). M.Sc. Thesis, Queen's Univ., Kingston, Ontario, Canada.

Youla, D. C. (1978). *IEEE Trans. Circuits Syst.* **CAS-25**, 694–702.

Zhou, Y., and Rushforth, C. K. (1982). *Appl. Opt.* **21**, 1249–1252.

CHAPTER 5

Application to Electron Spectroscopy for Chemical Analysis

Peter A. Jansson

Engineering Physics Laboratory, E. I. du Pont de Nemours and Company (Inc.)
Wilmington, Delaware

I.	Introduction	136
II.	Distortions Present in ESCA Spectra	138
	A. Electron Inelastic Scattering	139
	B. X-Ray Excitation Broadening	140
	C. Other Broadening	140
III.	Correction of Distortions	141
	A. Inelastic Scattering Base Line	141
	B. X-Ray Excitation Broadening	143
	C. Instrumental Considerations	144
IV.	Applications	145
	A. Silver Doublet	145
	B. Artificial Doublets	145
	C. Silicon	148
	D. Sample Charging in Polyester Film	149
V.	Concluding Remarks	150
	References	151

LIST OF SYMBOLS

A_1, A_2	relative intensities of the x-ray $K\alpha$ doublet
f	fraction of electrons scattered inelastically
$H(x)$	1 when $x > 0$; when $x \leq 0$; additional background in Chapter 1
$H_-(x)$	$H(-x) = 1 - H(x)$
$(i)_j$	data $i(x)$ sampled with energy spacing Δx
i_l	lth data element, the element at the high-energy end of the scan range
i_L	number of electrons uniformly scattered from energies above limit x_L
$i(x), i$	observed ESCA spectrum; the data
$i_S(x), i_S$	observed spectrum contribution due to inelastically scattered electrons
l	number of data elements in a digitized scan
$(\hat{o}_B)_j$	discrete version of estimate of function $o_B(x)$ sampled with energy spacing Δx

135

DECONVOLUTION:
WITH APPLICATIONS IN SPECTROSCOPY

$\hat{o}_i^{(k)}$	kth estimate of the true spectrum as described in Chapter 1, here identified as estimate of $o_R(x)$
$o(x), o$	spectrum devoid of broadening due to natural and instrumental effects
$o_B(x), o_B, i_R$	ESCA spectrum as observed less inelastic-scattering broadening
$o_R(x), o_R$	spectrum showing both natural and energy analyzer broadening
$s_B(x), s_B$	$\delta(x) + s_S$, the combined base-line spread function due to electrons inelastically scattered in sample plus those unscattered
$s_E(x), s_E$	spread function due to electrical filtering
$s_M(x), s_M$	spread function due to energy analyzer broadening
$s_N(x), s_N$	spread function due to natural broadening of ESCA spectral line
$s_R(x), s_R$	$s_X(x) \otimes s_E(x)$
$s_S(x), s_S$	spread function due to inelastic scattering
$s_X(x), s_X$	spread function due to broadening by exciting x-ray line
x	electron kinetic energy
x_1, x_2	relative energies of the x-ray $K\alpha$ doublet components
x_L	upper energy limit of a spectrometer scan
$\delta(x), \delta$	Dirac δ function or impulse
Δx_X	half-width at half maximum (HWHM) of each of the x-ray $K\alpha$ doublet components
κ	relaxation function
κ_0	constant in relaxation function

I. INTRODUCTION

The pioneering efforts of K. Siegbahn and associates in the 1950s led to the development of a new type of spectroscopy uniquely suited to the study of surfaces (Siegbahn et al., 1967). The technique, in Siegbahn's own nomenclature, is that of electron spectroscopy for chemical analysis (ESCA), which can reveal information about both chemical bonding and elemental composition within about 50 Å of the surface of a sample. It is the x-ray-excited variety of electron spectroscopy; other modes of excitation exist (Herglotz, 1982). ESCA has proved valuable in practical studies of material properties, catalysis, and adhesion, as well as fundamental investigations into the nature of surface chemical bonds. Herglotz and Suchan (1975) have reviewed ESCA methods, instruments, results, and limitations for workers in surface research. The current state of the art in this field has been summarized in a series edited by Brundle and Baker (1977, 1978, 1979, 1981).

In a typical ESCA experiment, a sample is placed in an evacuated chamber and bombarded with x rays, as illustrated schematically in Fig. 1. The x rays dislodge inner-shell electrons, which leave the sample with kinetic energy determined by the difference of the incident x-ray photon energy and the binding energy of the electron before the interaction. The energy analysis of these escaping electrons gives the binding-energy spectrum of the electrons belonging to the various atoms that make up the sample surface.

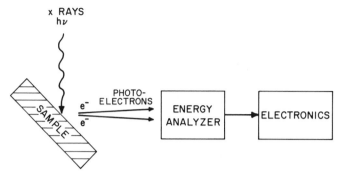

Fig. 1   ESCA spectrometer.

In addition to the principal influence of the atomic energy levels, the binding energy of inner-shell electrons is affected to a lesser degree by the outer electrons that participate in chemical bonding. The various atomic lines in the spectrum therefore appear slightly shifted in energy. The degree of shift depends on the nature of the binding of the atom from which the electron was ejected. This shift is typically small, often not larger than the breadth of the exciting x-ray line, energy analyzer broadening, and other effects.

Detailed analysis of the chemical shifts demands higher resolution than a basic ESCA spectrometer can provide. The needed resolution improvement may be achieved experimentally by monochromatizing the x rays and/or by high-resolution energy analysis of the electrons. Either way, because of the reduced slit widths required, penalties are paid in decreased signal-to-noise ratio and longer observation time. Furthermore, the problem of sample charging (Section IV.D) is typically more severe when monochromatized x rays are used (Brundle and Baker, 1981).

An elegant instrumental technique for partially circumventing the signal-to-noise versus resolution trade-off has been found (Siegbahn *et al.*, 1972). In this technique, x rays are dispersed in energy across the sample. The ejected electrons pass into an energy analyzer the dispersion of which exactly cancels the dispersion of the x rays. The result is the acquisition of a high-resolution spectrum. The added cost of the required equipment, however, combined with the sample-charging disadvantage and the increasing availability of low-cost computation, serves to provide a keen motivation for the development of a computational means of improving resolution.

All known methods for computational resolution improvement, however, involve signal-to-noise versus resolution trade-offs. Clean data having good signal-to-noise ratios are required if useful resolution improvement is to be obtained. The computational approach was facilitated for us by the availability of a unique spectrometer that provides a very high signal-to-noise ratio while retaining moderate to good resolution.

The work reported in this chapter was conducted in 1971, at the Engineering Physics Laboratory of E. I. du Pont de Nemours and Company. This research was performed as part of the ESCA instrument development program active at that time, and was presented by Jansson and Davies (1974). An example of a deconvolved polyester spectrum obtained in the early work appeared in the review article by Herglotz and Suchan (1975).

II. DISTORTIONS PRESENT IN ESCA SPECTRA

The resolution of overlapping spectral peaks depends on their separations, intensities, and widths. Whereas separation and intensity are predominantly functions of the sample, peak width is strongly influenced by the instrument's design. The observed line is a convolution of the natural line, a function characteristic of inelastically scattered electrons that produces a skewed base line, and the instrument function. The instrument function is, in turn, the convolution of the x-ray excitation line shape, the broadening inherent in the electron energy analyzer, and the effect of electrical filtering. This description is summarized in Table I.

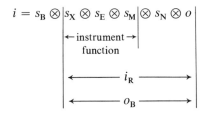

$$ i = s_B \otimes \left| s_X \otimes s_E \otimes s_M \right| \otimes s_N \otimes o $$

TABLE I

Spread Function Component Definitions

$s_S(x)$	spread function due to inelastic scattering
$s_N(x)$	spread function due to natural broadening of spectral line observed
$s_X(x)$	spread function due to broadening by exciting x-ray line
$s_B(x)$	$\delta + s_S$, the combined base-line spread function due to electrons inelastically scattered in sample plus those unscattered
$s_M(x)$	spread function due to energy analyzer broadening
$s_E(x)$	spread function due to electrical filtering

Two contributions to broadening can be thoroughly and easily characterized—the excitation line and electrical filtering. They are therefore good candidates for deconvolution. Although a third contribution—inelastic scattering—is a function of sample composition and morphology, it can be well approximated by a simple model. This model serves as the basis of an effective procedure for correcting the resulting uneven base line. Spectral resolution can be significantly improved by mathematically removing the broadening and distortion produced by these three effects.

A. Electron Inelastic Scattering

Many photoelectrons are not able to emerge from the sample without suffering energy loss by inelastic collisions. At a given kinetic energy, in addition to electrons from an atomic energy-level transition at that energy, one observes some electrons from higher-energy transitions that have partially lost their kinetic energy because of inelastic scattering. The general shape of these energy-loss curves is well known. For the sake of simplicity, we chose to model the scattering contribution at energy x as

$$i_S(x) = f \int_x^\infty o_B(x') \, dx', \tag{1}$$

where $o_B(x')$ is the number of unscattered electrons at energy x', and f is a fraction of the electrons scattered uniformly (in this approximation) to lower energy. We further observe that this integral may be converted to a convolution by introducing the Heaviside step function in abscissa-reversed form:

$$i_S(x) = f \int_{-\infty}^\infty H_-(x - x') o_B(x') \, dx', \tag{2}$$

where $H_-(x) = H(-x) = 1 - H(x)$. When expressed as a convolution, the number of unscattered electrons at a given energy appears as a simple Dirac δ function. For the total contribution at x we then have

$$i(x) = o_B(x) + i_S(x) = \int_{-\infty}^\infty [\delta(x - x') + f H_-(x - x')] o_B(x') \, dx'. \tag{3}$$

The kernel of this equation may be identified by the characteristic base-line response function

$$s_B(x) = \delta(x) + f H_-(x), \tag{4}$$

and Eq. (3) may be abbreviated

$$i = s_B \otimes o_B. \tag{5}$$

B. X-Ray Excitation Broadening

The Lorentzian shape of x-ray emission lines is well founded in quantum theory and has been substantiated experimentally (Hoyt, 1932). Siegbahn *et al.* (1967) discuss the aluminum anode x-ray source as applied to ESCA. Beatham and Orchard (1976) list doublet separations and half-widths derived from the literature and optimized by computer simulation. Källne and Åberg (1975) and Senemaud (1968) also provide values.

We may write the spread-function contribution due to the x-ray source as the following Lorentzian doublet:

$$s_X(x) = \sum_{j=1}^{2} \frac{A_j}{1 + [(x - x_j)/\Delta x_X]^2}. \tag{6}$$

In this expression the A_j are the relative intensities and the x_j the relative energy coordinates of the components. The quantity Δx_X is the component half-width at half maximum (HWHM).

C. Other Broadening

1. *Electron Energy Analyzer*

The precise nature of broadening due to the energy analyzer depends on analyzer design and may not be easy to compute, even in the ideal case. Measurement may also be difficult. The determination of the broadening function $s_M(x)$ may be attempted with the aid of narrow spectral lines or spectral lines of known shape, or by supplying the analyzer with electrons from a thermionic source (Lee, 1973).

The energy analyzer used in the present investigation is a unique high-étendue unit developed by Lee (1973) that facilitates the high signal-to-noise ratio desirable for deconvolution. The effect of broadening due to the energy analyzer was minor when compared with the x-ray excitation broadening affecting the present experiments.

2. *Electrical Filters*

As discussed in Chapter 1, any linear analog filter network may be used. Its performance may always be described by convolution with a filter function, here called s_E. Many modern digital filters may be thus described. The instrument employed in the present work used a simple single-stage RC filter of the type analyzed in Section II.D of Chapter 2.

3. *Natural Line*

The principal inherent broadening of the ESCA line is related to lifetime effects in a manner similar to that discussed in Section I.A of Chapter 2.

III. CORRECTION OF DISTORTIONS

In this investigation, only electron inelastic scattering and x-ray line broadening were chosen for substantial correction. There were two principal reasons: first, these broadening mechanisms account for the largest part of the distortion, and, second their contributions are easily determined or approximated.

A. Inelastic Scattering Base Line

Because the function given by Eq. (4) neither obliterates nor strongly suppresses the high Fourier frequencies in the data, we would expect a linear method to perform relatively well. A simple iterative approach based on the direct method of Section I of Chapter 3 does, in fact, prove effective.

From Eqs. (1) and (3) we may write the total contribution of the observed electron flux as

$$i(x) = o_B(x) + f \int_x^\infty o_B(x')\, dx'. \tag{7}$$

For data taken over a limited scan range it is convenient to break up the integral:

$$i(x) = o_B(x) + i_L + f \int_x^{x_L} o_B(x')\, dx'. \tag{8}$$

In this equation x_L is the high-energy scan limit, and we have defined the number of electrons uniformly scattered from energy above x_L as

$$i_L = f \int_{x_L}^\infty o_B(x')\, dx'. \tag{9}$$

Solving for the number of unscattered electrons and converting the resulting

equation to a discrete form suitable for use with digitized data, we obtain
the spectrum as it would appear in the absence of scattering:

$$(o_B)_j = \begin{cases} (i)_j - i_L - f\,\Delta x \sum_{k=j+1}^{l} (o_B)_k, & j = l-1, l-2, \ldots, 0, \\ (i)_j - i_L, & j = l, \end{cases} \qquad (10)$$

where indices j and k denote the jth and kth kinetic energy intervals of width
Δx, respectively, and l is the index corresponding to the energy of the upper
scan limit x_L.

Equation (10) may be used as a recurrence formula to obtain an estimate
of $(o_B)_j$ that depends on having previously obtained estimates of $(o_B)_{j'}$ for
$j' = j + 1, \ldots, l$. The quantity f must be chosen correctly for this procedure
to work. One possible method of determining f begins with choosing a
section of spectrum that begins and ends in a "base-line" region devoid of
spectral structure, so that $(o_B)_l = (o_B)_0 = 0$. An assumed value of f is used
to compute the estimate $(\hat{o}_B)_j$ based on Eq. (10). The last value, $(\hat{o}_B)_0$, will be
either underestimated or overestimated, depending on the initial assumption
for f. This information may be used in a simple Newton's method update of
f and repeated until convergence is achieved $[(\hat{o}_B)_0 \approx 0]$.

For $j = l$, we have $(\hat{o}_B)_l = (i)_l - i_L$. For scans started in the "base-line"
region we find $(i)_l = i_L$. It may sometimes be convenient to subtract the value
$(i)_l$ from the entire data set before further processing so that the data may
be treated as if $i_L = 0$.

It is also possible (and instructive) to express Eq. (10) in a nonrecurrence
form containing the data values $(i)_k$. After some algebra, we obtain

$$(o_B)_j = (i)_j - i_L - f\,\Delta x \sum_{k=j+1}^{l} [(i)_k - i_L](1 - f\,\Delta x)^{k-j-1}, \qquad (11)$$

$$j = 1, \ldots, l-1.$$

The quantity f may be determined and Eq. (11) applied in the same manner
as for Eq. (10). This time, of course, the $(\hat{o}_B)_j$ may be computed in either
ascending or descending sequence. With both Eqs. (10) and (11) it may be
desirable to use the average of a group of points to fix values for the end
points $(i)_0$ and $(i)_l$. Only three to five iterations proved to be necessary in our
applications (Jansson and Davies, 1974).

A widely used method described by Shirley (1972) is similar, but it employs instead a summation over the observed electron flux $(i)_k$, which includes contributions from electrons already scattered. Our own iterative
method is based on the assumption of uniform scattering of a fixed fraction
of the electrons that would be observed in the complete absence of scat-

tering. Our formula reduces to the Shirley formula when the quantity $|(1 - f \Delta x)^l - 1|$ is small. This occurs when $f \Delta x$ is small.

Other workers have carried out Van Cittert deconvolution of the x-ray line contribution without treating inelastic scattering (Wertheim, 1975; Wertheim and Hüfner, 1975), or they have used Van Cittert's method to remove the scattering contribution (Madden and Houston, 1976). Wertheim and Hüfner (1975) have called into question the use of the "traditional" background correction methods that are based on the assumption of electron scattering. They present supporting arguments and data. Prior removal of the base line is, however, necessary to take full advantage of the physical-realizability constraints imposed by the method that we have used to remove the x-ray broadening.

We recognize that the present simple model for inelastic scattering does not account for certain details such as structure in the energy-loss spectrum due to plasmons (Davis and Lagally, 1981). Nevertheless, its simplicity and effectiveness bear out its practice.

B. X-Ray Excitation Broadening

A review of deconvolution methods applied to ESCA (Carley and Joyner, 1979) shows that Van Cittert's method has played a big role. Because the Lorentzian nature of the broadening does not completely obliterate the high Fourier frequencies as does the sinc-squared spreading encountered in optical spectroscopy (its transform is the band-limiting rect function), useful restorations are indeed possible through use of such linear methods. Rendina and Larson (1975), for example, have used a multiple filter approach. Additional detail is given in Section IV.E of Chapter 3.

Constraining the solution to be positive can nevertheless provide the added benefits of reduced sensitivity to noise and improved resolution. Gold's iterative ratio method (Chapter 1, Section IV.A), for example, has been used successfully by a number of workers, including MacNeil and Dixon (1977) and Delwicke et al. (1980). MacNeil and Delwicke have compared it with the standard Van Cittert method, which is linear.

Other constrained methods have also been applied. Beatham and Orchard (1976) experimented with Biraud's method but experienced only limited success. Vasquez et al. (1981) found that maximum entropy is capable of yielding excellent results on simulated ESCA data. The authors, who used Burg's method, cite its freedom from need for trial-and-error optimization. They did, however, have to develop methods of dealing with problems of instability and lack of an order-selecting criterion.

For the present work, we chose the constrained method described by Jansson (1968) and Jansson *et al.* (1968, 1970). See also Section V.A of Chapter 4 and supporting material in Chapter III. This method has also been applied to ESCA spectra by McLachlan *et al.* (1974). In our adaptation (Jansson and Davies, 1974) the procedure was identical to that used in the original application to infrared spectra except that the data were pre-smoothed three times instead of once, and the variable relaxation factor was modified to accommodate the lack of an upper bound. Referring to Eqs. (15) and (16) of Section V.A.2 of Chapter 4, we set $\kappa = 2\hat{o}_n^{(k)}\kappa_0$ for $\hat{o}_n^{(k)} \leq \frac{1}{2}$ and $\kappa = \kappa_0 \exp[\frac{1}{2} - \hat{o}_n^{(k)}]$ for $\hat{o}^{(k)} > \frac{1}{2}$. This function is seen to apply the positivity constraint in a manner similar to that previously employed but eliminates the upper bound in favor of an exponential falloff. We also experimented with $\kappa = \kappa_0$ for $\hat{o}_n^{(k)} > \frac{1}{2}$, and found it to be equally effective. As in the infrared application, *only 10 iterations were needed.*

C. Instrumental Considerations

We applied this method to the removal of broadening by $s_X(x)$, which describes the x-ray line broadening. In generating $s_X(x)$, we were guided by the Kα doublet intensity, separation, and width data of Senemaud (1968). By using component half-widths somewhat larger than the literature values, we partially compensated for the energy analyzer broadening $s_M(x)$ as well.

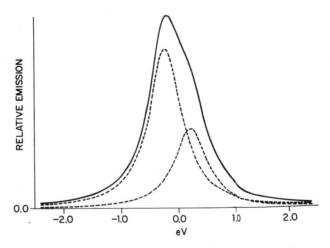

Fig. 2 Aluminum x-ray Kα doublet, widened to partially compensate for energy analyzer broadening. Dotted curves, components; solid curve, their sum.

In the present experiment, we took the aluminum Kα doublet components to be separated by 0.430 eV and to have a half-width of 0.35 eV each. The components were given a 2:1 intensity ratio and scaled to normalize the area under the combined peak. Figure 2 shows the resulting widened $s_X(x)$. We decided to use the same procedure employed in the earlier infrared studies, retaining the original parameters, that is, sample rate, number of passes, type of smoothing, and so on. For the procedure to work properly in this form, it was critical that the data be scaled to lie between 0 and 1 and that sample spacing relative to $s_X(x)$ be approximately the same as that used in the original infrared studies. Here we used 52 points per electron volt.

As noted in the Introduction, resolution correction requires data of superior quality. In the present study we were fortunate to have access to a spectrometer that provided the needed high counting rate without resolution degradation. Two design features combined to provide this capability: a high-intensity aluminum Kα x-ray source and an energy analyzer of uniquely high étendue. The source has been described by Davies and Herglotz (1969) and the analyzer by Lee (1973).

IV. APPLICATIONS

A. Silver Doublet

The silver $3d_{3/2}$–$3d_{5/2}$ doublet at approximately 367- and 373-eV binding energy was chosen for an initial demonstration of the method. Figure 3 compares raw data obtained from our spectrometer with the result after base-line correction and deconvolution. It reveals a nearly threefold improvement in resolution. There is a conspicuous absence of the noise and spurious peaks that are typical of most methods. The slight peak asymmetry is a consequence of the point-successive nature of the iteration scheme. The resulting linewidth represents the remaining broadening due to the natural line plus the finite bandpass of the electron energy analyzer, the small amount of electrical filtering employed, and the function used to presmooth the data.

B. Artificial Doublets

We observed the silver doublet again, this time with a 27-Hz square-wave modulation applied to the retarding potential of the energy analyzer. The spectrometer thus was made to alternate rapidly between two different scan

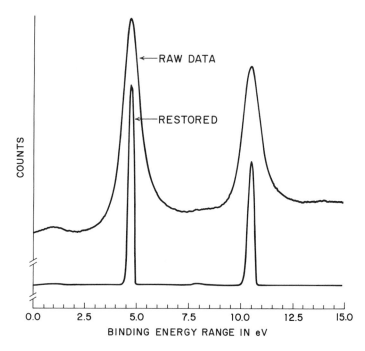

Fig. 3 Silver $3d_{3/2}-3d_{5/2}$ doublet.

ranges, the energy separation between ranges being dictated by the peak-to-peak amplitude of the square wave. Our control over the duty cycle of the square wave was used to adjust the relative intensities of the two spectra being superimposed. The concept applied here may be understood by noting that the observation time spent in each of two states of the square wave is proportional to the number of electrons counted in that state. The energy separation of the superimposed spectra was controlled by variation of the square-wave amplitude. By this means, artificial doublets were created for the purpose of testing deconvolution. The data recorded were otherwise exactly as they would be rendered in any normal spectrometer scan.

Figure 4 shows both data and resulting deconvolution when the silver doublet is observed with a 1-V peak-to-peak square-wave modulation having 2:1 duty-cycle ratio. Each of the doublet components is itself doubled. This component doubling is apparent in the raw data only as a modest broadening but becomes quite obvious after deconvolution. Figure 5 shows a similar run employing a 0.75-V square wave. It is apparent from these runs that the intensity ratios of deconvolved components of strongly blended lines should be taken as being only approximate.

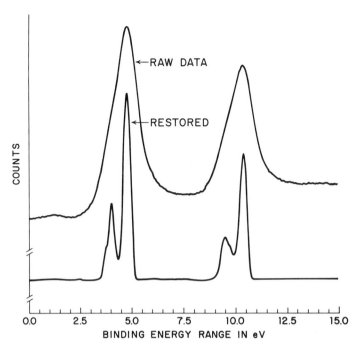

Fig. 4 Artificial doublets: silver spectrum modulated by a 1-V peak-to-peak square wave.

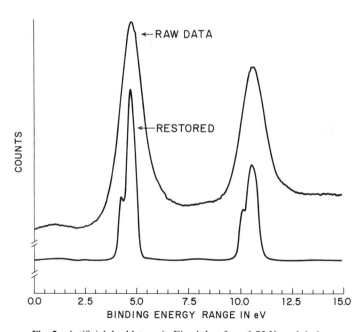

Fig. 5 Artificial doublets as in Fig. 4, but for a 0.75-V modulation.

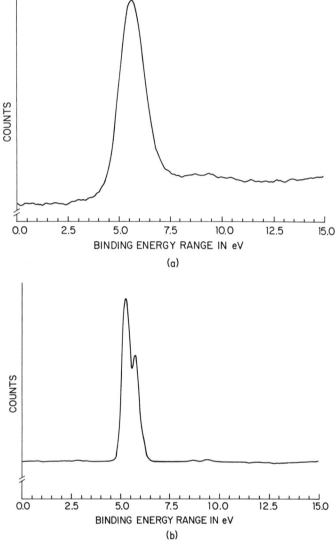

Fig. 6 $2p_{1/2}$ and $2p_{3/2}$ lines of silicon: (a) raw data; (b) result of deconvolution.

C. Silicon

Unaided by deconvolution, only instruments employing monochromatized x rays are able to resolve the $2p_{1/2}$ and $2p_{3/2}$ lines in the spectrum of silicon. Figure 6(a) displays our raw data before deconvolution; there is no obvious evidence of the presence of two components. Yet, the deconvolved spectrum in Fig. 6(b) shows both components clearly. They are separated by less than 0.6 eV. The resolution obtained compares favorably with that obtained by other methods.

D. Sample Charging in Polyester Film

The carbon $1s_{1/2}$ transition was observed in a sample of polyester film at near 1195 eV. The raw data indicate the presence of the three different carbon sites in the polyester molecule. The deconvolved data reveal additional spectral structure (Fig. 7). The additional structure appears in the form of a threefold repetition of the same curve. We easily reproduced the

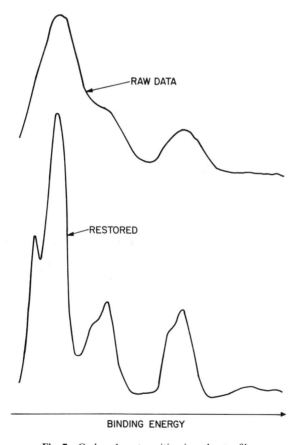

Fig. 7 Carbon $1s_{1/2}$ transition in polyester film.

raw data, but each time the experiment was run, the deconvolution produced a threefold repetition of a different curve.

These results are readily explained (Brundle and Baker, 1981) as being a consequence of the nonuniform electrostatic charging of the sample. The sample loses electrons continuously during the observation process. Owing to the nonconducting nature of the sample, the resulting charge buildup is spatially distributed in a nonuniform manner. Electrons leaving the sample from different areas of the surface move through slightly different potential fields. The observed repeated curve can thus be explained as the potential energy distribution of the surface of the sample. Photoelectrons from each of the characteristic carbon sites are subjected to identical energy shifts. The raw data are therefore seen to result from the convolution of the desired spectrum $o(x)$ with the potential energy distribution.

The foregoing observations suggest that further deconvolution may be possible. Specifically, we could attempt to remove sample-charging effects by using an observed isolated line as a measure of the potential distribution on the surface. Like all computer processing discussed in this chapter, this correction would be applied to data acquired in the normal way and would involve no alteration of the experimental apparatus or sample conditions, as would the electron flood gun method of minimizing such charging effects (Huchital and McKeon, 1972; Wagner et al., 1979).

V. CONCLUDING REMARKS

We have shown that a well-known simplified model of inelastic scattering can be described as a convolution and have applied an iterative base-line correction method based on it. The method proved effective in preprocessing spectra before deconvolution of x-ray line broadening. A deconvolution method constrained to produce only physically realizable solutions was effective in removing this broadening. We tested the method by modulating the potential of conducting samples to create artificial doublets. Deconvolution of spectra from nonconducting samples has explicitly revealed the presence of a potential distribution due to electrostatic charging.

A FORTRAN program implementing this analysis has been run off line in Univac 1108 and Digital Equipment Company PDP-10 computers. The same program, slightly modified, has been run on line in a PDP-11/20 with extended arithmetic option. It corrected the base line and deconvolved a 780-point spectrum in about 2.5 min. The modification included a conversion from floating point to integer arithmetic and assembler language coding of a few statements comprising the convolution loop.

Execution time can be significantly reduced without loss of effectiveness by data sample density reduction and corresponding alteration of the smoothing polynomial. Further improvements can be realized by using modern high-speed microcomputers. Constrained deconvolution times of well under one minute for equivalent spectra should be possible with compact and inexpensive equipment.

ACKNOWLEDGMENTS

The author thanks his colleagues J. D. Lee, for highly valued advice and contributions to stimulating discussions, and H. K. Herglotz, for his guidance as technical leader of the ESCA program at the Du Pont Engineering Physics Laboratory and for the impetus he gave to the research reported here. For their essential contributions, the author is especially indebted to his close collaborators in the present work, R. D. Davies, who spearheaded the deconvolution effort, especially the experimental aspects, and H. L. Grams, for computer programming and valuable insights. The present chapter is, in fact, largely based on the investigations of R. D. Davies. H. L. Suchan is thanked for his substantial engineering contribution in the design of the apparatus. The author is also grateful to his colleagues for their valuable criticisms and suggestions concerning the manuscript for this chapter.

REFERENCES

Beatham, N., and Orchard, A. F. (1976). *J. Electron Spectrosc. Relat. Phenom.* **9**, 129–148.
Brundle, C. R., and Baker, A. D., eds. (1977, 1978, 1979, 1981). "Electron Spectroscopy: Theory, Techniques and Applications," Vols. 1–4. Academic Press, New York.
Carley, A. F., and Joyner, R. W. (1979). *J. Electron Spectrosc. Relat. Phenom.* **16**, 1–23.
Davies, R. D., and Herglotz, H. K. (1969). *Adv. X-Ray Anal.* **12**, 496–505.
Davis, G. D., and Lagally, M. G. (1981). *J. Vac. Sci. Technol.* **18**, 727–731.
Delwiche, J., Hubin-Franskin, M. J., Caprace, G., and Natalis, P. (1980). *J. Electron Spectrosc. Relat. Phenom.* **21**, 205–218.
Herglotz, H. K. (1982). *In* "Surface Treatments for Improved Performance and Properties" (J. J. Burke and V. Weiss, eds.), pp. 19–49. Plenum, New York.
Herglotz, H. K., and Suchan, H. L. (1975). *Adv. Colloid Interface Sci.* **5**, 79–103.
Hoyt, A. (1932). *Phys. Rev.* **40**, 477–483.
Huchital, D. A., and McKeon, R. T. (1972). *Appl. Phys. Lett.* **20**, 158–159.
Jansson, P. A. (1968). Ph.D. Dissertation, Florida State Univ., Tallahassee.
Jansson, P. A., and Davies, R. D. (1974). *J. Opt. Soc. Am.* **64**, 1372.
Jansson, P. A., Hunt, R. H., and Plyler, E. K. (1968). *J. Opt. Soc. Am.* **58**, 1665–1666.
Jansson, P. A., Hunt, R. H., and Plyler, E. K. (1970). *J. Opt. Soc. Am.* **60**, 596–599.
Källne, E., and Åberg, T. (1975). *X-Ray Spectrom.* **4**, 26–27.
Lee, J. D. (1973). *Rev. Sci. Instrum.* **44**, 893–898.
McLachlan, A. D., Jenkin, J. G., Liesegang, J., and Leckey, R. C. G. (1974). *J. Electron Spectrosc. Relat. Phenon.* **3**, 207–216.
MacNeil, K. A. G., and Dixon, R. N. (1977). *J. Electron Spectrosc. Relat. Phenom.* **11**, 315–331.

Madden, H. H., and Houston, J. E. (1976). *J. Appl. Phys.* **47**, 3071–3082.

Rendina, J., and Larson, P. (1975). Paper Number 66, The Pittsburgh Conference on Analytical Chemistry and Applied Spectroscopy, Cleveland Convention Center, Cleveland, Ohio, March 3–7 (abstract only).

Senemaud, C. (1968). Thèsis, docteur ès Sciences Physiques, l'Université de Paris.

Shirley, D. A. (1972). *Phys. Rev.* **B5**, 4709–4714.

Siegbahn, K., Nordling, C., Fahlman, A., Nordberg, R., Hamrin, K., Hedman, J., Johansson, G., Bergmark, T., Karlsson, S., Lindgren, I., and Lindberg, B. (1967). "ESCA-Atomic, Molecular and Solid State Structure Studied by Means of Electron Spectroscopy." Almquist and Wiksells, Uppsala.

Siegbahn, K., Hammond, D., Fellner-Feldegg, H., and Barnett, E. F. (1972). *Science* **176**, 245–252.

Vasquez, R. P., Klein, J. D., Barton, J. J., and Grunthaner, F. J. (1981). *J. Electron Spectrosc. Relat. Phenom.* **23**, 63–81.

Wagner, C. D., Riggs, W. M., Davis, L. E., Moulder, J. F., and Muilenberg, G. E. eds. (1979). "Handbook of X-Ray Photoelectron Spectroscopy." Perkin-Elmer, Eden Prairie, Minnesota.

Wertheim, G. K. (1975). *J. Electron Spectrosc. Relat. Phenom.* **6**, 239–251.

Wertheim, G. K., and Hüfner, S. (1975). *Phys. Rev. Lett.* **35**, 53–56.

CHAPTER **6**

Instrumental Considerations

William E. Blass
George W. Halsey

Department of Physics and Astronomy, The University of Tennessee
Knoxville, Tennessee

I.	Introduction	155
II.	Resolution–Acquisition-Time Trade-Offs	156
III.	Acquiring the Data: Dispersive Spectrophotometers	157
	A. System Model	157
	B. Analytical Characterization of the System Model	160
	C. General Data-Acquisition Considerations	163
	D. Acquisition with Continuous Scanning	170
	E. Acquisition with Step Scanning	171
	F. Calibration of Spectra	171
IV.	Details of Spectroscopic Experiments	173
	A. Impact of Noise on Deconvolution	173
	B. Impact of Bouguer–Lambert Law	174
	C. Impact of Instrumental Response Function	177
V.	Preparing to Deconvolve a Data Set	179
	A. Smoothing	180
	B. Base-Line Considerations	181
VI.	Deconvolving the Data: Constrained Deconvolution	182
	A. Deconvolution Algorithm	183
	B. The Nonlinearity of Deconvolution	184
	References	185

LIST OF SYMBOLS

$A(f)$	Fourier transform of signal processing function $A(t)$
$A(t)$	signal processing function
$A_M(x)$	measured absorptance
$\hat{A}_T(x)$	estimate of true absorptance $A_T(x)$
BW	full bandwidth of tuned amplifier
$B_T(x)$	flux at zero sample pressure
$\text{chop}(x)$	chopping function
d	grating spacing
$d\bar{v}/d\theta_d$	variation in monochromator wave number as the diffraction angle is varied
$d\bar{v}/d\theta_g$	variation in monochromator "center" wave number as the grating angle is varied

DECONVOLUTION:
WITH APPLICATIONS IN SPECTROSCOPY

$D_T(x)$	stray background flux
$e^{(k)}$	error function, also called root-mean-square error associated with kth iteration
f_0	center frequency of tuned amplifier
f_c	focal length of principal collimating mirror
f_{ch}	chopping frequency
gate(t/T)	modulation function related to rect(t); gate$(t/T) = 1$ when t lies between 0 and T, and 0 otherwise
Hg:Ge	mercury-doped germanium detector
$H(t)$	1 when $x > 0$, 0 when $x \le 0$
InSb	indium antimonide detector
$i(x)$	observed signal
j	integer-sequence-ordering variable
k	grating constant
m	number of passes
M	molecular weight
$M(t)$	signal modulation function (e.g., the change in center passband frequency of spectrometer as a function of time)
n	diffraction-order number
$n(t)$	noise voltage
N_j	number of counts in photon channel j
NEP	noise-equivalent power
$o(x)$	true signal as a function of x, here equivalent to time as a sequence-ordering variable
$\hat{o}^{(k)}(x)$	kth estimate of $o(x)$
P	power falling on detector
P_1, P_2	arbitrary power levels
$P(x)$	absorption coefficient
$P_V(x)$	absorption coefficient for a Voigt profile line
Q	Q of a tuned amplifier ($\approx f_0/$BW)
$r_{max}^{(k)}$	constant coefficient of relaxation function $r^{(k)}[\hat{o}^{(k-1)}(x)]$
$r^{(k)}[\hat{o}^{(k-1)}(x)], r^{(k)}[y]$	relaxation function used at kth iteration
$r(x)$	response function
R	responsivity of a detector in volts per watt
R_D	detector resistance
$(R_D)_0$	R_D at zero signal
R_L	load resistance
R_{new}	data-acquisition rate with resolution degraded by multiplicative factor γ
R_{old}	data-acquisition rate prior to degradation
RAM	random access memory
RC	resistance–capacitance
$s(x)$	response function or instrumental spread function
$S(f)$	spectral density function
S/N	signal-to-noise ratio
t	time
T	absolute temperature
T_c	period of chop(x)
TDL	tunable diode laser
$U_T(x)$	transmittance; transmitted spectral flux
$v(t), V_{in}(t)$	input voltage of RC filter

V	voltage
V_0	voltage across detector at zero signal: $R_D = (R_D)_0$
$V_C(t)$, $V_{out}(t)$	output voltage of RC filter
$V_{sig}(\alpha)$	output signal of bias circuit
w_g	ruled width of grating
W	slit width
x	independent variable
X	pressure times path length of sample
X_1, X_2	arbitrary quantities
α	parameter embodying all of the dependence of R_D on incident photon flux
γ	resolution degradation factor
δ_d	$\theta_g - \theta_d$
δ_i	$\theta_i - \theta_g$
Δt_{csam}	continuous scan sampling time interval
$\Delta\theta_d$	angle subtended by exit-slit width at principal collimating mirror
$\Delta\theta_{ssam}$	grating angle interval between data samples; if Ω_{step} is the change in grating angle for one step, $\Delta\theta_{ssam}$ is an integer multiple of Ω_{step}
$\Delta\bar{v}_D$	full width at half maximum of Doppler profile
$\Delta\bar{v}_{min}$	limit of resolution
$\Delta\bar{v}_{obs}$	full width at half maximum of observed line: this usage is different than in other chapters, where $\Delta\bar{v}$ is given a somewhat different meaning
$\Delta\bar{v}_{res}$	full width at half maximum of response function
$\Delta\bar{v}_{ssw}$	spectral slit width in cm^{-1}; note that $\Delta\bar{v}_{ssw}$ is a full width at half maximum equivalent (see text)
$\epsilon(\lambda)$	intermediate symbolic parameter
$\zeta(\omega)$	phase of $\Xi(\omega)$
η	signal-to-noise improvement factor
θ	angle between grating normal and optical axis of collimating mirror
θ_d	diffraction angle
θ_i	angle of incidence
θ_g	grating angle
λ	wavelength
$\bar{v}$	wave number $(= 1/\lambda)$
$\xi(t)$	transfer function of filter
$\Xi(\omega)$	transform of $\xi(t)$
τ	amplifier time constant

INTRODUCTION

Application of deconvolution is at the same time seductively attractive and potentially dangerous. Deconvolution is attractive because the researcher can obtain resolution beyond that achievable without deconvolution. In addition, when resolution enhancement is not a goal, deconvolution can lead to richly enhanced acquisition rates. Deconvolution is dangerous because numerical processing may radically alter the character of experimental data and hence the conclusions drawn from them.

To grasp the flavor of this caveat, consider this: the adjustment range of an instrument's physical parameters is limited. Furthermore, many modifications, because they are time consuming and expensive, are carefully considered before implementation and fully tested afterwards. On the other hand, because deconvolution is numerical, radical changes are readily made and often not as thoroughly tested as are physical changes to the instrument. Very subtle effects are possible.

Lest the reader become disenchanted, consider the following: although the path to utilization of deconvolution contains pitfalls for the unwary, the process is testable at each step. Careful users come to view the process (and requisite computer system) as an extension of the observing device.

II. RESOLUTION–ACQUISITION-TIME TRADE-OFFS

Deconvolution of spectra, such as infrared absorption spectra, provides researchers with a tool that they can use to carry out a particular experiment. It provides an extra measure of flexibility in the design of experiments and in the observation process. In dispersive infrared spectroscopic systems, Blass and Halsey (1981) have shown that effective resolution–acquisition-time trade-offs may be made, owing to the fact that dispersive infrared spectroscopy is usually detector noise limited. Acquisition rates are therefore optical throughput dependent, which is equivalent to saying that acquisition rates are resolution dependent. Blass and Halsey (1981) show that, for a constant signal-to-noise ratio,

$$R_{new} = \gamma^5 R_{old},$$ (1)

where γ is the factor by which the resolution has been degraded, R_{old} the old acquisition rate, and R_{new} the enhanced rate. For example, degrading the resolution by a factor of 2 increases the allowable acquisition rate by a factor of 32. Deconvolution can readily recover the factor of 2. Expensive scientific equipment may therefore be used more effectively. It also permits experiments that would otherwise be impractical owing to excessive observation time. Degrading resolution by a factor of 3, for example, allows an increase in the acquisition rate by a factor of 243.

When an improved signal-to-noise ratio is required, Eq. (1) becomes

$$R_{new} = (\gamma^5/\eta^2)R_{old}, \tag{2}$$

where η is the required improvement factor for the signal-to-noise ratio. Not all forms of spectroscopy allow such productive resolution–acquisition-rate trade-offs. Where they are possible, deconvolution becomes an even more powerful experimental tool.

III. ACQUIRING THE DATA: DISPERSIVE SPECTROPHOTOMETERS

A large proportion of spectral data is acquired by dispersive spectrophotometry. The discussion that follows is restricted to instruments that use a diffraction grating as the principal dispersive element. The sense of the following also applies to systems that use a prism. In general, we treat systems using photosensitive detectors and fixed-position slits. Scanning is achieved by rotation of the diffraction grating.

To place the discussions of this and the following chapter in a clear context, we describe a specific system model. Application discussions may be seen in the light of a specific instrument. To use deconvolution techniques effectively, the instrument as well as the data acquired must be analytically well characterized.

A. System Model

For clarity and precision in the following discussion, the 5-m Littrow spectrometer at the University of Tennessee at Knoxville will serve as the prototype (Jennings, 1974). The optical diagram is presented in Fig. 1. The system is used as an absorption spectrometer; most of the discussion also applies to acquisition of emission spectra.

The source is a resistively heated carbon rod operating at 2600 K (Boyd *et al.*, 1974). From the source, the radiation follows the mirror path in sequence. As the optical beam enters the round vacuum tank the beam begins traversing a prism spectrometer. The need for the prism predisperser arises from the low density of rulings of the echelle gratings used in the grating monochromator (31.6 and 79/mm). The incident beam must pass through this optical bandpass filter prior to detection because of the limited free spectral range of the grating monochromator. The particular prism

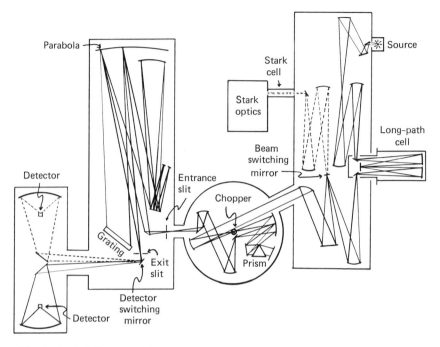

Fig. 1 Optical diagram of the prototype system model. The drawing is not to scale and is not to be considered an optical ray diagram. The principal dispersing element is a coarse echelle ruled grating 20 × 40 cm wide. Theoretical double-pass resolution at four normal slits is approximately 0.095 cm^{-1}; actual achievable resolution is approximately 0.009 cm^{-1}.

predisperser currently used is a Wadsworth–Littrow prism monochromator (Wadsworth, 1894; James and Sternberg, 1969; Stewart, 1970).

As the beam leaves the prism predisperser, it is focused on the entrance slit of the grating monochromator. The slit is curved, has variable width, and opens symmetrically about the chief ray (optical center line of system). The monochromator itself is of the off-axis Littrow variety (James and Sternberg, 1969; Stewart, 1970; Jennings, 1974) and uses a double-pass system described by McCubbin (1961). The double-pass aspect of the system doubles the optical retardation of the incident wave front and theoretically doubles the resolution of the instrument. The principal collimating mirror is a 5-m-focal-length, 102-cm-diam parabola.

From the exit slit, the beam is directed to either of two detectors by the movable mirror behind the exit slit. Typical detectors are InSb, a photovoltaic indium antimonide detector operated at 77 K, and He:Ge, a photoconductive mercury-doped germanium detector maintained at ≈15 K by a closed-cycle helium refrigerator

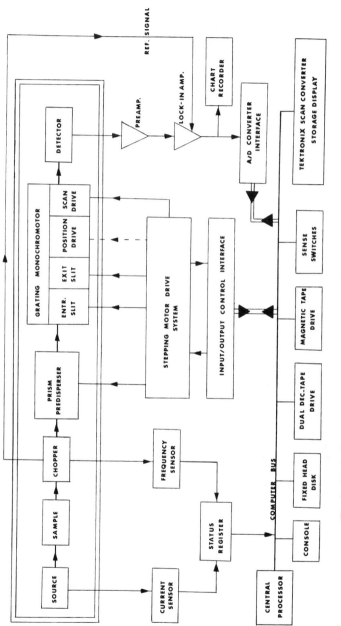

Fig. 2 Block diagram of the spectrometric system at the University of Tennessee.

The rotation of the grating is achieved by using a tangent arm drive. The tangent arm system is driven by a 200-step/revolution stepping motor. One step corresponds to approximately 1.45×10^{-7} rad (8.3×10^{-6} deg). Note particularly that the tangent drive results in an approximately linear relation between grating angle and drive-shaft rotation angle (Blass, 1976b). A block diagram of the entire system is shown in Fig. 2.

B. Analytical Characterization of the System Model

The grating equation for both single- and double-pass Littrow monochromators is given by

$$\bar{v} = nk \csc \theta_g, \qquad (3)$$

where $\bar{v} = 1/\lambda$ is the wave number, $[\bar{v}] = \text{cm}^{-1}$ (λ is the wavelength, $[\lambda] = \text{cm}$); $n = 1, 2, 3, \ldots$ is the grating order; θ_g is the grating angle, that is, the angle between the collimating mirror axis and the grating normal; and $k = 1/2d$ is a grating constant, $[k] = \text{cm}^{-1}$ (d is the grating spacing, $[d] = \text{cm}$). Equation (3) is correct for a true Littrow in single pass and for a double-pass Littrow when the inversion of the single-pass exit image to form the double-pass entrance image is carried out symmetrically about the collimation axis (Jennings, 1974).

The following definitions are required: θ_g is the angle between a ray parallel to the collimator axis and the grating normal, the angle of incidence $\theta_i \equiv \theta_g + \delta_i$, and the diffraction angle $\theta_d \equiv \theta_g - \delta_d$. In the symmetrical Littrow case $\delta_i = -\delta_d$, and Eq. (3) follows from the general grating equation

$$\lambda = (2/nk)(\sin \theta_i + \sin \theta_d). \qquad (4)$$

For the dispersion one finds for single pass (Jennings, 1974)

$$d\bar{v}/d\theta_d = -(\bar{v}/2) \cot \theta_g, \qquad \theta_g, \theta_i = \text{const}, \qquad (5)$$

$$d\bar{v}/d\theta_g = -\bar{v} \cot \theta_g, \qquad \theta_d, \theta_i = \text{const}, \qquad (6)$$

and for double pass

$$d\bar{v}/d\theta_d = -(\bar{v}/4) \cot \theta_g, \qquad \theta_g, \theta_i = \text{const}, \qquad (7)$$

$$d\bar{v}/d\theta_g = -\bar{v} \cot \theta_g, \qquad \theta_d, \theta_i = \text{const}. \qquad (8)$$

From Eqs. (6) and (8) one verifies that the angular separation of two specific wave numbers is the same in single and double pass. Equations (6) and (8) are used to determine scan rates or scanning observation intervals as discussed later.

To determine the spectral slit width, it is necessary to use Eqs. (5) and (7).

Conceptually this can be viewed as the variation in the wave number passed through the monochromator as one moves from one edge of the exit slit to the other. The monochromator is assumed to be set at some center wave number given by Eq. (3).

Using Eqs. (5) and (7) and assuming $\delta_i = \delta_d \approx 0$, we obtain for the spectral slit width ($\Delta\bar{v}$, edge to edge of the slits)

$$\Delta\bar{v}_{ssw} = \left|\partial\bar{v}/\partial\theta_d\right| \Delta\theta_d, \qquad \theta_g, \theta_i = \text{const}, \tag{9}$$

$$\Delta\bar{v}_{ssw} = (\bar{v}W/2mf_c) \cot\theta_g, \tag{10}$$

where W is the slit width, given by

$$W = f_c \, \Delta\theta_d, \tag{11}$$

m is the number of passes, and f_c is the focal length of the principal collimating mirror. The quantity $\Delta\theta_d$ is the angle subtended by the slit width at the principal collimating mirror. Equations (5) and (7) have been used to simplify Eq. (9).

The theoretical Rayleigh limit of resolution is given by

$$\Delta\bar{v}_{min} = (2mw_g \cos\theta_g)^{-1}, \tag{12}$$

where w_g is the ruled width of the grating. The calculated spectral slit width of a well-designed, properly aligned system should be close to the observed resolution of the instrument. Resolution may be approximated by the full width at half maximum (FWHM) of spectral features expected to be considerably narrower in wave number than $\Delta\bar{v}_{ssw}$. Alternatively, for high-dispersion systems, the FWHM of a known Doppler-limited single transition can be used as a measure of $\Delta\bar{v}_{res}$, the instrumental resolution (see Chapter 2, Sections II.G.1–II.G.3). Note that, throughout Chapters 6 and 7, $\Delta\bar{v}$ with any subscript is the *full width* at *half maximum* (FWHM) of the function specified.

The Doppler width (FWHM) is given by Townes and Schalow, 1955)

$$\Delta\bar{v}_D = 7.162 \times 10^{-7}\bar{v}\sqrt{T/M}, \tag{13}$$

where T is the absolute temperature and M the molecular weight of the gas phase sample.

The properties of a gaussian result in the relationship

$$(\Delta\bar{v}_{obs})^2 = (\Delta\bar{v}_{res})^2 + (\Delta\bar{v}_D)^2, \tag{14}$$

where the instrument function is taken to be well approximated by a gaussian of full width (FWHM) equal to $\Delta\bar{v}_{res}$. Strictly speaking, Eq. (14) holds only for emission spectra. In absorption spectra, the transmitted flux is given by Eq. (9) of Chapter 2. The measured absorptance is given by Eq. (50) using

Eq. (46) of Chapter 2. The measured absorptance profile for a Gaussian absorption coefficient of half width at half maximum (HWHM) Δx_{Dop} and Gaussian response function of half width (HWHM) Δx_{res} was calculated. The resulting HWHM was measured, as was the apparent peak absorptance. In Fig. 3 the percentage error, as $100[1 - \Delta x_{\text{obs}}(\text{meas})/\Delta x_{\text{obs}}(\text{calc})]$, is plotted against $\Delta x_{\text{Dop}}/\Delta x_{\text{res}}$. Figure 3 also presents the apparent peak absorptance versus $\Delta x_{\text{Dop}}/\Delta x_{\text{res}}$ for line strengths corresponding to true peak absorptances of 0.1 to 0.9. Note that all HWHM quantities (in the notation of

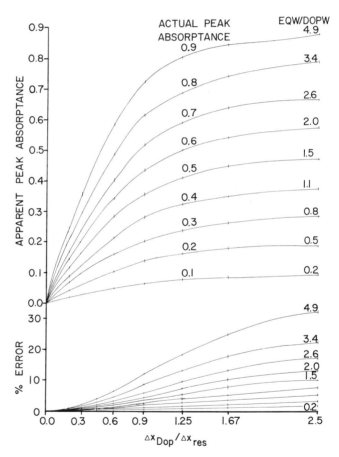

Fig. 3 Upper traces, apparent peak absorptance vs $\Delta x_{\text{Dop}}/\Delta x_{\text{res}}$, the Doppler width per unit resolution. Each trace is identified by the actual peak absorptance. Lower traces, percentage error incurred when $(\Delta x_{\text{Dop}}^2 + \Delta x_{\text{res}}^2)^{1/2}$ is used to approximate Δx_{obs} for an absorption line vs $\Delta x_{\text{Dop}}/\Delta x_{\text{res}}$. The curves are labeled with the appropriate equivalent width per unit Doppler width as EQW/DOPW.

Chapter 2) appear as ratios and are thus equivalent to ratios of the corresponding FWHM quantities.

For a more accurate representation of the instrumental distortion effects see Section II.G of Chapter 2. The approximation given by Eq. (14) is convenient and for high-dispersion instruments observing weakly absorbing spectral lines; it is often accurate enough for a number of experimental uses.

C. General Data-Acquisition Considerations

There are two different methods of changing the grating angle θ_g. One method continuously changes θ_g and samples the data at sequential, equal time intervals Δt_{csam}. The alternative approach samples the data at specific angular intervals $\theta_g + j\,\Delta\theta_{ssam}$, where j is an integer-sequence-ordering variable and $\Delta\theta_{ssam}$ is the grating angle interval between samples. In both cases, the sampled data are converted from analog to machine-readable form and stored on some transportable medium such as magnetic tape or a floppy disk. [There is no reason why simple fast random access memory (RAM) storage is precluded except that all processing of the data—such as deconvolution—must be performed before the RAM can be used again to acquire data from the instrument. Such an approach is often not cost effective.]

The processes involved in data acquisition and the considerations relevant to obtaining appropriate data for deconvolution are nearly identical for both continuous scanning and step scanning instruments. Where no specific distinction is noted, no differences are significant at the current state of the deconvolution art. Sampling, aliasing, and other details of the data-acquisition process can be analyzed using the methods discussed in Chapter 1.

1. Signal-to-Noise Ratios

To increase resolution of recorded data effectively by using deconvolution, appropriately high signal-to-noise ratios are necessary. Depending on the spectral region involved, a factor that usually dominates the selection of detectors used, very different situations regarding signal-to-noise ratios may prevail.

a. Infrared Detection   In the infrared region of the spectrum a variety of detectors are available (Stewart, 1970; Blass and Nielsen, 1974). The limiting source of noise is generally the detector, regardless of type, in infrared spectrophotometry using well-designed instrumentation. Most detectors in routine use in high-resolution spectroscopy are either photoconductive or photovoltaic, and detection is achieved by measuring either a voltage or a current as a function of incident photon energy density. Noise in infrared spectrophotometry is generally assumed to be band-limited Gaussian white noise. (Other types of detection, including thermal detectors, are adequately modeled in this fashion.)

Because infrared spectrophotometry is detector noise limited, a profitable trade-off is possible between signal-to-noise ratio and resolution. Such a trade-off may be useful when deconvolution is used to recover resolution "lost" by opening monochromator slits to increase acquisition rates. In any system where degrading the resolution improves the signal-to-noise ratio nonlinearly, a potentially useful trade-off is possible.

b. Visible Detection   Visible spectrophotometry generally makes use of photomultiplier detectors and photon-counting techniques. In this region, source noise is often the dominant noise source. Because the noise is assumed to be Poisson distributed, the signal-to-noise ratio may be represented as

$$S/N \propto N_j/\sqrt{N_j}, \tag{15}$$

where N_j represents the photon count in channel j. Data taken with equal counts in each channel are most desirable for deconvolution. Unfortunately, this places severe time restrictions on acquisition both in emission and in absorption spectrophotometry and is not often used. Equal-time-per-channel counting results in relatively poor signal-to-noise ratios in the small-signal regions in emission spectroscopy and in the vicinity of absorption lines in absorption spectroscopy.

c. Signal-to-Noise Calculations   In a concise treatment of signal-to-noise ratios in Fourier spectroscopy compared with dispersive spectroscopy, Treffers (1977) shows that

$$S/N = (P/NEP)\sqrt{2} \int_{-\infty}^{\infty} A(t)M(t) \, dt \bigg/ \left[\int_{-\infty}^{\infty} A(t)^2 \, dt \right]^{1/2}, \tag{16}$$

where the noise is assumed to be white noise, P is the signal power falling on the input aperture of the modulator represented by $M(t)$, NEP is the noise-equivalent power of the detector, and $A(t)$ is the signal processing function described later.

Treffer's model system consists of signal power P (in watts) falling on the input aperture of the modulator (i.e., the spectrophotometer) that modulates the input power as a function of time by a factor $M(t)$ such that $0 \le M(t) \le 1$. The modulation function $M(t)$ is *not* the modulation of the signal due to chopping but modulation of the signal due to scanning. The modulated signal falls on a detector with responsivity R (in volts per watt) (Kruse *et al.*, 1962; Stewart, 1970) and flat frequency response. The idealized instantaneous output voltage $PRM(t)$ (in volts) has noise $n(t)$ (in volts) added to it, and the resulting model signal is linearly transformed, producing the processed and integrated signal

$$S = PR \int_{-\infty}^{\infty} A(t)M(t)\, dt. \tag{17}$$

It is assumed that the noise voltage $n(t)$ is the result of a real stationary process (Davenport and Root, 1958) with zero mean. Because it can be shown that the spectral density function $S(f)$ is the Fourier transform of the autocorrelation function of the noise, it follows that the rms noise is given by

$$N = \left[\int_{-\infty}^{\infty} S(f)|A(f)|^2 \, df \right]^{1/2}, \tag{18}$$

where $A(f)$ is the Fourier transform of the signal processing function $A(t)$.

Assuming that the noise is white noise, that is, $S(f) = S_0$, Eq. (18) is simplified and the signal-to-noise ratio is determined from Eqs. (17) and (18). Equation (18) is simplified by the definition of the noise-equivalent power (in watts per $Hz^{1/2}$)

$$NEP = (2S)^{1/2}/R, \tag{19}$$

yielding Eq. (16).

We define a function gate (t/T), which is equal to 1 when t is in the range from 0 to T, and 0 otherwise. The quantity T is thus the length of time the function is nonzero. Using the result given by Eq. (16), Treffers shows the following.

(1) For a photometer with no chopping such that

$$M(t) = A(t) = \text{gate}(t/T) = \begin{cases} 1 & \text{if} \ \ 0 \le t \le T \\ 0 & \text{otherwise,} \end{cases}$$

$$S/N = (P/NEP)(2T)^{1/2}. \tag{20}$$

(2) For a square-wave-chopped photometer with up–down integration such that

$$M(t) = A(t) = \text{gate}(t/T)$$

half the time and

$$M(t) = 0 \quad \text{and} \quad A(t) = -\text{gate}(t/T)$$

half the time,

$$S/N = (P/NEP)(T/2)^{1/2}. \tag{21}$$

(3)   For a square-wave-chopped photometer with phase-sensitive detection following a bandpass filter centered on the chopping frequency such that

$$M(t) = \begin{cases} \text{gate}(t/T) & \text{when} \quad A(t) > 0 \\ 0 & \text{when} \quad A(t) < 0 \end{cases}$$

and

$$A(t) = \cos 2\pi f t \ \text{gate}(t/T),$$

$$S/N = (P/\text{NEP})(4\pi/\pi^2)^{1/2}. \tag{22}$$

These results are valid for consideration of the signal-to-noise ratios of the several signal processing modes. The processor may be expanded to include other elements in the systems and the signal-to-noise ratio calculated on the basis of Eq. (16).

2. Analog Signal Processing

The detector output signal is generated by a current preamplifier for photovoltaic detectors, such as InSb, and by a simple detector bias circuit shown in Fig. 4 for photoconductive detectors, such as PbS and Hg:Ge. The voltage signal derived from the bias circuit is normally preamplified and forwarded to a phase-sensitive synchronous detector usually embodied in a lock-in amplifier (Stewart, 1970; Blass, 1976b).

The behavior of the bias circuit of Fig. 4 is of some interest. Let $(R_D)_0$ be the equivalent detector resistance at zero signal. The quantity R_L is the load resistance (preferably wire-wound or metal film; see Stewart, 1970) and V_0 is the background detector voltage that corresponds to a detector resistance $R_D = (R_D)_0$. Then the output signal of the bias circuit, $V_{sig}(\alpha)$, is given by

$$V_{sig}(\alpha) = V R_D(\alpha)/[R_L + R_D(\alpha)], \tag{23}$$

where α represents the net parametric dependence of R_D on incident photon flux. Because the radiation is modulated by the spectrophotometer, generally

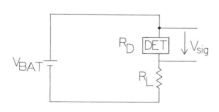

Fig. 4 Simple photoconductive detector bias circuit. R_D is the detector resistance, $(R_D)_0$ the equivalent detector resistance at zero signal (i.e., background flux only on detector). R_L is the load resistance, V_0 the signal at $(R_D)_0$, V_{sig} the signal voltage at R_D. $V_0 = V_{bat}(R_D)_0/[R_L + (R_D)_0]$, $V_{sig}(t) = V_{bat}R_D/(R_L + R_D)$, and V_{bat} is the battery voltage.

by a mechanical chopper, the quantity of interest, for the signal sensed across
the detector, is

$$\frac{\partial^2 V_{sig}}{\partial R_L \, \partial \alpha} = \frac{V(R_D - R_L)}{(R_D + R_L)^3} \frac{\partial R_D}{\partial \alpha}, \tag{24}$$

which indicates that $(\partial V_{sig}/\partial \alpha) \, \delta \alpha$, the ac component of V_{sig}, is a maximum
for $R_L = R_D$. It is readily verified that if the signal is sensed across the load
resistance, the condition $R_L = R_D$ produces a maximum $\partial V/\partial \alpha$ in that case
also. In either case, the ac component of the signal voltage is given by

$$V_{sig} = (V_{sig})_0 - (V/4R_D)(\partial R_D/\partial \alpha) \, \delta \alpha. \tag{25}$$

If it is assumed that (i) the signal is sensed at a fixed wavelength and (ii) the
chopping function is given by

$$\text{chop}(x) = \begin{cases} 1 & \text{for} \quad T_c/4 \leq x \leq 3T_c/4 \\ 0 & \text{for} \quad 0 < x < T_c/4, \, 3T_c/4 < x < T_c, \end{cases} \tag{26}$$

where T_c is the period of the chopping function, so that

$$\delta \alpha = \gamma \, \text{chop}(x), \qquad \gamma = \text{const}, \tag{27}$$

(iii) the detector response time is negligibly short, and (iv) $\varepsilon(\lambda)$ is given by

$$\varepsilon(\lambda) = -(\partial R_D/\partial \alpha)\gamma, \tag{28}$$

then we obtain

$$V_{sig} = (V_{sig})_0 + [\varepsilon(\lambda)V/4R_D] \, \text{chop}(x). \tag{29}$$

This is the appropriate equation to Fourier transform if the frequency
content of the detector output is desired.

The typical synchronous detection or demodulation system follows a
tuned amplifier (tuned to optical chopping frequency f_{ch}) and is realized in
the form of a lock-in amplifier. The lock-in amplifier effectively filters out the
dc component of V_{sig} and produces an output proportional to the fundamental component (at f_{ch}) of the time-varying V_{sig}. As a part of the lock-in
signal processing chain, the user must select an amplifier time constant τ,
which is related to the bandwidth of the amplifier centered on f_{ch}. Although
the bandwidth depends on the specific lock-in amplifier, a reasonable assumption is that the signal energy ($\propto V^2$) is down by 3 dB (50%) at

$$f = \pm 1/2\pi\tau, \tag{30}$$

which is just the case for a simple single-section RC filter (see Fig. 3 of Chapter

2) with 6 dB rolloff per octave (20 dB per decade). For the single-section RC filter we have from Eqs. (41) and (42) of Chapter 2

$$V_C(t) = V_{out}(t) = \xi(t) \otimes v(t) = \xi(t) \otimes V_{in}(t), \tag{31}$$

where

$$\xi(t) = (1/\tau)H(t)e^{-t/\tau}, \tag{32}$$

$$\tau = RC, \tag{33}$$

and $H(t)$ is the Heaviside step function (see Chapter 1).

The transform of $\xi(t)$ is given by

$$\Xi(\omega) = |\Xi(\omega)|e^{j\zeta}, \tag{34}$$

where the gain function normalized to unity at $\omega = 0$ is given by

$$|\Xi(\omega)| = (1/\tau)[\omega^2 + (1/\tau)^2]^{-1/2} \tag{35}$$

and the phase is

$$\zeta(\omega) = \tan^{-1}(\omega\tau). \tag{36}$$

The 3-dB points are found by setting

$$(V_{out}/V_{in})^2 = \tfrac{1}{2} \tag{37}$$

and solving Eq. (35) for f as a function of τ; the result is given by Eq. (30), the absolute value of which is the half-bandwidth. The decibel is a logarithmic ratio measure. The quantity X_1 is said to be 3 dB over X_2 when X_1 is twice X_2. That is, X_1 is $10 \log_{10}(X_1/X_2)$ dB greater than X_2. Because power is a function of the square of voltage, power ratios in terms of voltages are given by

$$(P_1/P_2) = 20 \log_{10}(V_1/V_2) \text{ dB.} \tag{38}$$

Some lock-in amplifiers have switch-selectable (6 dB)–(12 dB) rolloff filters, and the specific bandwidth properties should be obtained from the manufacturer's specifications.

It is also true that some lock-in amplifiers as well as tuned preamplifiers allow the user to select the Q of the tuned amplifiers. The Q of an amplifier is given approximately by $Q \approx f_0/\text{BW}$, where f_0 is the center frequency and BW the full bandwidth. The system bandwidth must be carefully and properly selected when recording data to be deconvolved, and *all* factors that influence the system bandwidth must be carefully taken into account.

3. Digitization of Signals

The output of the lock-in amplifier is input to a sample-and-hold amplifier or directly to the analog-to-digital converter. This signal is converted to a digital and thus machine-readable form.

In considering the signal digitizing process, one must exercise a great deal of care to fully understand just how a specific system works. This understanding should include detailed characterization of both hardware *and* software functions. For example, in a continuous scanning system, "integration" over a period of several time constants will yield different results depending on whether a number of successive "sample-and-hold" samples are averaged or a number of successive samples from the signal line are fed directly into an analog-to-digital converter (ADC). In many cases, such differences are of little consequence. To be sure of this, however, one must consider *all* aspects of the acquisition process in detail and consider carefully the impact of the approximations that are made. ·

a. Sample-and-Hold Amplifiers Sample-and-hold amplifiers are operational amplifiers; the output tracks (follows) the input until a hold command is received. While the hold command is asserted, the output signal is equal to the input signal averaged over a window in time. The time at which the window begins is determined by the hold command. A typical window aperture might be 100 nsec.

Most sample-and-hold amplifiers are biased, unity gain, inverting amplifiers of a specific type—a fact that calls for some care in the subsequent processing of the digitized data. The sample-and-hold amplifier biased at 10 V converts, for example, 10 V to 0 V, 0 V to 10 V, and 3 V to 7 V. This example assumes that the input range of the amplifier is 0–10 V and that it is biased at 10 V, so the output signal is $10 \text{ V} - 1 \times V_{\text{in}}$. The need for careful and periodic adjustment of sample-and-hold amplifiers might best be communicated by reminding the reader that a sample-and-hold amplifier is essentially a dc amplifier with a very high slew rate.

b. Analog-to-Digital Conversion Analog-to-digital converters are generally of the *n*-bit variety if they are interfaced to computers, the term "*n* bit" meaning that the maximum number that can be generated as digitized output is $2^n - 1$. (The least significant bit has a value of 2^0 and the most significant a value of 2^{n-1}.) The range of input voltage that can be converted is specified for each ADC. Some are bipolar, some are unipolar, and all have a specific, ideally linear, relationship between input analog voltage and digitized representation. For example, a 12-bit ADC, unipolar with an input range of 0 to 10 V, may represent 0 V as $000\ 000\ 000\ 000_2$ or 0000_8, whereas

10 V becomes $111\ 111\ 111\ 111_2$ or 7777_8, and 2 V is represented by $001\ 100$ $110\ 011_2$ or 1463_8. For those who find such things mystical, to convert V_{10} into V_2 or V_8, calculate (for an n-bit converter)

$$\left[\frac{V_{10}}{(V_{AD\,max})_{10} - (V_{AD\,min})_{10}} \times 2^n - 1 \right]_{10} = \text{(converted value)}_{10} \quad (39)$$

and convert to base 2. V_{10}, V_2, and V_8 are voltages in base 10, 2, and 8, respectively; $(V_{AD\,max})_{10}$ and $(V_{AD\,min})_{10}$ are the maximum and minimum ADC input voltages. We may then readily convert to base 8. The binary representation goes over to octal by writing down the octal value of three binary digits at a time starting from the least significant bit on the right. The specific binary representation of a converted voltage depends on the specific device. Manufacturer specifications should be consulted.

Just as sample-and-hold amplifiers must be properly adjusted, so must ADCs; both range and linearity are generally adjustable. As with sample-and-hold amplifiers, periodic validation of adjustment should be carried out.

   c. Digital Sampling   The analog signal must be sampled at a frequency sufficiently high to avoid aliasing in the deconvolution process. Although the Jansson algorithm is carried out in signal space, there is nevertheless potential for aliasing. Note that the term "aliasing" describes a consequence of the sampling process, which may be considered in either signal or Fourier space (see Chapter 1). The sampling of a spectrum of Gaussian lines has been previously discussed in detail (Blass, 1976a; Blass and Halsey, 1981). These results are summarized in Section III.D.

   d. Digital Filtering   In addition to analog filtering in the lock-in amplifier, we use some modest digital filtering prior to recording the raw data. The value in using digital filtering is twofold: high-frequency noise suppression is improved over the analog result, and the bandwidth may be more precisely set than with a typical discrete switch-selected time constant on a lock-in amplifier.

   Jansson further discusses electrical and digital filtering in Section II.D of Chapter 2 and Section III.C.5 of Chapter 3.

D. Acquisition with Continuous Scanning

   Blass (1976a) and Blass and Halsey (1981) discuss data acquisition for a continuous scanning spectrometer in detail. The principal concept is that as a system scans a spectral line at some rate, the resulting time-varying signal will have a distribution of frequency components in the Fourier domain.

The sampling theorem tells us that we must sample the signal at a rate equal to or greater than twice the highest-frequency component in the Fourier transform of the signal. For an actual spectrum of absorption lines, this maximum frequency depends on the selected scanning rate in a direct way. In addition, the bandpass of the electronics must be established in such a way that the maximum signal frequency will be minimally attenuated.

If there were no noise in the world, these factors would control the required sampling rate. However, because noise exists and noise can be aliased as well as data, the maximum noise frequency passed by the electronics system actually determines the required sampling rate.

The truth of this is readily seen in the following hypothetical example. Assume that a spectrum of Gaussian lines is to be scanned at a rate such that the maximum Fourier component is 100 Hz. We might then establish the electronic bandpass such that a 100-Hz component is attenuated less than 3 dB. Without noise we would sample at 200 Hz. However, significant noise signals exist out to at least six times the passband frequency of 100 Hz, which means that we must sample at 1200 Hz to avoid aliasing the noise as we deconvolve. This is an extremely conservative approach, and one might well sample less frequently without difficulty.

The net result of these considerations is that, for a spectrum of Gaussian lines, one should sample 10 times per resolution element ($\approx$ FWHM of isolated lines) and that the scan rate should be adjusted to yield a scan rate of 10 times constants to scan one resolution element (Blass, 1976a).

E. Acquisition with Step Scanning

Although one can treat data acquisition with discrete or step scanning quite apart from the continuous scanning case, we generally opt for treating our data acquisition as if we were scanning continuously. This is reasonable because the stepping time consumed by step scanning is negligible. We therefor acquire data as described in the preceding section. The only significant difference is that we stop and digitally integrate at one grating position for each sample acquired. The integration time is approximately one time constant (determined by the system bandwidth) and, as in the continuous case, we acquire 10 samples per resolution element.

F. Calibration of Spectra

Wavelength or wave-number calibration of infrared absorption spectra is normally accomplished by simultaneously recording the spectrum along with a spectrum of a simple molecule having precisely measured transition frequencies.† There are a number of "variations on a theme" in this process, but generally it is necessary to deconvolve both the desired absorption spectrum *and* the calibration spectrum. The stability of observed transition frequencies under deconvolution has been investigated (Willson, 1973; Blass and Halsey, 1981). Using "point-simultaneous" constrained nonlinear deconvolution (see Chapter 3, Section III.C.1 and III.C.2), Blass and Halsey (1981) found line-center frequency stability to be excellent. (Willson, 1973, found a stability problem in "point-successive" deconvolution.)

Recent work on a production basis involving 5-μm spectra of $^{12}CD_3F$ has verified line-center frequency stability under deconvolution. Large numbers of experimental records of $^{12}CD_3F$, simultaneously recorded with CO in the sample, were calibrated before and after deconvolution (point-simultaneous methods, Jansson algorithm). No systematic differences were detected in comparisons of the before and after frequencies of nonblended absorption lines. That is, the variance was consistent with the optomechanical precision of the spectrometer and the mean deviation summed to approximately zero, validating the frequency calibration of the deconvolved data.

There are some potential calibration problems with deconvolution. The principal potential problems stem from the use of calibration lines in a grating order different from that in which the target spectrum is recorded. The dispersion $\Delta\bar{\nu}/\Delta\theta_g$ is a function of the wave number $\bar{\nu}$ passed by the monochromator [see Eqs. (5)–(8)] and thus, for calibration lines of significantly different $\bar{\nu}$, the instrument response function has a significantly different width (in either time or steps as we scan the spectrum). Especially in such cases, the recorded data file should be split into two files: one for calibration lines and one for target spectrum lines. The gaps in each file should be zero filled. The two files should then be deconvolved in precisely the same way (i.e., identical parametetrization of the process in terms of relaxation parameter and interation number) using an instrument response function appropriate to each spectrum file. To date, we have not accumulated a sufficiently extensive data base in such cases to make any definitive comments. However, we see no serious problem with the approach; the process may be tested in the same way that the deconvolution of the 5-μm $^{12}CD_3F$ data was tested.

In the $^{12}CD_3F$ tests, there were several examples of a 10% variation in $\bar{\nu}$

---

† For simplicity of expression, the wave number associated with a given transition will often be referred to as a frequency. This is not without some physical basis, because $\bar{\nu} = cf$, where $\bar{\nu}$ is in reciprocal centimeters, c in centimeters per second, and f in hertz; that is, we often think of wave number as scaled frequency, with the proportionality constant being equal to the speed of light.

over the range of the target spectrum. In these cases we used a response function appropriate to the target spectrum average $\bar{\nu}$ for all lines and experienced no difficulties. However, if one were using 5-μm (2000-cm^{-1}) CO lines to calibrate 12-μm C_2H_6, for example, it is likely that not only would the frequency dependence of the transfer function be involved but also the slit widths used for the 5- and 12-μm observations would be different. The splitting of the data to be deconvolved into two files, and so forth, would be mandatory in this case.

IV. DETAILS OF SPECTROSCOPIC EXPERIMENTS

In this section we offer some comments based to a large extent on eight years of the study, application, and production use of deconvolution with high-resolution infrared absorption spectra. In Chapter 7, a number of specific examples and test cases are presented that illustrate many of the following comments.

A. Impact of Noise on Deconvolution

Consider the following scenario. You have an experiment that you wish to perform—perhaps you intend to obtain the high-resolution absorption spectrum of ethane at 12 μm. With the average dispersive spectrometer, this is a demanding task and also a time-consuming experimental run. As you set up the run, you adjust experimental and instrumental parameters to maximize the signal and minimize the run time. You are faced with a number of competing and contradictory goals.

The decisions to be made if the resulting data are to be deconvolved differ considerably from those if the data are to be measured as recorded. For example, the most significant single cause of the failure of the deconvolution process to produce physically acceptable results is *excessive noise* in the recorded data. Typically the spectroscopist seeks to maximize resolution and, to some extent, minimize the run time. For direct measurement of recorded data, this is often an acceptable decision. In many cases, the person measuring the data is capable of making informed judgments of high quality at signal-to-noise ratios as low as 20:1. Such is not the case when data are to be deconvolved. For resolution enhancements of a factor of 2.5 to 3, a minimum signal-to-noise ratio for band-limited Gaussian noise is 75:1 to 100:1 (Blass and Halsey, 1981).

As a rule of thumb, we have found that, in the range of resolution enhancements from 2 to 5, the minimum required signal-to-noise ratio can be approximately represented by an empirical relationship

$$(S/N)_{min} \approx 6 \exp(\Delta\bar{\nu}_{meas}/\Delta\bar{\nu}_{decon}). \tag{40}$$

Equation (40) is presented as an approximate guideline, and no physical significance is implied by the form or factors used. A generated data test showing the impact of various signal-to-noise ratios is available (see Blass and Halsey, 1981, Fig. 24); further tests and comments will be found in the next chapter.

The effects of inadequate signal-to-noise ratio on deconvolution range from divergence of the iterative Jansson method to the generation of spurious weak features that appear to be absorption lines. We have never experienced divergence of the Jansson method used as described (Blass and Halsey, 1981) but that is due in large part to our not attempting to deconvolve really noisy data and to the approach that we take in the deconvolution iteration process. Although this is discussed later, we remark that our approach is to use moderate underrelaxation in the initial iterations. This seems to aid in the stabilization of the process (see Blass and Halsey, 1981, pp. 43–47).

One further comment regarding noise: in absorption spectral data, it is the signal-to-noise ratio that affects the quality of the results, not the "peak-height"-to-noise or information-to-noise ratio. This statement assumes that the data to be deconvolved are principally 10–30% absorbing and that the signal-to-noise ratio satisfies the requirements of Eq. (40). That this is a reasonable observation follows from the physically meaningful constraints that are imposed and the deconvolution process as discussed in Chapters 4 and 7.

B. Impact of the Bouguer–Lambert Law

As shown in Section I.E.2 of Chapter 2, the transmitted spectral flux U is given by Eq. (9), and in the case of a pressure-broadened line

$$U = U_0 \exp(-\text{natural} \otimes \text{collision} \otimes \text{Doppler}). \tag{41}$$

From Section II.E.1 of Chapter 2 we find that the measured absorptance (where x is identified with $\bar{\nu}$, the frequency in reciprocal centimeters) is given by

$$A_M(x) = r(x) \otimes \left[1 - \frac{U_T(x) - D_T(x)}{B_T(x) - D_T(x)} \right] \tag{42}$$

using Eqs. (49) and (52) from Chapter 2. The quantity U_T is the transmitted spectral flux and is given by Eq. (41); the quantity B_T is the maximum transmitted flux for zero sample pressure and may be identified with U_0 in Eq. (41). If we assume that $D_T(x)$, the stray background flux, is negligible, as well it might in a well-designed and maintained system, then Eq. (42) becomes

$$A_M(x) = r(x) \otimes [1 - \exp(-\text{natural} \otimes \text{collision} \otimes \text{Doppler})]. \quad (43)$$

Figure 5 illustrates the measured data values as well as the data converted to absorptance and transmittance. There are a number of implications of Eq. (43) for deconvolution, and we shall discuss each of these in turn.

1. Apparent versus Actual Absorption

Rewriting Eq. (43) in the sense of Eq. (9) of Chapter 2, we find

$$A_M(x) = r(x) \otimes [1 - \exp(-PX)], \quad (44)$$

where P is the spectral absorption coefficient and X the pressure times the path length. For the present we ignore the fact that PX may be represented as a convolution as indicated in Eq. (43), and focus on the fact that $P(x)X$ is a typical "bell"-shaped function that gives rise to an $A_M(x)$ as illustrated in Fig. 5. Remember that as the pressure times path length increases, the peak absorption $[A_M(x)]_{max}$ does not increase linearly as a function of X, and that for a fixed X the area under the $A_M(x)$ curve is a constant independent of the half-width of the instrument function $r(x)$ (or, loosely, the instrumental resolution).

Consider an example. Assume that we are looking at a spectral line and trying to decide how to observe the line for deconvolution giving width reduction by a factor of 3. Further assume that pressure broadening is

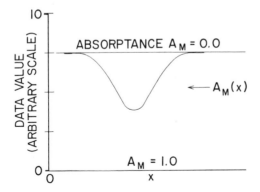

Fig. 5 Absorption line as recorded with measured data values related to absorptance and transmittance.

negligible and that the Doppler full width is 0.005 cm^{-1}. We are going to adjust the pressure path for good deconvolved results. It should be easy, but it is not always obvious. There is a large difference between the choice of X for spectrometers with resolutions of 0.008 and 0.050 cm^{-1}. In the former case, deconvolving to Doppler width increases resolution only by a factor of 1.9, whereas in the latter case a full factor of 3 achieves only a FWHM of 0.017. Thus the 0.050-cm^{-1} instrument will "see" an increase in peak height $[A_M(x)]_{max}$ that exceeds the increase for the 0.008-cm^{-1} instrument simply because a greater narrowing of the observed line is achievable in the latter case.

There is, however, a further and more subtle effect. A 0.050-cm^{-1} instrument will make a 99%-absorbing 0.005-cm^{-1} Doppler line look like a well-behaved 5%-absorbing line, and even mild deconvolution reveals obviously saturated spectral lines. Examples of this effect may be found in the literature on deconvolution (see, e.g., Blass and Halsey, 1981, Fig. 22b, third and fourth traces, and Fig. 2 of Chapter 4).

A rough rule of thumb can be set forth to help avoid having your deconvolved spectral lines exhibit saturation effects. Reduce the pressure path (and thus apparent absorption) until you are sure that it is actually too low; then cut it in half again. As an alternative, one could easily generate a dictionary of simulations based on the ratio of the FWHM of the true line to the equivalent FWHM characterizing specific instrumental resolution. In our laboratory we have simply developed experience in evaluating the requirements. Our instrumental resolution of 0.009 cm^{-1} makes that fairly easy.

2. Deconvolving beyond the Spread Function

There have been a number of discussions about "deconvolving beyond the Doppler limit" (see Pliva *et al.*, 1980) but we prefer to go back a step further and simply discuss deconvolving beyond removal of the instrumental spread function. Jansson discusses deconvolving inherent broadening in Section I.E.2 of Chapter 2.

We rewrite Eq. (44) as

$$A_M(x) = r(x) \otimes \{1 - \exp[-P(x)X]\}, \tag{45}$$

where generally $P(x)$ is given by $P_V(x)$ [Chapter 2, Eq. (9)],

$$P_V(x) = P' \frac{\mu}{\pi} \int_{-\infty}^{\infty} \frac{\exp(-\zeta^2)}{\mu^2 - (x - x_0 - \zeta)^2} \, d\zeta, \tag{46}$$

the Voigt function discussed in Chapter 1. Though Eq. (45) looks complicated in the light of Eq. (46), remember that Eq. (45) can be represented as

$$A_M(x) = r(x) \otimes \{1 - \exp[P_N(x) \otimes P_P(x) \otimes P_D(x)X]\}. \qquad (47)$$

Examining Eq. (47), we see two possible ways to go beyond a modified (i.e., fully deconvolved) $A_M(x)$, that is, beyond our best estimate $\hat{A}_T(x)$ of the true absorptance A_T:

$$\hat{A}_T(x) \approx \{1 - \exp[P_N(x) \otimes P_P(x) \otimes P_D(x)X]\}. \qquad (48)$$

The most obvious approach has not been attempted to our knowledge—namely, deconvolve $A_M(x)$ until the effect of $r(x)$ has been removed, subtract both sides of Eq. (48) from 1, take the natural logarithm of the resulting equation [and data set: $1 - \hat{A}_T(x)$], and proceed. (As Jansson mentions in a slightly different context in Chapter 2, the impact of noise has not been studied in this case.) The modified data set is now of the form

$$\ln[1 - \hat{A}_T(x)] = P_N(x) \otimes P_P(x) \otimes P_D(x)X. \qquad (49)$$

As mentioned before, we know of no one who has attempted this approach. It is obvious that extremely good signal-to-noise ratios would be required.

The second approach is to expand Eq. (47) for small values of the experimental argument. This is discussed briefly by Jansson in Section I.E.2 of Chapter 2. For very-high-resolution data where very short paths and low pressures are used (i.e., where X is small) this approach is potentially feasible though signal-to-noise ratio requirements would be very demanding for weak apparent absorptions.

For tunable-diode-laser(TDL) studies where $r(x)$ is much narrower than even a modestly pressure-broadened line, one might either attempt the former method characterized by Eq. (49) or perhaps simply ignore (cautiously) the impact of $r(x)$. From the foregoing discussion, however, it should be obvious that the considerations and potential problems involved in going beyond removal of the effect of $r(x)$ are in no way trivial. Nevertheless, we have successfully deconvolved some data beyond the limit of the removal of $r(x)$. In many instances we have only taken 0.012-cm^{-1} data for a 0.005-cm^{-1} Doppler spectrum to 0.004 cm^{-1} and have encountered no serious problems. (Quantities refer to FWHM of the peaks.) In other instances, we have deconvolved 0.002-cm^{-1} TDL spectra of C_2H_6 at 12 μm to 0.0009 cm^{-1}, which is roughly one-half the Doppler limit. An example is shown in Chapter 7.

C. Impact of Instrumental Response Function

Jansson discusses the determination of the instrumental response function as well as its analytical characterization in Chapter 2.

We have found, for dispersive infrared spectrophotometry at extremely high resolution, that a Gaussian function works extremely well for critical line-position work. Most of our deconvolution experience of a production nature (as opposed to tests and trials) involves spectra showing predominantly Doppler broadening with a Doppler width FWHM on the order of 0.005 cm^{-1}. Even though we have verified that the double-pass 5-m Littrow system achieves double-pass theoretical resolution at four normal (4N) slits (Todd and McCubbin, 1977), we generally run the system at 2.5 to 3.5 times the Doppler width or 0.012 to 0.018 cm^{-1} resolution. This results in observed FWHM values of 0.0130 to 0.0187 cm^{-1} for single transitions. Our general procedure is to make several measurements of the FWHM of isolated lines and use a gaussian of approximately the same (or slightly smaller) width as an instrument function. For our system, this appears to be completely satisfactory for line-position work. M. A. Dakhil of our laboratory, in the process of taking data on $^{12}CD_3F$ in the 5-μm region, has run numerous spectra under the typical conditions just described. Using ^{12}CO, ^{13}CO, $^{12}CO_2$, and $^{13}CO_2$ absorption lines as standards he has calibrated the CD$_3$F spectra. Prediction of well-known frequencies observed in the spectra led to a standard deviation of errors in the range of 3×10^{-4} to 10^{-3} cm^{-1} prior to deconvolution. Following deconvolution (including the calibration lines) the standard deviations are as good or better than those of the calibration fits. Extensive comparisons of measured wave numbers in nondeconvolved and deconvolved spectra show no systematic variations. Standard deviations of $^{12}CD_3F$ lines based on comparisons are consistent with the standard deviations of the calibration fits. We believe that the residual errors are due to mechanical limitations of the grating drive mechanism. The magnitudes of the variances are consistent with this assumption.

It is also necessary to pay attention to the variation of the instrumental dispersion over the range of the data to be deconvolved. We have not experienced any practical problems with dispersion variations of up to 10% from one end to another of a 150-cm^{-1} spectrum using a fixed-width (in data points) Gaussian response function—although the final deconvolved resolution was not constant over the range of the data. We have considered linearizing infrared dispersive spectrometer data in wave number prior to deconvolution, but have not implemented such procedures at this time. For situations where the dispersion varies significantly over the range of the data, either linearization should be considered or the spectrum broken into smaller segments and deconvolved piecewise using appropriate response functions.

For TDL spectra we have in fact linearized the data by using simultaneously recorded Ge étalon fringes. For such spectra this is often an absolute necessity because of the "dispersion variation" in the recorded data.

If one were to attempt any critical-intensity or line-shape-related deconvolution, we would seriously suggest that a great deal of effort be expended to determine the detailed nature of the instrumental response function. Effects due to neglect of response-function asymmetry have been discussed in some detail (Blass and Halsey, 1981, pp. 107–112).

In Chapter 2, Jansson describes the determination of the system response function for a dispersive spectrometer system. We have made a number of such determinations using very-low-pressure samples of, for example, CO in the 5-μm region. As discussed by Jansson, one records the data and then removes the Doppler profile using deconvolution, yielding the system response function.

Our system is set up with a stepping motor drive on the grating tangent arm, and our tests show repeatability of position to better than $\pm \frac{1}{10}$ step (an average tolerance equal to 8×10^{-6} cm^{-1} at 1000 cm^{-1}). We are therefore able to scan a very weakly absorbing CO line as many as 150 times and use the average as the input data for deconvolution. In this fashion, we avoid the problems concerning the signal-to-noise ratio mentioned by Jansson in Section II.G.1 of Chapter 2.

Another useful approach is to determine the profile of the very weakly absorbing single Doppler-limited line to be used as the system response function. Instead of removing the Doppler profile by deconvolution, however, one might convolve the observed data with the theoretical Doppler profile and use the observed line directly as the effective response function for the modified data set. This technique has been tested and works well. The general approach might be especially useful where a pressure-broadened line has to be used to obtain the response function. Instead of deconvolving the theoretical Voigt profile, the data might well be convolved with the theoretical Voigt profile and the directly observed "response function" used without modification. This approach has one very positive feature, in that it does not add noise to the response function as does deconvolution (to some degree, deconvolution enhances residual noise).

V. PREPARING TO DECONVOLVE A DATA SET

Preparation of data for deconvolution must begin prior to the acquisition of the first data point. Once the resolution of the system is set (in a dispersive spectrometer this is equivalent to setting the slit width), the density of data points per resolution element must be chosen as discussed in Sections III.D and III.E. There are some subtle factors that must be taken into account. For example, for continuous scanning, approximately 10 data points per resolution element are recommended (Blass, 1976a) to capture all of the information required by the data *and* the noise. On the other hand, the data-point density must be great enough to characterize the spectral lines *after* deconvolution. Because a gaussian requires only 2 data points per FWHM, 10 points per resolution element is a sufficient density in the raw data for deconvolution by a factor of up to 5. Of course, when only 2 points per FWHM remain, interpolation is necessary to render the spectrum visually useful.

Under less restrictive noise/bandwidth considerations, one might drop the density to 6 points per resolution element at the risk of some minor noise aliasing. However, deconvolution by a factor of 3 would leave only two points per FWHM of a Gaussian spectrum—a number sufficient to characterize the spectrum, although display and measurement are difficult.

In the latter event, one may use a sinc-function interpolation algorithm to increase the data-point density in the deconvolved spectrum (Willson and Edwards, 1976). A density-doubling routine is especially easy to implement. A word of warning, however—do not attempt to interpolate prior to deconvolution. For any reasonable-length interpolation filter, sufficient additional high-frequency components are added to the spectrum to destroy the validity of the deconvolution process. After deconvolution a density doubler of length 50 works well for measurement and display if 4 or 5 data points remain per FWHM. If the remaining number is 2 or 3, then a longer filter will be required for a reasonable representation (on the order of 200 to 500 points long).

Returning to the mainstream discussion of data preparation, we note that, for a 6-dB-per-octave-rolloff RC filter network in a lock-in amplifier, the continuous scan rate amounts to approximately one time constant per data point or 10 time constants per resolution element (Blass, 1976a). Some time is saved if only six data points are taken per resolution element. We have tried acquiring in this fashion, with no visible negative effects.

In our operations, we often establish step-scan parameters as if we were continuously scanning. Because continuous scanning requirements are more demanding, no problems arise using this approach.

Remember, also, that the signal-to-noise ratio must be high enough to accommodate the desired deconvolution (see Section IV.A). As noted, inadequate signal-to-noise ratio is the major cause of unsatisfactory deconvolution results.

A. Smoothing

Once the data have been recorded in machine-readable form, they may be further processed to improve the signal-to-noise characteristics of the record. Note, however, that an improvement of the signal-to-noise ratio will most often be at the expense of resolution, and thus smoothing the data record is not without problems. In Section III.C.5 of Chapter 3, Jansson comments on band limiting and polynomial filters. Examples of smoothing using polynomial filters are presented in Chapter 7.

We have carried out simulations using polynomial least-squares filters of the type described by Savitzky and Golay (1964) to determine the impact of such smoothing on apparent resolution. For quadratic filters, a filter length of one-fourth of the linewidth (at FWHM) does not seriously degrade the apparent resolution of two Gaussian lines in very close proximity.

As a rule of thumb, we generally restrict the filter length used to one-fourth of the FWHM of an observed Doppler-broadened line. Willson and Edwards (1976) have carried out numerous simulations and provide insight into the impact of smoothing on spectral data records.

In Chapter 7, the high-frequency attenuation characteristics of polynomial filters are discusssed. These considerations are relevant to questions regarding the desirability of a single application of a filter of given length as opposed to multiple application of shorter filters.

Because the most serious problem arising in the deconvolution of spectra is that of noise, detailed attention to smoothing in a fashion consistent with the uniform attenuation of high-frequency noise will result in the best possible deconvolution results.

B. Base-Line Considerations

Owing to the nature of typical raw absorption data, extreme care must be used in preparing the data for deconvolution. Referring to the ordinate (data values) in Fig. 5, one sees that 0% absorptance corresponds to an arbitrary voltage at the output of the typical lock-in amplifier. A typical data record will have been converted by subtracting the lock-in output (Fig. 5) from, say, a value of 10 V, thus inverting ("tipping over") the spectral line and resulting in the appearance of a line as an emission line. However, the 0% absorptance level now corresponds to some voltage (data value) of, say, 2 V. In addition, this 0% absorptance level is generally a slowly varying function of wave number.

Anyone proceeding to deconvolve the raw, "tipped-over" data set will be disappointed (see Pliva et al., 1980; e.g., Section 3). It is easy to see why this is so when constraints are placed on deconvolution as in the Jansson method (Blass and Halsey, 1981). The most effective constraint placed on the deconvolution process in the Jansson algorithm is that 0% absorptance is a lower limit in the deconvolved estimate. That is, the deconvolved absorption spectrum is not allowed to exhibit an emission signal. If the observed data include a bias (i.e., offset) that does not represent actual absorption, then the

0% absorptance constraint will take effect at the apparent 0% absorptance level, and not at the true 0% absorptance level. This results in numerous spectral artifacts. Examples are presented in Chapter 7.

Processing the data file to remove the artificial absorption signal is relatively easy in the event that the spectrum is not too densely populated with lines and 0% absorptance levels may be found across the range of the recorded spectrum. We find that a simple point-to-point base-line subtraction method is simplest and gives the best results. This requires some accurate method to display and measure data values in the file. In our case, we use a graphic display with light pen to acquire a base-line subtraction data set. Other, less direct methods should work also. Sensitivity tests are not difficult to design and execute.

In the event that one is dealing with very dense spectra, the best method of establishing 0% absorptance levels is the inclusion of measured 0% absorptance levels periodically in the data file.

VI. DECONVOLVING THE DATA: CONSTRAINED DECONVOLUTION

The actual deconvolution of a data set is formally straightforward. Let $\hat{o}^{(k)}(x)$ be the kth iterative estimate of the actual spectrum $o(x)$, where x is nominally "time" viewed as a sequence-ordering variable. Further, let $i(x)$ be the actual observed spectrum that has been instrumentally convolved with the observing system response function $s(x)$. The observed data set $i(x)$ is assumed to be related to $o(x)$ by the convolution integral equation

$$i(x) = \int s(x - x')o(x')\,dx' \tag{50}$$

or, equivalently,

$$i = s \otimes o. \tag{51}$$

Let us note that $\hat{o}^{(k)}(5)$ is the amplitude of $\hat{o}^{(k)}(x)$ at $x = x_5$. Let $r^{(k)}[\hat{o}^{(k-1)}(x)]$ be a weighting function used in determining the kth estimate $\hat{o}^{(k)}(x)$ of $o(x)$. We shall have occasion to abbreviate $r^{(k)}[\hat{o}^{(k-1)}(x)]$ as $r^{(k)}[y]$.

The constrained deconvolution algorithm is then specified by

$$\hat{o}^{(k)}(x) = \hat{o}^{(k-1)}(x) + r^{(k)}[\hat{o}^{(k-1)}(x)][i(x) - \hat{o}^{(k-1)}(x) \otimes s(x)]. \tag{52}$$

The $\hat{o}^{(0)}(x)$ estimate is generally taken to be $i(x)$.

A. Deconvolution Algorithm

Step by step, the constrained deconvolution algorithm may be stated as follows.

(1) Set $k = 1$, $\hat{o}^{(0)}(x) = i(x)$.
(2) Form

$$\hat{o}^{(k)}(x) = \hat{o}^{(k-1)}(x) + r^{(k)}[\hat{o}^{(k-1)}(x)][i(x) - \sum \hat{o}^{(k-1)}(x')s(x - x')],$$

where the sum over x' is over the range of nonzero values of $s(x - x')$.

(3) Invoke termination–convergence tests. If satisfied, the current $\hat{o}^{(k)}(x)$ is the estimate of $o(x)$ and the procedure is completed. If not satisfied, set $k = k + 1$ and return to step (2).

Convergence may be monitored by forming the root-mean-square error of the current estimate as

$$e^{(k)} = \left\{ \frac{1}{N} \sum_x [i(x) - \hat{o}^{(k-1)}(x) \otimes s(x)]^2 \right\}^{1/2}. \tag{53}$$

Note that $e^{(k)}$ is not an infallible measure of convergence.

1. Relaxation Functions

The key to the success of constrained deconvolution is the relaxation function $r[x]$. For absorption spectra the data are scaled to lie in the range from 0 to 1. For example, Jansson (1970) used

$$r^{(k)}[y] = r^{(k)}[\hat{o}^{(k-1)}(x)] = r^{(k)}_{\max}[1 - 2|\hat{o}^{(k-1)}(x) - \tfrac{1}{2}|], \tag{54}$$

where $r^{(k)}_{\max}$ is a constant for the kth iteration. Blass and Halsey (1981) after Willson (1973) use

$$r^{(k)}[y] = r^{(k)}[\hat{o}^{(k-1)}(x)] = r^{(k)}_{\max}\{\hat{o}^{(k-1)}(x)[1 - \hat{o}^{(k-1)}(x)]\}^p, \tag{55}$$

where p is normally set equal to 1. As Frieden (1975) points out, one may modify Eq. (54) to specify limits other than 0 and 1 by using

$$r^{(k)}[y] = r^{(k)}[\hat{o}^{(k-1)}(x)] = C[1 - 2(B - A)^{-1}|\hat{o}^{(k-1)}(x) - (A + B)/2|], \tag{56}$$

where A is the allowed minimum and B the allowed maximum data value.

In Chapter 7 several tests are presented using a Gaussian relaxation function such that

$$r^{(k)}[y] = r^{(k)}[\hat{o}^{(k-1)}(x)] = r^{(k)}_{\max}\{\exp[-(y - 0.5)^2/0.11]\}, \tag{57}$$

which yields $r^{(k)}[0] = r^{(k)}[1] = 0.1$ and $r^{(k)}[0.5] = 1.0$.

Another function used in Chapter 7 is described by

$$r^{(k)}[y] = 1, \qquad 0 \le y \le 1,$$
$$\hat{o}^{(k)}(x) = 1, \qquad y > 1, \tag{58}$$
$$\hat{o}^{(k)}(x) = 0, \qquad y < 0,$$

which is a form of clipping and not a relaxation function in the usual sense of the term.

B. The Nonlinearity of Deconvolution

The constrained deconvolution algorithm produces estimates that cannot be obtained from the data by simple linear inverse filtering. This is most readily seen using the Blass–Halsey weight function as an example.

Using Eq. (55) for $r[y]$, we obtain for step (1) in the iterative algorithm of Section VI.A

$$\hat{o}^{(1)}(x) = i(x) + i(x)[1 - i(x)]\{i(x) - r[y] \otimes i(x)\} \tag{59}$$
$$= i(x) + [i(x)]^2 - [i(x)]^3 - i(x)\{r[x] \otimes i(x)\}$$
$$+ [i(x)]^2\{r[x] \otimes i(x)\}. \tag{60}$$

One readily sees that on an analytical level the algorithm rapidly becomes complicated. This is true even if one simply constrains the data to be non-negative, that is,

$$r^{(k)}[y] = \hat{o}^{(k-1)}(x),$$

whence

$$\hat{o}^{(1)} = i(x) + i(x)\{i(x) - r[y] \otimes i(x)\} \tag{61}$$
$$= i(x) + [i(x)]^2 - i(x)\{r[y] \otimes i(x)\}. \tag{62}$$

Two conclusions follow readily. First, the constrained deconvolution algorithm is decidedly nonlinear in the observed data. Second, no easy analytical interpretation of the effects of a particular relaxation function may be obtained by considering the Fourier transform of $\hat{o}^{(k)}$ cast in terms of $i(x)$ and $r[i(x)]$.

In the next chapter, we present the results of tests carried out with the relaxation functions described here. Simulated spectra generated by computer as well as actual experimental spectra are used in the tests.

REFERENCES

Blass, W. E. (1976a). *Appl. Spectrosc.* **30**, 287–289.
Blass, W. E. (1976b). *Appl. Spectrosc. Rev.* **11**, 57–123.
Blass, W. E., and Halsey, G. W. (1981). "Deconvolution of Absorption Spectra." Academic Press, New York.
Blass, W. E., and Nielsen, A. H. (1974). Chap. 2.2 *in* "Methods of Experimental Physics" (D. Williams, ed.), Vol. 3A, 2nd Ed., pp. 126–192. Academic Press, New York.
Boyd, W. J., Jennings, D. E., Blass, W. E., and Gailar, N. M. (1974). *Rev. Sci. Instrum.* **45**, 1286–1288.
Davenport, W. B., Jr., and Root, W. L. (1958). "An Introduction to the Theory of Random Signals and Noise." McGraw-Hill, New York.
Frieden, B. R. (1975). *In* "Picture Processing and Digital Filtering" (T. S. Huang, ed.), p. 177. Springer-Verlag, Berlin and New York.
James, J. F., and Sternberg, R. S. (1969). "The Design of Optical Spectrometers." Chapman and Hall, London.
Jansson, P. A. (1970). *J. Opt. Soc. Am.* **60**, 184.
Jennings, D. E. (1974). Ph.D. Dissertation, University of Tennessee, Knoxville.
Kruse, P. W., McGlauchlin, L. D., McQuistan, R. B. (1962). "Elements of Infrared Technology: Generation, Transmission, and Detection." Wiley, New York.
McCubbin, T. K., Jr. (1961). *J. Opt. Soc. Am.* **51**, 887.
Pliva, J., Pine, A. S., and Willson, P. D. (1980). *Appl. Opt.* **19**, 1833.
Savitzky, A., and Golay, M. J. E. (1964). *Anal. Chem.* **36**, 1627.
Stewart, J. E. (1970). "Infrared Spectroscopy." Dekker, New York.
Todd, T., and McCubbin, T. K. (1977). *Appl. Spectrosc.* **31**, 326.
Townes, C. H., and Schawlow, A. (1955). "Microwave Spectroscopy." McGraw-Hill, New York.
Treffers, R. R. (1977). *Appl. Opt.* **16**, 3103–3106.
Wadsworth, F. O. (1894). *Philos. Mag.* **38**, 337–351.
Willson, P. D. (1973). Ph.D. Thesis, Michigan State University, East Lansing.
Willson, P. D., and Edwards, T. H. (1976). *Appl. Spectrosc. Rev.* **12**, 1.

CHAPTER 7

Deconvolution Examples

George W. Halsey
William E. Blass

Department of Physics and Astronomy, The University of Tennessee
Knoxville, Tennessee

I.	Introduction	188
II.	Examining the Deconvolution Process	188
	A. Deconvolving Noise-Free Data	189
	B. The Deconvolution Process	194
III.	Deconvolution, Noise, and Smoothing	195
	A. Deconvolving Spectra with Broadband White Noise Added	195
	B. Deconvolving Spectra with Band-Limited White Noise Added	196
IV.	Relaxation Methods	201
	A. Comparison of Several Constraints	201
	B. Maximizing the Rate of Convergence	205
V.	Operating Considerations	206
	A. Impact of Response-Function Errors	207
	B. Interpreting Root-Mean-Square-Error Plots	208
	C. Impact of Errors in Determining Minimum and Maximum Signal Levels	210
	D. Impact of Analog-to-Digital Conversion Accuracy	210
VI.	Deconvolution of Other Types of Spectra	211
	A. Deconvolution of Interferometric Spectra	211
	B. Deconvolution of Pressure-Broadened Infrared and Raman Spectra	213
VII.	Deconvolution Examples	215
	A. Deconvolution of Grating Spectra	215
	B. Deconvolution of Tunable-Diode-Laser Spectra	217
	C. Deconvolution of Interferometric Absorption Spectra	219
	D. Deconvolution of Pure Rotational Laser Raman Spectra	221
	E. Deconvolution of γ-Ray Spectra	222
	F. Conclusions and Beginnings	224
	References	225

LIST OF SYMBOLS

$e^{(k)}$	root-mean-square error defined in Eq. (53) of Chapter 6
$E^{(k)}(x)$	error spectrum $o(x) - \hat{o}^{(k)}(x)$
$i, i(x)$	observed spectrum
$o, o(x)$	true spectrum, not "distorted" by observation
$\hat{o}^{(k)}(x)$	kth estimate of $o(x)$
PSD	power spectral density

DECONVOLUTION:
WITH APPLICATIONS IN SPECTROSCOPY

$r_{max}^{(k)}$	constant coefficient of relaxation function $r[\hat{o}^{(k)}(x)]$
$r[\hat{o}^{(k)}(x)]$	weighting or relaxation function used in constrained deconvolution
RMSE	root-mean-square error
$s, s(x)$	response function
$\mathrm{sinc}(x)$	$(\sin \pi x)/\pi x$
$\mathrm{sinc}^2(x)$	$(\sin^2 \pi x)/(\pi x)^2$
x	sequence-ordering variable such as time or wave number

I. INTRODUCTION

Since 1973 we have been exploring deconvolution, first as a fascinating novelty and recently as a valuable tool. We originally approached deconvolution with a large measure of skepticism. In the process of testing and evaluating the practical limit of deconvolution, we have gained a certain measure of confidence in the results. Most of our work has centered on infrared absorption spectroscopy using both grating and tunable-diode-laser spectrometers. Many of our conclusions can be directly applied to other forms of spectroscopy. In this chapter we present examples of deconvolution of computer-generated data and real spectra; we hope that these examples will provide useful insight into the practical limits of deconvolution. We investigate key questions pertaining to noise, smoothing, base-line variations, appropriateness of the response function, and several relaxation functions.

By far the easiest and most informative way to examine the deconvolution process is to use a computer in place of a spectrometer. By simulating a spectrum rather than using an actual spectrometer to record a spectrum, we have complete control over such factors as resolution, line shape, and signal-to-noise ratio. Also, we know what the perfectly deconvolved spectrum would look like. If the deconvolution process generates extra lines, it will be immediately apparent when the original spectrum (before convolution) and the deconvolved spectrum are compared. Another obvious advantage of using a simulation process is that the performance of a large number of spectrometers operating under all possible conditions can be evaluated.

II. EXAMINING THE DECONVOLUTION PROCESS

Mathematically speaking, deconvolution refers to the method or methods used to solve the convolution integral equation

$$i(t) = \int s(t - x)o(x) \, dx, \tag{1}$$

where i is the measured spectrum, s the system spread function or response function, and o the ideal spectrum, free of any instrumental effects. For many practical reasons solving Eq. (1) for o is impossible for a real observed spectrum i. A more practical answer to the question, *What is deconvolution?* is that deconvolution is the process of finding the best estimate of o for a particular measured spectrum i and response function estimate s, including utilization of any a priori information available about o.

There are many reasons why deconvolution algorithms produce unsatisfactory results. In the deconvolution of actual spectral data, the presence of noise is usually the limiting factor. For the purpose of examining the deconvolution process, we begin with noiseless data, which, of course, can be realized only in a simulation process. When other aspects of deconvolution, such as errors in the system response function or errors in base-line removal, are examined, noiseless data are used. The presence of noise together with base-line or system transfer function errors will, of course, produce less valuable results.

For all the tests detailed in this chapter, deconvolution was carried out for 100 iterations using a relaxation function of the form

$$r^{(k)}[\hat{o}^{(k-1)}(x)] = r_{\max}^{(k)}\hat{o}^{(k-1)}(x)[1 - \hat{o}^{(k-1)}(x)],$$

with $r_{\max}^{(k)} = 2.0$ unless otherwise specified. This was done for the purpose of consistency in the test process, and 100 iterations was estimated to be the largest number of iterations that would ever be necessary. For Gaussian response function and practical deconvolution applications, most of the resolution improvement occurs in the first 10 iterations.

A. Deconvolving Noise-Free Data

For most real spectrometers, the response function has a gaussianlike profile. As a result, observation of a spectrum using such an instrument results in attenuation of the higher spectral frequencies.† Practical deconvolution therefore involves enhancing the higher spectral frequencies in the recorded spectrum while maintaining a physically meaningful and useful representation of o. This point is made clearer upon examination of the spectra presented in Figs. 1 and 2. These figures represent the history of two

---

† In this chapter we assume that spectral data records (e.g., an absorption spectrum) are a function of time. Time is to be viewed as a sequence-ordering variable. All references to frequency are references to spectral frequency. Specifically, we have the convolution integral $i(t) = \int s(t - x)o(x)\,dx$ and the Fourier transform of $i(t)$, $I(\omega) = (1/\sqrt{2\pi})\int_{-\infty}^{\infty} i(t)e^{-j\omega t}\,dt$, where $s(t - x)$ is the system response function, $o(x)$ the true spectrum, $i(t)$ the observed spectrum, t "time" as a sequence-ordering variable, $I(\omega)$ the Fourier transform of $i(t)$, and ω the Fourier frequency in radians per unit time.

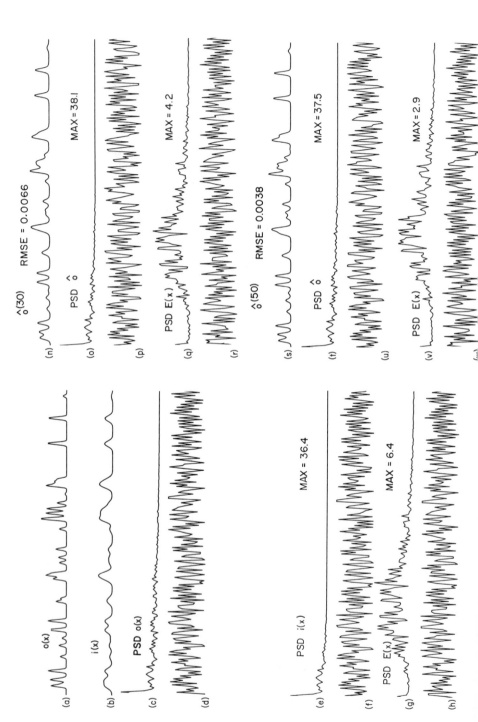

(a) o(x)

(b) i(x)

(c) **PSD** o(x)

(d)

(e) PSD i(x)   MAX = 36.4

(f)

(g) PSD E(x)   MAX = 6.4

(h)

(n) ô(30)   RMSE = 0.0066

(o) PSD ô

(p)   MAX = 38.1

(q) PSD E(x)   MAX = 4.2

(r)

(s) ô(50)   RMSE = 0.0038

(t) PSD ô

(u)   MAX = 37.5

(v) PSD E(x)   MAX = 2.9

(w)

190

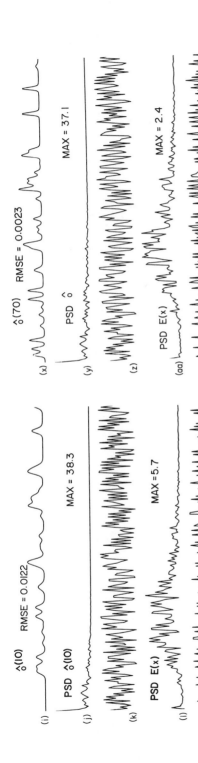

Fig. 1 Deconvolution of simulated *noiseless* data using the Jansson weighting scheme. Trace (a) is the original spectrum $o(x)$, trace (b) the convolved spectrum $i(x)$. Traces (c) and (d) are the power and phase spectra of $o(x)$, traces (e) and (f) the power and phase spectra of $i(x)$, traces (g) and (h) the power and phase spectra of the error spectrum $E(x)$. Traces (i)–(m) are the deconvolution result, the power and phase spectra of the deconvolution result, and the power and phase spectra of the error spectrum, respectively, after 10 iterations with $r_{max}^{(k)} = 1.0$. Traces (n)–(r) are the same results after 20 additional iterations with $r_{max}^{(k)} = 2.0$. Traces (s)–(w) are the same results after 20 additional iterations with $r_{max}^{(k)} = 3.5$. Traces (x)–(bb) are the same results after 20 additional iterations with $r_{max}^{(k)} = 5.0$.

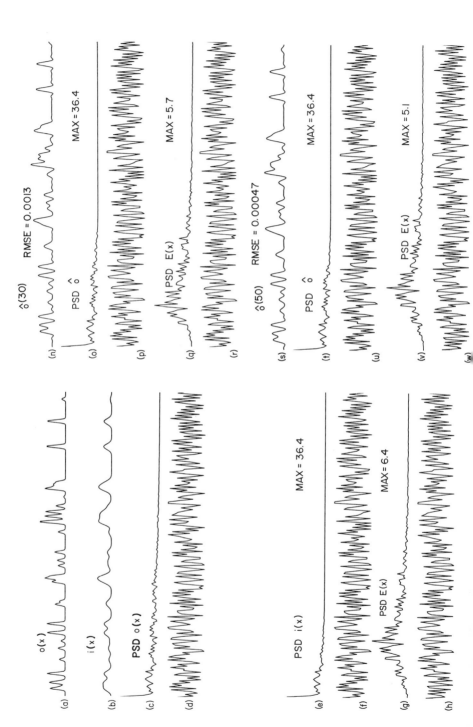

192

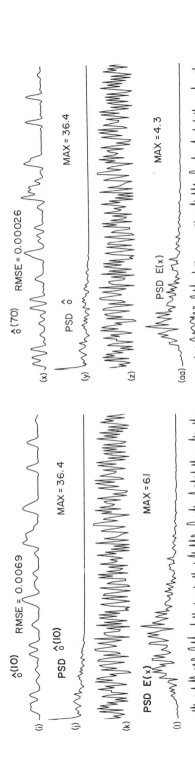

Fig. 2 Deconvolution of simulated *noiseless* data with no weighting. Trace (a) is the original spectrum $o(x)$, trace (b) convoluted spectrum $i(x)$. Traces (c) and (d) are the power and phase spectra of $o(x)$, traces (e) and (f) the power and phase spectra of $i(x)$, traces (g) and (h) the power and phase spectra of the error spectrum $E(x)$. Traces (i)–(m) are the deconvolution result, the power and phase spectra of the deconvolution result, and the power and phase spectra of the error spectrum, respectively, after 10 iterations with $r_{max}^{(k)} = 1.0$. Traces (n)–(r) are the same results after 20 additional iterations with $r_{max}^{(k)} = 2.0$. Traces (s)–(w) are the same results after 20 additional iterations with $r_{max}^{(k)} = 3.5$. Traces (x)–(bb) are the same results after 20 additional iterations with $r_{max}^{(k)} = 5.0$.

simulated deconvolutions. In both figures, trace (a) is labeled the original spectrum. This is the o of Eq. (1), and in these cases is a medium-density absorption spectrum having a natural line shape that is a Gaussian profile with a full width at half maximum (FWHM) of three data points. Trace (b) of Figs. 1 and 2 is labeled the convolved spectrum, the result of convolving the original spectrum with a Gaussian response function having a FWHM of eight data points. In the absence of noise, base-line offsets, and other practical problems associated with a real spectrometer, this would be the spectrum available to a researcher and corresponds to $i(t)$ in Eq. (1).

It is obvious upon examination of traces (a) and (b) of Figs. 1 and 2 that an apparent loss of information has occurred. The loss occurred in the process of convolving the original spectrum with the response function. Although the loss of information that accompanies the loss of resolution may be at first apparent, it, in fact, did not occur. The explanation will take some time and we ask the readers for their patience. Traces (c) and (e) of Figs. 1 and 2 represent the power spectral densities (PSDs) of the original and convolved spectra, respectively. The PSDs are plotted on the same scale, and it is easy to see that the convolution process has attenuated the higher frequencies. Traces (d) and (f) of Figs. 1 and 2 represent the phase spectra of the original and convolved spectra, respectively. From a simplified point of view, the PSD contains information regarding the line profiles in the spectrum, whereas the phase spectrum contains only information regarding the positions and intensities of the spectral features. Examining traces (c) and (f) carefully, one sees that the phase spectra are identical for frequencies that are significantly nonzero in the original spectrum. Thus the information is still there but is in a form less useful to the spectroscopist. The recovery of the attenuated frequencies that returns the spectrum to the more useful form is the goal of deconvolution. Figures 1 and 2 detail the recovery of these frequencies.

B. The Deconvolution Process

In solving Eq. (1) we follow an iterative process that has been refined in the last several years (Jansson, 1970; Jansson et al., 1970; Willson, 1973; Pliva et al., 1980; Blass and Halsey, 1981). In this iterative process, the zeroth approximation to o, the ideal spectrum, is taken as i, the measured spectrum, and the iterative process can be represented as

$$\hat{o}^{(0)} = i, \tag{2}$$

$$\hat{o}^{(k)} = \hat{o}^{(k-1)} + r^{(k)}[\hat{o}^{(k-1)}][i - s \otimes \hat{o}^{(k-1)}], \tag{3}$$

where $\otimes$ represents the convolution operation and $r^{(k)}[\hat{o}^{(k-1)}]$ is a weighting function or relaxation parameter. At each iteration, an error spectrum can be defined as

$$E^{(k)} = o - \hat{o}^{(k)}. \tag{4}$$

Traces (g) and (h) of Figs. 1 and 2 represent the PSD and phase spectrum of $E^{(0)}$, respectively.

The remaining traces of Figs. 1 and 2 show the progress of the deconvolution algorithm at various points in the deconvolution process. For Fig. 1 the relaxation scheme of Eq. (55) of Chapter 6 was used, whereas for Fig. 2 no weighting was done. For each deconvolution the progress is examined at four points by plotting the current deconvolved spectrum, its PSD and phase spectrum, and the PSD and phase spectrum of the error spectrum. In both deconvolution tests the final deconvolved spectrum has superior resolution when compared with the convolved spectrum. Thus the deconvolutions were technically successful. However, in the unweighted case shown in Fig. 2, negative points and extra lines were generated. In the unweighted case the root-mean-square correction $e^{(k)}$ [Chapter 6, Eq. (53)] was smaller, but the PSD of the error spectrum was greater. This demonstrates that $e^{(k)}$ may not be the best parameter with which to evaluate or monitor the deconvolution process. In fact, for real data with noise, $e^{(k)}$ cannot be reduced below a certain point without generating meaningless results. Relaxation methods will be dealt with in more detail in Section IV.

III. DECONVOLUTION, NOISE, AND SMOOTHING

In Section II, the deconvolution examples used noise-free simulated spectra. Any real spectrum will be corrupted by noise. The noise can be reduced by smoothing, but smoothing generally attenuates high spectral frequencies in data. There is an operational conflict, however, because it is these same high spectral frequencies that we wish to enhance by deconvolution. In this section, the effects of noise on deconvolution are demonstrated and several smoothing techniques are evaluated.

A. Deconvolving Spectra with Broadband White Noise Added

In Fig. 3 we use the same simulated spectrum as in Figs. 1 and 2. The original spectrum and convolved spectrum are shown in traces (a) and (b). Traces (c)–(g) represent the result of deconvolving trace (b) 100 iterations.

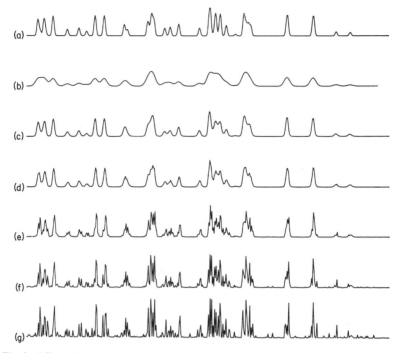

Fig. 3   Effects of noise on deconvolution. Trace (a) is the original spectrum $o(x)$, trace (b) the convolved spectrum $i(x)$ displayed with no noise added. Traces (c)–(g) are the deconvolution results for *broadband white-noise* signal-to-noise ratios of $\infty:1$, $580:1$, $115:1$, $60:1$, $30:1$, respectively.

Prior to deconvolution, the convolved spectrum was corrupted with additive noise to form rms signal-to-noise ratios of $\infty:1$, $580:1$, $115:1$, $60:1$, and $30:1$ for traces (c)–(g), respectively. The noise generated is similar to white noise (or detector noise) after passing through an amplification system with a negligible time constant. A typical spectrum would be recorded with the higher noise frequencies already attenuated, and thus our example may not represent a realistic situation. It should also be pointed out that the simulated noise differs from trace to trace in Fig. 3 only in amplitude and is thus not truly random.

B. Deconvolving Spectra with Band-Limited White Noise Added

It can be concluded from the results shown in Fig. 3 that without some form of electronic or digital smoothing the minimum required signal-to-noise ratio for successful deconvolution might be as high as 200:1. In Fig. 4,

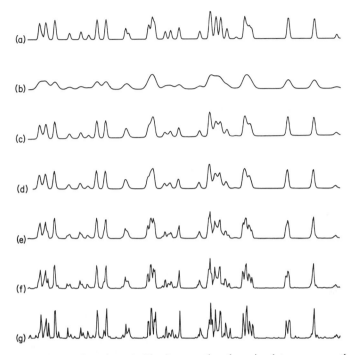

Fig. 4 Same data and results as in Fig. 3 except that the noisy data were smoothed with a quadratic five-point smoothing filter.

the results of deconvolving the same noisy data are shown, except that a mild form of digital smoothing was performed prior to deconvolution. The smoothing was done using a five-point quadratic least-squares technique described by Savitzky and Golay (1964). With this very mild smoothing the results are improved to the point that a signal-to-noise ratio of 100:1 might be adequate for most deconvolution purposes. This process is repeated using a more severe form of smoothing, a 13-point quartic least-squares smoothing, and the results are displayed in Fig. 5. Here it appears that data with a signal-to-noise ratio of only 60:1 might be suitable for deconvolution if the data are presmoothed in this manner.

1. *Testing Smoothing Techniques*

In testing a smoothing technique, two criteria should be considered. Obviously, a smoothing technique should reduce the magnitude of the noise and the impact of noise on the deconvolved spectrum. Second, a smoothing technique should not seriously affect the deconvolution process or the deconvolved spectrum. That is, if the smoothing is too severe, it will further

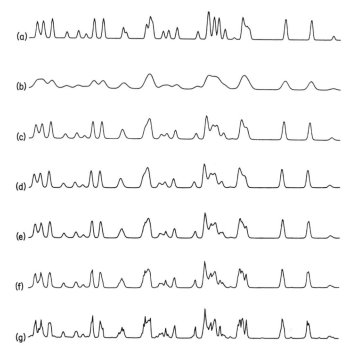

Fig. 5   Same data and results as in Fig. 3 except that the noisy data were smoothed with a quartic 13-point smoothing filter.

attenuate the spectral frequencies that the deconvolution process must enhance, slowing down the convergence rate and limiting the resolution improvement that can be achieved. Clearly the two smoothing techniques tested here do attenuate the noise, but are they too severe? This question can be answered in two ways. First, by comparing the deconvolved spectra for the noiseless case with and without smoothing, any oversmoothing would be apparent. In examining traces (c) of Figs. 3–5 together, no oversmoothing effects are observed, except for a very slight loss of resolution for the 13-point quartic smooth. A second way to judge the effects of smoothing on deconvolution is to compare the PSDs of the original and convolved spectra with the PSD of the smoothing profile. This is done in Fig. 6. Traces (a) and (e) represent the PSDs of the original and convolved spectra of Figs. 3–5 respectively. Trace (b) is the PSD of the five-point quadratic smooth profile and trace (c) the PSD of the 13-point quartic smooth profile. Trace (b) is essentially unity for frequencies where the PSD of the original spectrum is significant, whereas trace (c) indicates that the 13-point quartic smooth may attenuate some frequencies that are important in the original spectrum.

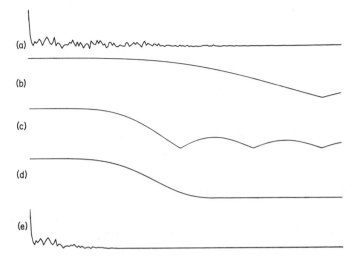

Fig. 6 Trace (a) is the power spectral density of the original spectrum $o(x)$, trace (e) the power spectral density of the convolved spectrum $i(x)$. Traces (b), (c), and (d) are the power spectra of the following smoothing filters: 5-point quadratic, 13-point quartic, and multismooth (a convolution of a 5-point quadratic, a 7-point quadratic, an 11-point quartic, and a 13-point quartic.

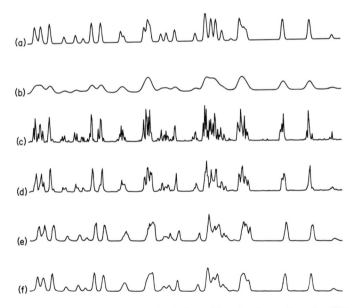

Fig. 7 Deconvolution results for simulated data with a signal-to-noise ratio of 60:1. Trace (a) is the original spectrum $o(x)$, trace (b) the convolved spectrum $i(x)$. Traces (c)–(f) are the results of deconvolution with no smoothing, 5-point quadratic smoothing, 13-point quartic smoothing, and multismoothing, respectively.

2. In Search of the Ideal Smoothing Filter

Examination of traces (b) and (c) of Fig. 6 reveals a problem with these two smoothing techniques. Ideally the PSD of a smoothing profile should be approximately 1 for frequencies significant in the spectrum to be smoothed and gradually approach zero for higher frequencies. The PSDs shown in traces (b) and (c) have windows at high frequencies that allow noise to pass through these filters. Trace (d) shows the PSD of a smoothing profile much closer to the ideal. It represents a smoothing profile (later called multismooth) formed by using a series of smoothing filters, specifically, a 5-point quadratic, a 7-point quadratic, an 11-point quartic, and a 13-point quartic least-squares smoothing filter. The deconvolution results for this smoothing technique are shown in Fig. 7, along with the results for no smoothing, 5-point quadratic smoothing, and 13-point quartic smoothing, all starting with a signal-to-noise ratio of 60:1 in the convolved spectrum. In Fig. 8, the root-mean-square error (RMSE) is plotted against iteration number for each of the four smoothing techniques. Trace (a) is the RMSE plot for the no-smoothing case. Comparison with the other RMSE plots indicates that the multi-smoothing technique is very effective.

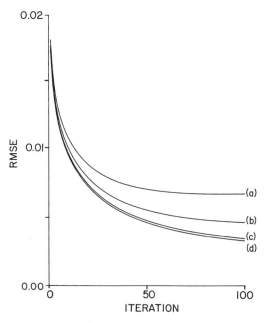

Fig. 8　Root-mean-square-error (RMSE) plots for the results of Fig. 7. Traces (a)–(d) are for no smoothing, 5-point quadratic smoothing, 13-point quartic smoothing, and multismoothing, respectively, corresponding to the results shown in traces (c)–(f) of Fig. 7.

IV. RELAXATION METHODS

Much of the evolution of deconvolution from a fascinating novelty to a realistic spectroscopic tool can be attributed to developments in the application of constraints to the deconvolution process (Jansson *et al.*, 1970; Blass and Halsey, 1981). It has been shown that if the corrections to the deconvolved spectrum are weighted the results can be markedly improved. In Figs. 1 and 2, two deconvolutions were detailed. These two deconvolution tests used the same simulated spectrum, but one (Fig. 1) employed a relaxation method similar to that described by Jansson *et al.* (1970). With this relaxation method, a physically meaningful deconvolved spectrum was found that was in good agreement with the original spectrum in that simulation.

A. Comparison of Several Constraints

In this section, three relaxation methods are tested and compared: Jansson weighting, clipping, and Gaussian weighting characterized by weighting functions given in Chapter 6 by Eqs. (55), (58), and (57), respectively.

In Fig. 9, results for these relaxation methods are compared with an unconstrained deconvolution. Traces (a) and (b) are the same simulated infrared absorption spectra used earlier. Traces (c)–(f) represent the results of deconvolving with four different weighting or relaxation schemes. Trace (c) is the deconvolution result using no weighting and exhibits extra lines and negative values that are not physically meaningful for an absorption spectrum. Trace (d) is the deconvolution result using a modified form of the Jansson weighting. Here the defects seen in trace (c) are absent but the resolution improvement suffers. Trace (e) represents the result of clipping as a method of constraint. In clipping, when a point in the spectrum exceeds a limiting value, it is set equal to that limiting value. Here negative points are set equal to 0 and points greater than 1 are set equal to 1 because the spectrum represents an absorption spectrum. This method always produces a spectrum within the constraint limits but, as in this example, it can produce a curious line shape. The fourth weighting function is a gaussian centered at 50% absorptance and falling to $\frac{1}{10}$ at 0 and 100% absorption. The results for Gaussian weighting are shown in trace (f). This method does not assure nonnegativity, but the negative lobes and sidelobes are less than in the nonweighted case and the convergence is faster than in the Jansson weighting scheme. The RMSEs of these four test runs are plotted in Fig. 10.

Figure 11 duplicates the tests shown in Fig. 9, except that the absorber density in the simulated spectrum is increased by a factor of 7. Here the

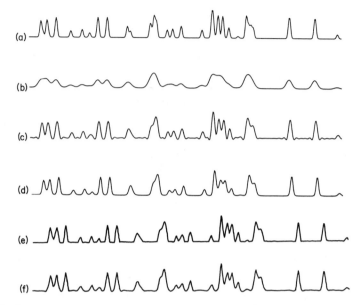

Fig. 9   Deconvolution results for various relaxation schemes: (a) $o(x)$, (b) $i(x)$, (c) $\hat{o}^{(k)}(x)$ using no relaxation (d) $\hat{o}^{(k)}(x)$ using Jansson-type relaxation, (e) $\hat{o}^{(k)}(x)$ using clipping, (f) $\hat{o}^{(k)}(x)$ using Gaussian weighting.

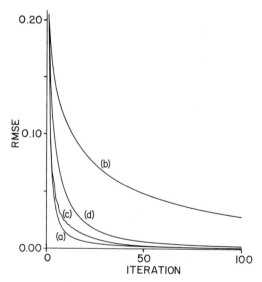

Fig. 10   Root-mean-square-error (RMSE) plots for the deconvolutions of Fig. 9: (a) no relaxation, (b) Jansson-type relaxation, (c) clipping, (d) Gaussian weighting.

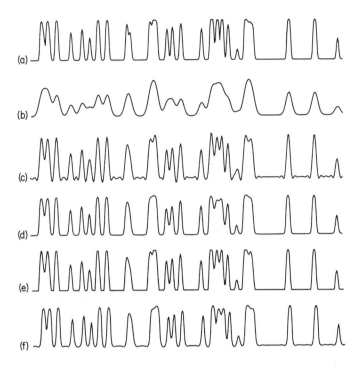

Fig. 11 Deconvolution results as in Fig. 9 except that the simulator absorber is seven times stronger; i.e., the absorptance is seven times that for Fig. 9.

sidelobes produced in the nonweighted deconvolution are even more pronounced. The nonconstrained deconvolution also produces several absorption lines showing absorption greater than 100%. Again the Jansson weighting scheme produces slightly less resolution, but superior line shapes, no extra lines, no negative points, and no superabsorptance points (absorptance greater than 1). The RMSEs for the deconvolution runs of Fig. 11 are shown in Fig. 12.

All of the constraining methods outlined here have some merit. The test results shown here are for noiseless data, where the base line and response function were known exactly. In the presence of typical noise, the problems inherent in the nonweighting scheme seem to worsen. The clipping method exhibits a tendency to produce artificial lines and to lose weak lines in very dense spectra. The Jansson weighting algorithm seems to work for a wider variety of conditions and is especially good where the desired line shapes are Gaussian or near Gaussian.

The principal drawback of the Jansson weighting technique is the relatively slow convergence rate. This can be partially overcome by increasing the

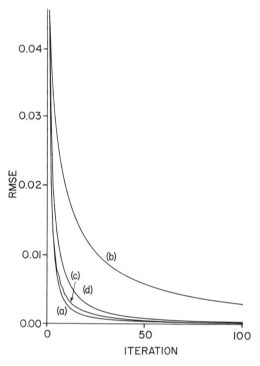

Fig. 12   Root-mean-square-error (RMSE) plots for the deconvolutions of Fig. 11: (a) no relaxation, (b) Jansson-type relaxation, (c) clipping, (d) Gaussian weighting.

magnitude of the weighting function $r^{(k)}[\hat{o}^{(k-1)}(x)]$ in Eq. (3). However, there is a limit beyond which convergence is no longer assured. In practice, we generally use a weighting function of the form

$$r^{(k)}[\hat{o}^{(k-1)}(x)] = r_{\max}^{(k)}\hat{o}^{(k-1)}(x)[1 - \hat{o}^{(k-1)}(x)] \tag{5}$$

for absorption spectra, where $r_{\max}^{(k)}$ is constant for the particular iteration. Typically values of $r_{\max}^{(k)}$ for deconvolving a high signal-to-noise spectrum would be

$$r_{\max}^{(k)} = \begin{cases} 0.5, & 1 \le k \le 10, \\ 1, & 10 < k \le 20, \\ 2, & 20 < k \le 30, \\ 3.5, & 30 < k \le 40, \\ 5, & 40 < k \le 60. \end{cases} \tag{6}$$

B. Maximizing the Rate of Convergence

It seems feasible to set $r_{max}^{(k)}$ dynamically based on the rate of decrease of the root-mean-square error. The goal is to achieve convergence in the fewest number of iterations while retaining the stability of the process.

Figure 13 summarizes results of six attempts at dynamic weighting using 100 iterations. The weighting schemes are as follows: (a) $r_{max}^{(k)} = 2$, (b) $r_{max}^{(k)} = \sqrt{RMSE_0/RMSE_{k-1}}$, (c) $r_{max}^{(k)} = RMSE_0/RMSE_{k-1}$, (d) $r_{max}^{(k)} = 2\sqrt{RMSE_0/RMSE_{k-1}}$, (e) $r_{max}^{(k)} = 2RMSE_0/RMSE_{k-1}$, and (f) $r_{max}^{(k)} = 2RMSE_0/RMSE_{k-1}$ truncated at 40 iterations.

The corresponding RMSEs are shown in Fig. 14. The results in cases (a) and (e) are nearly identical and produce the same RMSE after 100 iterations. Cases (c), (d), and (f) are also similar and are superior to cases (a) and (b). Case (e) becomes nonconvergent after 42 iterations, with apparent absorption going to -2500% at one point but remaining reasonable elsewhere. Case (c) also shows nonconvergence starting with iteration 95, but there seems to be no effect in the deconvolved spectrum. Perhaps the limiting value to the parameter $e^{(0)}/e^{(k-1)}$ is about 20 for data for a very high signal-to-noise ratio, which would suggest a termination criterion for cases (b)–(e).

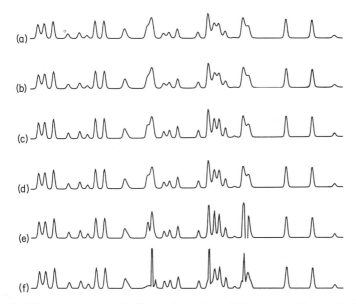

Fig. 13 Deconvolution results for a dynamic weighting test: (a) $\alpha = 2$, (b) $\alpha = \sqrt{RMSE_0/RMSE_{k-1}}$, (c) $\alpha = RMSE_0/RMSE_{k-1}$, (d) $\alpha = 2\sqrt{RMSE_0/RMSE_{k-1}}$, (e) $\alpha = 2RMSE_0/RMSE_{k-1}$, all for 100 iterations; (f) $\alpha = 2RMSE_0/RMSE_{k-1}$ for 40 iterations.

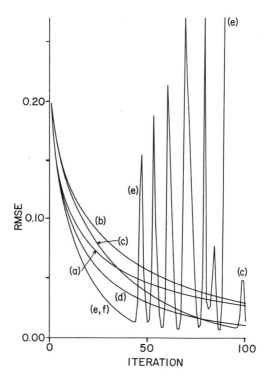

Fig. 14 Root-mean-square-error (RMSE) plots for cases (a)–(f) of Fig. 13. Traces (e) and (f) are superimposed; trace (f) is terminated at iteration 40.

V. OPERATING CONSIDERATIONS

When recording and processing a spectrum, several considerations must be kept in mind if deconvolution techniques are to produce optimum results. Noise is the foremost consideration and has been considered in some detail in Section III. Among the several other considerations of lesser importance that also have impact on the deconvolved spectrum is the appropriateness of the response function used in deconvolving a recorded spectrum as discussed in detail in Chapter 1. The accuracy with which the 0 and 100% absorption levels are determined can also affect the deconvolved spectrum. Also, because a spectrum must be digitized before it can be computer processed, the number of bits used in the digitization must also be considered.

A. Impact of Response-Function Errors

Because the response function of a real spectrometer changes to some extent when any of the operational parameters are changed, the question naturally arises, How well must the instrument function be known for deconvolution to be successful? One way to answer this question is to use a set of "wrong" response functions in deconvolving a simulated spectrum. Figure 15 shows the results of just such a test. Here the simulated spectrum is the same infrared absorption spectrum with three-point Gaussian line-widths convolved with an eight-point Gaussian response function. Traces (a)–(f) represent the results of deconvolving this spectrum with Gaussian response functions with widths of 7, 7.5, 8, 8.5, 9, and 9.5 points, respectively. The earlier question can now be answered, at least in part.

At first glance all six traces of Fig. 15 are identical, indicating that the width of the response function is less important than would probably be expected. There are some minor differences to be pointed out. If an isolated line is used to measure resolution by a direct determination of its full width at half maximum, then the 9.5-point response function performed best. In

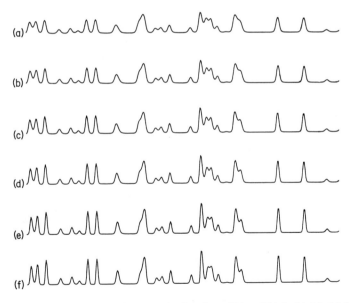

Fig. 15 Deconvolution results for response-function widths of (a) 7, (b) 7.5, (c) 8, (d) 8.5, (e) 9, and (f) 9.5 points. The true response-function width is 8 points. In traces (e) and (f) the resulting linewidths for single lines are narrower than in the original data.

fact, single lines in trace (f) are about 15% narrower than in the original spectrum because some of the original line broadening has been removed by deconvolution. However, in examining a close pair of lines as a resolution test, the best results are to be found in traces (c) and (d), corresponding to response-function widths of 8 and 8.5 points, respectively. Because narrowing an isolated line is of little practical importance and is generally not the motivation behind deconvolution, it can be concluded that the width of the response function should be within 5% of the proper value. However, if the correct response function is not readily determined, there is some justification for deconvolving with "a good guess" because even the extreme case shown in Fig. 15 is superior to trace (b), the spectrum without deconvolution.

B. Interpreting Root-Mean-Square-Error Plots

In Fig. 16 the RMSE is plotted for the six deconvolutions of Fig. 15. A few words of explanation are in order. An RMSE plot similar to those of Fig. 16 presents us with a practical method of monitoring the progress of a

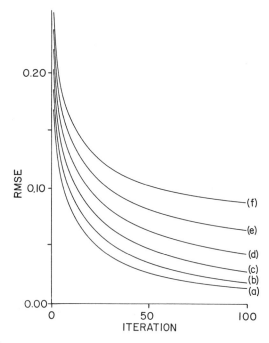

Fig. 16 Root-mean-square-error (RMSE) plots for the deconvolutions of Fig. 15. The response-function widths are (a)7, (b) 7.5, (c) 8, (d) 8.5, (e) 9, and (f) 9.5 points.

deconvolution. When the RMSE no longer decreases appreciably in several iterations there is no need to continue. The actual value of the RMSE is of little use, however, because it is dependent on the data being deconvolved. As seen in Fig. 16, the initial RMSE increases linearly with the width of the deconvolution response function, but this effect cannot be used to test the appropriateness of the response function, because the initial RMSE also depends on the spectral line density, true response function, noise level, and many other parameters. For a fixed response function, the initial RMSE varies as the square root of the spectral line density for a wide range of values of this parameter, decreasing slightly as the lines begin to become very overlapped. Even when the response functions used for the convolution and deconvolution are identical, the RMSE is affected by the response-function width, decreasing slightly as the width increases. Noise also affects the RMSE plot, having a small effect on the initial RMSE but greatly reducing the rate at which the RMSE decreases if the signal-to-noise ratio is much less than 50:1.

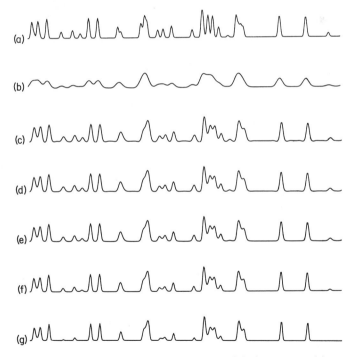

Fig. 17 Base-line variation effects. Trace (a) is the original spectrum $o(x)$, trace (b) the convoluted spectrum $i(x)$. Traces (c)–(g) are the deconvolution results after offsets of -5, -2, 0, 2, and 5%, respectively, have been applied to the convolved data; i.e., the base line used for trace (c) was placed at the "true absorptance" of -0.05 (on a scale of 0.0 to 1.0).

C. Impact of Errors in Determining Minimum and Maximum Signal Levels

Another consideration in the processing of data to be deconvolved is the accuracy of the 0 and 100% absorption levels (or the zero signal for non-absorption spectra). The constraints discussed in Section IV greatly increase the stability of the deconvolution process, but this effect would be diminished if the 0 and 100% levels were poorly determined. The accuracy required for successful deconvolution and the effects of base-line variations can be determined directly by the simulation method. Figure 17 shows the standard test spectrum and five deconvolved spectra. Prior to deconvolution, constant offsets were added to the convolved spectrum. Traces (c)–(g) of Fig. 17 represent the deconvolution results for offsets of -5, -2, 0, 2, and 5%, respectively. Here an offset is defined as $\pm x$ if the true zero signal or 0% absorption level corresponds to $\pm x$ in the digitized spectrum. After the offsets were added, negative points were reset equal to zero. The most striking effect of the base-line offset is the reduction of intensity for weak lines when the offset is negative. The $-5%$ offset causes the weakest lines to vanish, but has little effect on the intensity of stronger lines or the resulting resolution. For an offset of 5%, sidelobes develop as in the unconstrained case [trace (c) of Fig. 9], but are much smaller. Thus, if the 0% level cannot be directly determined within a 1% error, it is better to guess too low than too high when assuming a base line.

D. Impact of Analog-to-Digital Conversion Accuracy

Before any spectrum can be deconvolved it must be recorded in machine-readable form. This is generally accomplished with an analog-to-digital converter (ADC). These are many such devices on the market today, each with its own advantages and disadvantages. Because the recording of the output of a scanning spectrometer does not require an ADC of high speed, the primary consideration here is the number of binary digits or bits resulting from the conversion. At the University of Tennessee at Knoxville we use a standard 10-bit, 0–10-V ADC, where the amplifier output may be as small as 3 V for 100% transmittance. This translates to approximately $\frac{1}{3}%$ absorption for the least significant bit of the ADC. Data simulated for ADCs using 6, 8, 10, and 12 bits were deconvolved and the results are shown in traces (c)–(f) of Fig. 18, respectively. The 10-bit simulation [trace (e)] shows almost no digitization effects, and even the eight-bit simulation [trace (d)] would be acceptable with minimal smoothing.

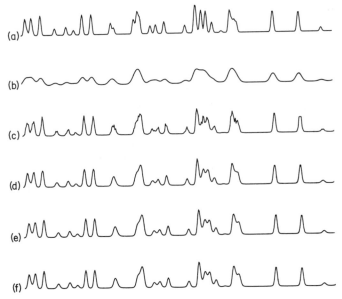

Fig. 18 Effects of digitization of the convolved spectrum $i(x)$ on deconvolution. Trace (a) is the original spectrum $o(x)$, trace (b) the convolved spectrum $i(x)$. Traces (c)–(f) are the deconvolved spectra after "digitizing" the convolved spectrum using 6, 8, 10, and 12 bits, respectively.

VI. DECONVOLUTION OF OTHER TYPES OF SPECTRA

The previous sections have dealt primarily with infrared absorption spectra, although the conclusions can in general be applied to other types of spectra. Here additional uses of deconvolution will be demonstrated. In the first example, a Fourier transform spectrum is simulated and several attempts to deconvolve this spectrum show limited success. In the second example, pressure-broadening effects in an infrared absorption spectrum and a Raman spectrum are simulated. An attempt at removing these effects by deconvolution shows some promise.

A. Deconvolution of Interferometric Spectra

The authors have witnessed many debates concerning the deconvolution of Fourier transform spectra. Some insist that it cannot work; others insist that it does, but that it should not; and others just assume that it works. In

Fig. 19, an unapodized spectrum [response function $(\sin \pi x)/\pi x = \text{sinc}(x)$] is shown in trace (b). For such a spectrum there will be sidelobes and "negative absorption" if the natural linewidths are narrower than the full width of the sinc-shaped response function. These are seen in Fig. 19, where the linewidth is three points and the response function width eight points. Here the phrase "instrument response function" may have a slightly different definition, but the meaning is clear. For such a response function, the direct deconvolution methods fall short.

The negative sidelobes of the unapodized spectrum render the constraints defined earlier useless. Trace (c) of Fig. 19 shows the result of unconstrained deconvolution. After only five iterations the algorithm begins to diverge and, as seen in Fig. 19, the spectrum is worse for our trouble. Trace (d) shows the result of zero clipping. Although there is some improvement and convergence does occur, the resulting spectrum appears real but is not a faithful representation of the original spectrum shown in trace (a), particularly where several lines overlap in the convolved spectrum. Trace (e) represents the same

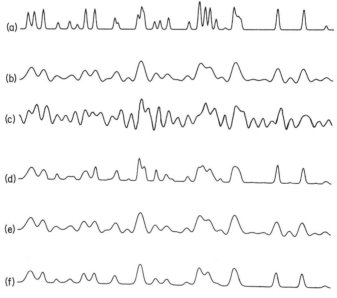

Fig. 19   Deconvolving $s(x) = \text{sinc}(x)$ and $s(x) = \text{sinc}^2(x)$ convolved data. Trace (a) is the original spectrum $o(x)$, trace (b) the result of convolving with an eight-point sinc function, trace (c) the result of unconstrained deconvolution using the same sinc function for just 10 iterations. Trace (d) is the result of 100 iterations using "zero clipping." Trace (e) is the original spectrum convolved with an $8\sqrt{2}$-point sinc-squared function, trace (f) the result deconvolving trace (e) with the same sinc-squared function for 100 iterations using the Jansson-type relaxation function $r^{(k)}[o^{(k-1)}(x)]$.

spectrum with triangular apodization used to generate a response function $s(x) = (\sin^2 \pi x)/\pi^2 x^2 = \mathrm{sinc}^2(x)$. Although some resolution is lost, the negative lobes are eliminated and the positive sidelobes greatly reduced. Such a spectrum can be deconvolved with the Jansson weighting technique even in the presence of the sidelobes (which are now all nonnegative). The result is shown in trace (f). This spectrum shows resolution equal to or slightly better than the unapodized spectrum shown in trace (b) and the sidelobes have been almost completely eliminated. For a very crowded spectrum this would appear to be a useful technique, particularly if there is interest in the weaker lines of the spectrum.

B. Deconvolution of Pressure-Broadened Infrared and Raman Spectra

In all the previous simulations we have assumed a narrow Gaussian line profile that results from Doppler broadening. This is often not the case, owing to the presence of pressure broadening that results in a Lorentzian line profile at sufficiently high pressures. This often means an unavoidable loss of resolution. To observe a spectrum when the transition intensities are low, it is often necessary to increase the sample gas pressure. In this case the pressure broadening can be sufficient to mask important spectral features. Conversely, when the pressure is low enough to allow these important spectral features to be resolved, they may be too weak to observe at achievable signal-to-noise ratios. To some extent this problem can be alleviated by using deconvolution to remove pressure-broadening effects (as well as instrumental effects). This is demonstrated for two cases: a simulated absorption spectrum and a simulated Raman spectrum.

For an absorption spectrum the effects of pressure broadening can be approximated by convolving the spectrum with a lorentzian of the proper width. (An accurate representation would require us to carry out the convolution on the absorption coefficient spectrum.) Thus a spectrum with an eight-point Lorentzian line profile convolved with an eight-point Gaussian response function can be deconvolved with a Voigt profile having Gaussian and Lorentzian components each eight points in width. This is shown in Fig. 20. Trace (a) represents an infrared absorption spectrum with a Gaussian line profile three points wide. Trace (b) is the same spectrum generated using an eight-point lorentzian for the line profile. Trace (c) is the convolution of trace (b) with an eight-point gaussian, and trace (d) is the result of deconvolving trace (c) with the appropriate Voigt profile. As can be seen in examining Fig. 20, most of the pressure-broadening effects have been eliminated, resulting in gaussianlike spectral lines with a FWHM of from four to five

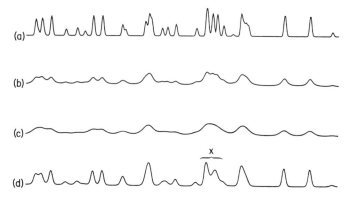

Fig. 20 Removing pressure-broadening effects with deconvolution. Trace (a) is a simulated absorption spectrum using a three-point Gaussian line shape. Trace (b) is the same infrared spectrum except that an eight-point lorentzian was used for the line shape. Trace (c) is the result of convolving trace (b) with an eight-point gaussian. Trace (d) is the result of deconvolving Trace (c) with a Voigt profile having Lorentzian and Gaussian widths of eight points. Feature marked "x" should be better resolved at the apparent deconvolved resolution.

points. The relative intensities and positions are well reproduced except for the feature marked "x" in trace (d), which should be partially resolved at the apparent resolution.

This scheme was also used to test pressure-broadening removal in a Raman spectrum. Here, no approximations are needed and any pressure-broadening effects can be considered as part of the instrument response function if the

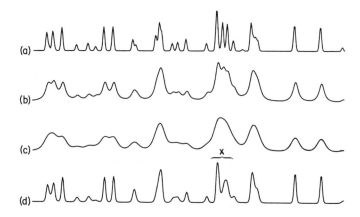

Fig. 21 Replication of Fig. 20 for Raman-type data. The constraint used was for negative suppression only. All plots are scaled so that the maximum signal is 1 in. This gives the *appearance* of nonconservation of area.

pressure broadening is the same for all lines in the spectrum. The results are shown in Fig. 21. The recovery of the original spectrum is quite good and the resulting line profile is approximately a gaussian with a FWHM of less than four points for most lines. Though the recovery is better in the Raman case, the features marked "x" in trace (d) of Fig. 21 should be better resolved if the resolution were as good as indicated by the widths of single lines.

VII. DECONVOLUTION EXAMPLES

We conclude this chapter by presenting several examples of deconvolution of real data. Most of these examples represent deconvolutions of data that were used as part of a spectral analysis rather than generated as deconvolution examples or tests. The examples include high-resolution grating spectra, tunable-diode-laser (TDL) spectra, a Fourier transform infrared spectrum (FTIR), laser Raman spectra, and a high-resolution γ-ray spectrum.

A. Deconvolution of Grating Spectra

For several years now we have routinely deconvolved spectra recorded on the grating spectrometer at the University of Tennessee at Knoxville. Two examples are presented here. The first is part of the v_4 band of CD_3F recorded in 13th order using a 31.6/mm grating. The spectrum was recorded once with a resolution of 0.010 cm^{-1} in a scan that covered most of the v_4 band. The spectrum was smoothed and base-line corrected prior to deconvolution, and a section of this data set (RQ_4 region) is shown in Fig. 22 as trace (a). The data set was deconvolved using the relaxation method described by Jansson with the weighting values similar to those recommended in Eq. (6). The resulting spectrum was then interpolated to double the data-point density, and the RQ_4 region is shown in trace (b) of Fig. 22. A calculated spectrum, using molecular parameters from a previous lower-resolution run ($\Delta \bar{v} = 0.028$ cm^{-1}), is shown in trace (c) using a resolution corresponding to the Doppler width ($\Delta \bar{v} = 0.004$ cm^{-1}). The analysis of v_4 is currently under way using this deconvolved data set, which should yield an improved parameter set.

The second example of deconvolved grating spectra is part of the v_3 band of $^{28}SiH_4$, and is shown in Fig. 23. Trace (a) was recorded at a pressure of 1 Torr of natural-abundance SiH_4 at a resolution of 0.020 cm^{-1} in 12th order of the 31.6/mm grating. This spectrum was recorded to study the v_3 spectrum of the less abundant isotopes ($^{29}SiH_4$ 4.70%; $^{30}SiH_4$: 3.09%) and was later deconvolved. Owing to the isotopic shift in the v_3 band, the

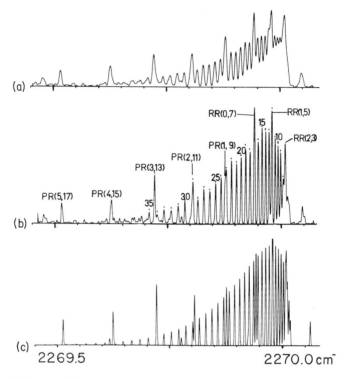

Fig. 22 Q branch of the RQ_4 subband of v_4 of CD_3F recorded on a grating spectrometer. Trace (a) is the raw data after smoothing and base-line subtraction. The resolution based on the widths of single lines is 0.010 cm^{-1}. Trace (b) is the result of deconvolving this data set with a gaussian of FWHM = 0.010 cm^{-1}. The resolution is 0.0045 cm^{-1}. Trace (c) is a calculated spectrum based on the analysis of an earlier spectrum with a resolution of 0.028 cm^{-1}. The resolution in trace (b) is 0.0045 cm^{-1} and in trace (e) is 0.0040 cm^{-1}.

interaction between v_3 and v_1 splits the J manifolds of $^{29}SiH_4$ and $^{30}SiH_4$ much more than those of $^{28}SiH_4$. The J manifolds of $^{28}SiH_4$ were then rescanned 12 times at higher resolution (0.012 cm^{-1}). These scans were averaged prior to further processing. The signal-to-noise ratio was approximately 60:1 for the individual scans and approximately 200:1 in the averaged spectrum. The averaged spectrum was not smoothed further, and the base-line-corrected spectrum for P_6 is shown in trace (b) of Fig. 23. A calculated spectrum, based on a preliminary analysis of v_1 and v_3, is shown in trace (c) using a resolution of 0.004 cm^{-1}, corresponding to the Doppler width. The deconvolved spectrum is shown in trace (d) of Fig. 23, and exhibits a linewidth of from 0.003 to 0.004 cm^{-1}. The instrument function used was a gaussian 0.012 cm^{-1} wide.

Fig. 23 P_6 of $^{28}SiH_4$ recorded on a grating spectrometer. Trace (a) is from a continuous scan of natural-abundance silane at a resolution of 0.020 cm^{-1}. Trace (b) is the same region recorded separately at a resolution of 0.012 cm^{-1} and is the average of 12 scans. Trace (c) is P_6 calculated at a resolution of 0.004 cm^{-1}, trace (d) the result of deconvolving the data shown in trace (b). The resulting resolution is between 0.003 and 0.004 cm^{-1}, and the relative intensities and positions are in good agreement with the calculated spectrum.

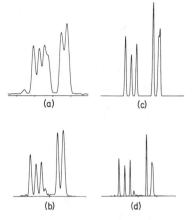

B. Deconvolution of Tunable-Diode-Laser Spectra

Tunable diode lasers have in recent years pushed the limits of resolution well beyond that achievable using conventional means. For the purposes of infrared molecular spectroscopy, the limiting factor is the Doppler broadening of spectral lines rather than the resolution when using a TDL spectrometer. Sophisticated techniques have been used to overcome the Doppler limit problem, such as cooling the sample under consideration and sampling through gas jets. Deconvolution presents a simpler way to limit the effects of Doppler broadening without investing in complicated and expensive instrumentation. The ν_9 band of ethane provides an interesting example.

The spectrum of ethane exhibits a splitting in all spectral lines at very high resolution. This effect is due to the torsional motion of the molecule that splits the energy levels. If the splittings in the upper and lower states of a transition differ, a splitting of the spectral line associated with that transition will occur. For the ν_9 band this splitting is approximately equal to the Doppler width at room temperature. The splitting increases with J and is enhanced in RQ$_0$ owing to the combined effects of a Coriolis interaction and l-doubling. For high values of J this splitting is resolved, but for lower values of J the splittings are unresolved even when using a tunable diode laser.

Figures 24 and 25 demonstrate the success of deconvolution beyond the Doppler limit. Trace (a) in both figures is the undeconvolved data and represents the result of signal averaging over a large number of scans to produce a signal-to-noise ratio of approximately 300:1. The spectra were base-line corrected using a 0% absorption corresponding to the apparent base line between lines and a 100% absorption level was assumed, which gave reasonable relative intensities for the range of J values observed. The

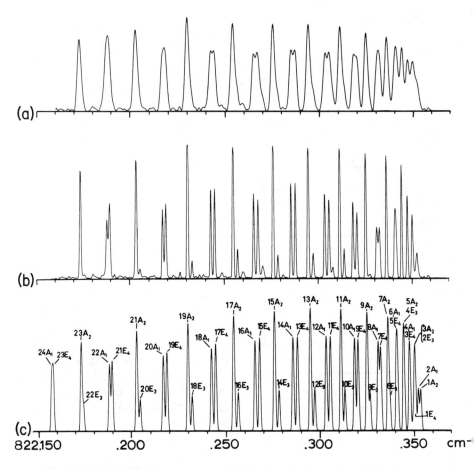

Fig. 24 Tunable-diode-laser spectrum of RQ_0 of v_9 of ethane. Trace (a) is the average of 250,000 scans and exhibits linewidths of 0.0022 cm^{-1} (the Doppler width is 0.0018 cm^{-1}). Trace (b) results from the deconvolution of the data in trace (a) using a gaussian with a FWHM of 0.0022 cm^{-1} as a response function. Trace (c) is the Q branch calculated using a model that includes torsional splitting effects $\Delta v = 1.95$ mk. Trace (c) is calculated for $\Delta v = 0.00075$ cm^{-1}, which is less than one-half the 300 K Doppler width.

spectra were then interpolated to a linear frequency scale of 0.00025 cm^{-1} per data point using a 3-in. germanium étalon producing a fringe spacing of 0.0161 cm^{-1}. The linewidths were nearly constant across the scans and approximately 0.0022 cm^{-1}. The spectra were then deconvolved using a Gaussian response function having a FWHM of 0.0022 cm^{-1}. The results are shown in trace (b) of both figures. Trace (c) is a calculated spectrum (using constants obtained from fitting the entire v_9 band) using linewidths

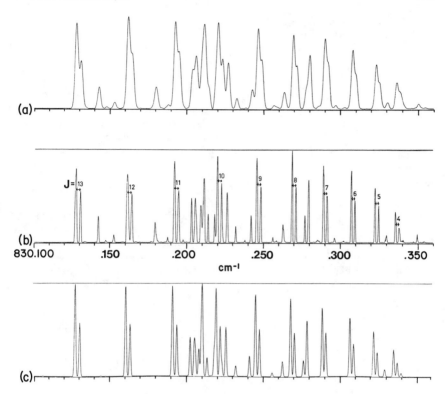

Fig. 25 Tunable-diode-laser spectrum of RQ_3 of v_9 of ethane. All details are the same as in Fig. 24.

of 0.00075 cm^{-1}, less than half the Doppler width. The agreement is very good, limited in this case by the linearization of the diode-laser scans.

C. Deconvolution of Interferometric Absorption Spectra

Figures 26 and 27 show the results of using deconvolution to remove the sidelobes present in a Fourier transform spectrum. These two spectral sections are taken from the v_2 absorption band of ammonia at about 848 and 828 cm^{-1}, respectively. The unapodized spectrum exhibits a resolution of 0.125 cm^{-1}, and is shown in trace (a). The spacing of the lines in Fig. 26 is such that the sidelobes when added together partially cancel, minimizing their effect. Sidelobes and the apparent negative absorption between the lines are both still present. The spacing of the lines in Fig. 27 is such that the sidelobes add constructively, accentuating their effect and producing

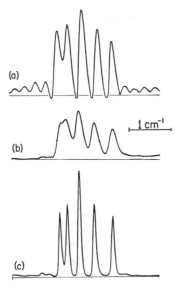

Fig. 26 Fourier transform spectrum of v_2 of ammonia. Trace (a) is a section of the infrared absorption spectrum of ammonia recorded on a Digilab Fourier transform spectrometer at a nominal resolution of 0.125 cm^{-1}. In this section of the spectrum near 848 cm^{-1} the sidelobes of the sinc response function partially cancel, but the spectrum exhibits "negative" absorption and some sidelobes. Trace (b) is the same section of the ammonia spectrum using triangular apodization to produce a sinc-squared transfer function. Trace (c) is the deconvolution of the sinc-squared data using a Jansson-type weight constraint.

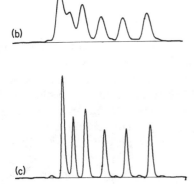

Fig. 27 Same as Fig. 26 except $\bar{v} \approx 828$ cm^{-1}. Sidelobes add constructively in the unapodized spectrum and produce very large sidelobes. The strongest feature is two unresolved lines.

spurious absorption lines between the true features. Trace (b) shows the same spectrum using triangular apodization. Here the sidelobes are greatly reduced and the negative absorption features are removed but the resolution is reduced. A single well-isolated line between these two sections was chosen for the instrument profile. The deconvolved spectrum presented in trace (c) still shows very weak sidelobes, but no negative absorption points. The resolution lost in the apodization has been recovered, the apparent linewidths of the deconvolved spectrum being about two-thirds of those in the un-apodized spectrum.

D. Deconvolution of Pure Rotational Laser Raman Spectra

Trace (a) of Fig. 28 represents part of the pure rotational laser Raman spectrum of O_2. This example is a single scan of the S_3 transition $J = 3$ to $J = 5$ in the ground state. This transition is split into three components

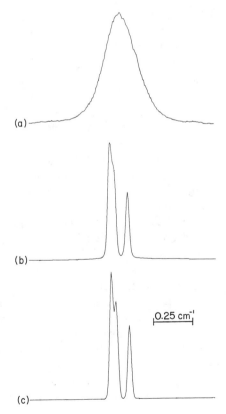

(a)

(b)

(c)

0.25 cm^{-1}

Fig. 28   Pure rotational spectrum of O_2. Trace (a) is the S_3 transition recorded at a pressure of 1.0 atm. Trace (b) is the result of deconvolving the S_3 profile with a Voigt profile to remove most of the pressure broadening, Doppler broadening, and instrument effects. Trace (c) was calculated using a 0.035-cm^{-1} Gaussian profile and calculated spin splittings. The traces are scaled to the same height.

because of the permanent dipole moment of the oxygen molecule in the ground state. This scan was recorded as part of a series of scans in a study of pressure broadening in oxygen. The pressure was 1.0 atm and induced a pressure broadening of approximately 0.1 cm^{-1}. In addition to pressure broadening (lorentzian), the profile was affected by Doppler broadening (0.042 cm^{-1}, gaussian) and the instrument function (0.035 cm^{-1}, quasi-gaussian). These effects combine to mask the splittings almost completely. The spectrum was recorded with 0.0051 cm^{-1} per data point, using 300 points to scan each line in the pure rotational spectrum. The peak in S_3 is approximately 2400 counts and there is a nearly constant background of 100 counts.

Prior to deconvolution, the background was subtracted and the data were smoothed with a 15-point quadratic least-squares polynomial followed by a 19-point quartic least-squares polynomial. The data were then scaled from 0 to 1. The S_3 profile was deconvolved using a weight constraint of the form

$$r^{(k)}[\hat{o}^{(k-1)}(x)] = \hat{o}^{(k-1)}(x)\exp[-\hat{o}^{(k-1)}(x)] \qquad (7)$$

for 75 iterations. The root-mean-square error was reduced by a factor of 5 and the resolution of the three components of S_3 is in fairly good agreement with the theory. The deconvolved spectrum is shown in trace (b) of Fig. 28, and a calculated spectrum with a 0.035-cm^{-1} Gaussian profile is shown in trace (c).

E. Deconvolution of γ-Ray Spectra

As a final deconvolution example, we present a high-resolution γ-ray spectrum. Figure 29 shows the γ-ray emission of Tm^{159} from 120 to 250 keV using a Tm^{159} source produced in a beam experiment. The response function here is in many ways similar to that of the tunable-diode-laser spectrum shown in Section VII.B. That is, the response function is partially due to Doppler broadening and is not constant over even a short section of the spectrum. Here the FWHM ranges from 7.0 points at the low-energy (left) end of the section shown to about 7.8 points at the high-energy (right) end, each data point corresponding to 0.25 keV. The sample under consideration is produced in such a way that there is a sizable background signal from many weak unresolvable γ-ray peaks and also background from the Compton scattering edges for higher-energy γ rays. The average background over this region is about 19,000 counts per channel and the highest peak exhibits about 25,000 counts per channel above background. Trace (a) of Fig. 29 represents the base-line-corrected data. Prior to deconvolution these data are smoothed using a seven-point quadratic and a nine-point quartic

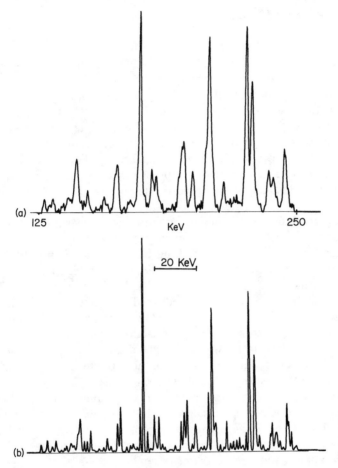

Fig. 29 Gamma-ray spectrum of Tm^{159}. Trace (a) is the recorded spectrum after the continuum has been subtracted. Trace (b) is the result of smoothing and deconvolving this spectrum with a Gaussian profile having a width corresponding to the narrowest features observed in the spectrum.

Savitzky–Golay smoothing profile. The response function used was a gaussian with a FWHM of seven points. The weighting scheme used was of the form

$$r[\hat{o}^{(k)}(x)] = \hat{o}^{(k)}(x) \exp[-\hat{o}^{(k)}(x)], \qquad (8)$$

which seems to maintain the relative intensities better than a weighting scheme with only a lower-limit constraint. The deconvolved result is shown in trace (b).

F. Conclusions and Beginnings

From our several years of testing and using deconvolution techniques, we have developed a favorable attitude toward constrained signal-space deconvolution. It may well be that our outlook is favorable because we have been in a position to test the method not only in simulation but also in practice. Some of the tests have involved degrading resolution to obtain data to deconvolve and compare with higher-resolution data. Other tests have involved synthesis of spectra based on analysis of nondeconvolved data and comparison with deconvolved data (compare, e.g., Figs. 23 and 24). Nevertheless, each time we change our approach or do something new, we carry out extensive testing to determine the impact of the changes and/or the implementation of new ideas.

There are, in fact, many unanswered questions regarding deconvolution. Even more certainly, there are many as yet unasked questions.

Remember also, that, as pointed out by Blass and Halsey (1981), deconvolution can be used to achieve resolution otherwise unavailable or to obtain a specific required resolution using hardware with a factor of 2 to 3 less in resolving power. To the dedicated experimental scientist this may sound like heresy. We are sure that in 1984 such reactions are not uncommon; in 1994 it is likely that most observational devices will have a deconvolution processor imbedded in them and that the deconvolution processor will, in fact, be thought of as an integral part of the experimental apparatus.

Today questions such as, Would you rather use deconvolution or build a higher-resolving-power instrument? are not uncommon. In fact, today they may actually be justified, considering what we have *not* learned about deconvolution. These same questions cannot be justified on the basis of a negative reaction to a numerical process that is no more mystical than the operation of any state-of-the-art instrument.

Constrained deconvolution is an experimental tool. Happily, it is a useful tool testable at each step in its development. Do not, however, assume that it can do no wrong. Utilization of state-of-the-art scientific instrumentation is a demanding task and is not for the faint of heart. Constrained deconvolution is a probable addition to many state-of-the-art instruments. Constrained deconvolution is rapidly becoming a part of data acquisition rather than a part of data analysis.

REFERENCES

Blass, W. E., and Halsey, G. W. (1981). "Deconvolution of Absorption Spectra." Academic Press, New York.

Jansson, P. A. (1970). *J. Opt. Soc. Am.* **60,** 184.

Jansson, P. A., Hunt, R. H., and Plyler, E. K. (1970). *J. Opt. Soc. Am.* **60,** 596.

Pliva, J., Pine, A. S., and Willson, P. D. (1980). *Appl. Opt.* **19,** 1833.

Savitzky, A., and Golay, M. J. E. (1964). *Anal. Chem.* **36,** 1627.

Willson, P. D. (1973). Ph.D. Thesis, Michigan State University, East Lansing.

CHAPTER **8**

Maximum-Likelihood Estimates of Spectra

B. Roy Frieden

Optical Sciences Center, The University of Arizona
Tucson, Arizona

I.	Introduction	229
II.	Orientation	230
III.	Physical Model for the Object	232
	A. Counting the Degrees of Freedom z	232
	B. Degeneracy Factor W	234
IV.	Most Probable Object: Definition	235
V.	Most Probable Object in the Presence of Data.	237
VI.	Object-Class Law	238
VII.	Case of a "White" Object; Maximum Entropy	239
VIII.	Scenarios for Knowing $p(q_1, \ldots, q_M)$	240
	A. Knowledge of Empirical Data	240
	B. Maximum Conviction without Empirical Data	244
	C. Zero Prior Knowledge; Maximum Ignorance	244
IX.	Estimators	246
	A. High-Conviction Cases	246
	B. Entropylike Estimators	247
	C. Estimators for Empirical Data; Impartial Conviction	248
X.	How to Handle Noise	250
XI.	Test Cases	252
	A. Application to Experimental Data	252
	B. Effect of Choice of Prior Object $\{Q_m\}$ and of z	254
XII.	Conclusion	258
	References	258

LIST OF SYMBOLS

A	slit area
c	speed of light
C	normalization constant
df	degrees of freedom
e_m	standard deviation of noise at mth point in image data
$\{\varepsilon_m\}$	noise in image data
h	Planck's constant
H	entropy, Jaynes form
H_1	entropy, Burg form

DECONVOLUTION:
WITH APPLICATIONS IN SPECTROSCOPY

$\{i_m\}$	image data
k	Boltzmann's constant
L	total number of photons in empirical object
m	general resolution cell in object
$\{m_m\}$	empirical object
M	number of resolution cells constituting object
ML	maximum likelihood
n_m	number of object photons in mth resolution cell
$\hat{n}_m$	estimated version of n_m
$\bar{n}_m$	average number of photons in mth resolution cell
n_m/z_m	occupancy ratio of mth resolution cell
N	total number of photons in object
o_m	object or true spectrum, mth value
$\hat{o}_m$	deconvolved, restored, or estimated (all equivalent) version of o_m
$p(q_1, \ldots, q_M)$	object-class probability law
P_N	probability of noise
$P_1(\{\varepsilon_m\} \mid \{n_m\})$	probability of noise conditional upon object
$P(n_1, \ldots, n_M)$	total probability of n_1 photons in cell $m = 1$ of object, and ..., and n_M photons in cell $m = M$ of object
$P_0(n_1, \ldots, n_M)$	probability of $\{n_m\}$ photons incident upon input slit
$P_0(n_1, \ldots, n_M \mid q_1, \ldots, q_M)$	$P_0(n_1, \ldots, n_M)$ conditional upon definite numbers $\{q_m\}$
$\mathscr{P}(n_1, \ldots, n_M, \varepsilon_1, \ldots, \varepsilon_M)$	joint probability of object and noise
q_m	probability that a randomly selected object photon will reside in cell m; relative intensity or "signal" intensity of cell m
$\{q_m\}$	prior object; prior spectrum
Q_m	prior object; prior spectrum; biases
R	distance between spectral source and entrance slit
rect	rectangular function
s_{mn}	sampled version $s(x_m - x_n)$
$s(x)$	point spread function of spectroscope
t	exposure time
T	temperature of a blackbody
w^2	area of spectral source
W	quantum-mechanical degeneracy factor
$W(n_1, \ldots, n_M)$	degeneracy as a function of $\{n_m\}$
x_m	generalized wavelength or wave number, mth value
z_m	number of quantum degrees of freedom in mth resolution cell
α_m	parameter used in forming $\{\hat{n}_m\}$ when empirical object given
δ	Dirac δ function
Δx	resolution "cell" length, set by user

$\Delta\lambda_m$	spectral wavelength range
Δv_m	spectral frequency range
$\Delta\tau_m$	coherence time for mth resolution cell
λ_m	wavelength, mth value
$\{\lambda_m\}$	Lagrange multipliers on image constraints
μ	Lagrange multiplier on normalization constraint
v	frequency (temporal)
$\bar{v}_m$	wave number, mth value
σ	standard deviation
σ_m	coherence area for mth resolution cell

I. INTRODUCTION

Every spectrum is obtained in a unique physical setting, defined by a unique set of *physical effects*. Source temperature, state of coherence, entrance-slit size, and so forth, all affect spectral shape. Therefore, the knowledge that the spectrum was formed under given circumstances should be used to form an *estimate* of the spectrum. In effect, the user hopes to "work backwards" from the given image data, through the physical effects, to the true spectrum. It follows that each estimation algorithm must be tailored to specific physical knowledge about the given spectrum. There should be no single best algorithm for use on all spectra.

Physical knowledge is part of a larger body of knowledge about the spectrum, called prior knowledge. (This is in the sense of prior to, or aside from, knowledge of the data.) A second component of prior knowledge is *statistical* in nature. The unknown spectrum may be imagined to belong to a "class" of spectra, all having the same statistical properties. For example, they may all have a small, but unknown, number of lines. Or the spectrum in question may have previously been estimated, but from observation of only a small number of photons. This would give us a random approximation to the true spectrum. Or the user may want to admit "maximum prior ignorance" about the probable spectrum present. This weakest form of prior knowledge will be discussed in detail. Finally, the reader may simply have some other form of prior knowledge at hand.

To date, the prior knowledge built into deconvolution algorithms has mainly been deterministic in nature, such as by use of positivity or boundedness. But a unified approach to estimating spectra should accommodate all possible physical and statistical prior knowledge about formation of the spectra. In particular, the approach should be physical, based on the Bose–Einstein nature of photons. Such an approach will be presented here. One general restoring principle will be derived, from which particular estimators follow, depending on what kind of prior knowledge is at hand. A maximum-likelihood† (ML) criterion will be the overall estimation criterion in use. This gives the estimates the following desirable property: if all the known deterministic and statistical conditions for forming the given spectrum were repeated over and over, the one spectrum that *actually occurred* the largest number of times would coincide with the ML estimate. That is, the estimate will be "right" more often than any other estimate. We suggest that this is about the most that one could ask of an estimate.

Depending on prior knowledge, the ML estimator will be either Jaynes maximum entropy, Kikuchi–Soffer maximum entropy, maximum weighted-Burg entropy, minimum "photon-site" entropy, minimum "empirical" entropy, some combination of these, or something altogether different. It behooves the user to specify most accurately and completely *his* state of prior knowledge, so as to tailor an ML estimator to the given problem. We shall show in detail how to do this.

II. ORIENTATION

This work will be restricted to the consideration of emission spectra. Absorption spectra and other kinds of spectra could be treated analogously but really need a separate physical development.

We shall assume that the spectrometer basically consists of an entrance slit of area A focused on an exit slit at $1:1$ magnification (Fig. 1). The recorded image is blurred because of diffraction and aberrations in the optics between the two slits and because of electronic blur in the readout circuit. The ultimate aim is to deconvolve or restore the spectral image so as to retrieve the spectral pattern at the entrance slit; the latter pattern is called the object.

The object consists of intensities o_m, $m = 1, \ldots, M$, or $\{o_m\}$ for short. These intensities are at wavelengths $\{\lambda_m\}$ or wave numbers $\{\bar{v}_m\}$; the choice is arbitrary. To keep the analysis general, we shall call these abscissas $\{x_m\}$. As is conventional, the deconvolved or restored version of the object is denoted by $\{\hat{o}_m\}$.

The subdivision of the $\{x_m\}$ abscissas is uniform,

$$x_m = m \, \Delta x, \qquad (1)$$

so that each x_m is centered on a resolution "cell" of length Δx (in either

---

† A Bayesian approach will be used, where all data and prior information are included. The approach is philosophically like maximum-a posteriori (MAP) estimation. Hence, our ML (or MAP) estimate is *maximum probable given the data and prior information.*

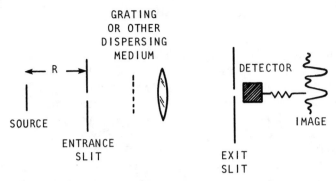

Fig. 1 Basic spectrometer layout. The source is extended and incoherent. Entrance and exit slits are at equal conjugates. The image is formed electronically after detection at the exit slit.

wavelength or wave-number units). The observer is assumed satisfied to resolve the spectrum at this ultimate level of resolution.

At a given x_m, the energy per photon is of course $h\nu_m$, where h is Planck's constant and ν frequency. Hence, if n_m denotes the number of object photons that radiate from the mth cell during a fixed exposure time, o_m obeys

$$o_m = n_m h\nu_m. \tag{2}$$

Obviously, if n_m is estimated, this relation allows o_m to be estimated as well (see Fig. 2).

Going one step further, suppose that the object consists of N radiated photons in all, that is,

$$\sum_{m=1}^{M} n_m = N. \tag{3}$$

Quantity N is assumed to be (approximately) known by conservation of energy from object to image.

In summary, then, we shall seek estimates $\{\hat{n}_m\}$ obeying normalization condition (3) and recorded image data. Once the $\{\hat{n}_m\}$ are known, Eq. (2) may be used to arrive at the object estimate $\{\hat{o}_m\}$.

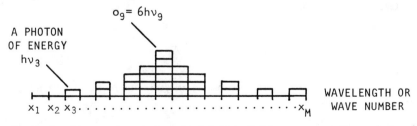

Fig. 2 An object spectrum to be estimated. Note that here $n_1 = n_2 = 0$, $n_3 = 1$, $n_4 = 0$, $n_5 = 2, \ldots$.

III. PHYSICAL MODEL FOR THE OBJECT

Imagine photons to be streaming from the entrance slit of area A toward the exit slit. These of course include all wavelengths of the source. Picture next just photons of one wavelength (or wave number) x_m as they flow from the entrance slit. Because these are quantum-mechanical entities, *they cannot occupy continuously* all positions in space during their flow. Instead, they may occupy only finite positions in space, called degrees of freedom (df) or modes. These are shown schematically as cubes in Fig. 3. Note that there are but a finite number z_m of such cubes and that we must subscript z because the number of modes will vary from one wavelength to another.

Each photon that flows from the entrance slit may exist only in such a mode. However, more than one may crowd in. This is a property of Bose–Einstein particles, to which photons belong as a class. Photons are also known to be microscopically indistinguishable, so that a given configuration of them within the modes cannot be distinguished from the same configuration where some of the photons have interchanged mode positions. As we shall see, the ML solution that we seek will depend vitally on this Bose–Einstein aspect of photon statistics.

A. Counting the Degrees of Freedom z†

Consider one resolution cell x_m of the spectrum. The number of df, z_m, is defined as the number of "independent" volumes that exist over the volume swept out by photons leaving the entrance slit during one exposure time t (see Fig. 3). An "independent" volume is the three-dimensional space over which a photon is coherent. Hence, if it has a coherence length $c\,\Delta\tau_m$, with c the speed of light and $\Delta\tau_m$ the coherence time, and if it has a coherence area σ_m, the coherence volume is simply their product, or $c\,\Delta\tau_m\,\sigma_m$. On the other hand, the total volume swept out by a wave front of photons all leaving the aperture of area A during exposure time t is ctA. Hence the number of coherence volumes is

$$\frac{ctA}{c\,\Delta\tau_m\,\sigma_m} \equiv z_m = \frac{t}{\Delta\tau_m}\,\frac{A}{\sigma_m}. \tag{4}$$

Of these factors, A and t are of course directly measurable and need no further consideration.

† Kikuchi and Soffer, 1977.

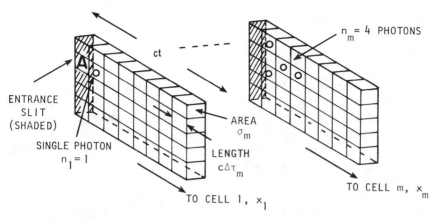

Fig. 3 Degrees of freedom (small cube) for light from the entrance slit (shaded area A) at wavelengths x_1 and x_m.

The coherence time $\Delta\tau_m$ relates to abscissa x_m and resolution interval Δx as follows. By the Heisenberg uncertainty principle, $\Delta\tau_m$ goes inversely with spectral purity, or

$$\Delta\tau_m \approx 1/\Delta\nu_m, \tag{5}$$

where $\Delta\nu_m$ is the frequency range consistent with x_m and Δx. By $\lambda\nu = c$,

$$\Delta\nu_m = (c/\lambda_m^2)\,\Delta\lambda_m,$$

so that by Eq. (5) we have

$$\Delta\tau_m \approx \lambda_m^2/c\,\Delta\lambda_m. \tag{6}$$

Then, if x_m represents wavelength λ_m, we conclude that

$$\Delta\tau_m \approx x_m^2/c\,\Delta x. \tag{7a}$$

But if x_m instead represents wave number $\bar{\nu}_m \equiv 1/\lambda_m$, we obtain

$$\Delta\tau_m \approx 1/c\,\Delta x. \tag{7b}$$

In either case [Eq. (7a) or (7b)], $\Delta\tau_m$ is related to observables.

The coherence area σ_m is measured at the entrance slit, because this is where the object is imagined to exist. From Fig. 1, this slit is distance R from the spectral source, which is incoherent and of area w^2. Then the Van Cittert–Zernike theorem (Born and Wolf, 1959) gives

$$\sigma_m \approx R^2\lambda_m^2/w^2. \tag{8}$$

Then, if x_m represents a wavelength, we obtain

$$\sigma_m \approx R^2 x_m^2/w^2, \tag{9a}$$

whereas if it represents a wave number,

$$\sigma_m \approx R^2/x_m^2 w^2 \tag{9b}$$

results.

The preceding is meant to be merely a representative calculation of σ_m. If instead a lens is interposed between source and entrance slit, σ_m will obey a different expression. The essential point is that σ_m is a computable quantity, whatever the setup.

The number of degrees of freedom, z_m, for a given resolution cell x_m may then be computed by combining Eq. (4) with Eq. (7a) or (7b) and Eq. (9a) or (9b).

It is sometimes not necessary to know the absolute value of z_m but rather the order of magnitude of *ratio* n_m/z_m, which might be called the occupancy ratio of the degrees of freedom by photons. This may be estimated as follows. Suppose that we knew that the spectrum was like that of a blackbody at temperature T. We could then use the well-known relation for the *average number* of photons per mode (degree of freedom),

$$\bar{n}_m/z_m = [\exp(h\nu_m/kT) - 1]^{-1}. \tag{10}$$

Parameters h and k are Planck's and Boltzmann's constants, respectively. This mean ratio can often be regarded as a useful approximation to the ratio n_m/z_m. In particular, it shows that for viewing the spectrum of the sun $(T = 6000$ K$)$ in the visible region the ratio is small,

$$n_m/z_m \ll 1, \tag{11a}$$

whereas at wavelengths exceeding 3 μm the ratio is large,

$$n_m/z_m \gg 1. \tag{11b}$$

Other kinds of sources are considered in Kikuchi and Soffer (1977).

B. Degeneracy Factor W

Figure 3 shows $n_1 = 1$ photon for cell 1 located at a particular df position. It also shows $n_m = 4$ photons located at four particular positions for cell m. In how many ways may one photon locate in one particular df position of cell 1 and four photons locate in four particular positions of cell m? Because we cannot label the photons in any way (they are indistinguishable), there is only one way in which these placements may be made.

By contrast, suppose the $n_1 = 1$ photon may be *at any* df position of cell 1. Now, in how many ways can this photon be placed within the cell? Because it may occupy any of z_1 positions, there are z_1 such ways. More generally, the number of ways n_m photons may locate in cell m is given by the Bose–Einstein factor (Kikuchi and Soffer, 1977)

$$(n_m + z_m - 1)!/n_m!(z_m - 1)!.$$

This is the number of ways n_m entities may be distributed among z_m positions, where any number may occupy a given position but where interchanges of the entities do not count as new ways.

Finally, we ask how many ways W can n_1 photons locate in cell 1, *and ...,* *and n_M* photons locate in cell M. Because these photon placements are independent from cell to cell, the answer is a product of the single-cell answers:

$$W(n_1, \ldots, n_M) = \prod_{m=1}^{M} \frac{(n_m + z_m - 1)!}{n_m!(z_m - 1)!}. \tag{12}$$

This is the net Bose–Einstein factor. It will be used in the following to form Bayesian estimators (see, e.g., Frieden, 1983) of the object $\{n_m\}$.

IV. MOST PROBABLE OBJECT: DEFINITION

The deconvolved or restored object that we seek is *the most probable number-count set* $\{n_m\}$. This is called the maximum-likelihood (ML) estimate of the object. It obeys simply

$$P(n_1, \ldots, n_M) = \text{maximum} \tag{13}$$

through choice of $\{n_m\}$. What is probability P?

We shall first find a component $P_0(n_1, \ldots, n_M)$ of P. Quantity P_0 describes the a priori probability of the $\{n_m\}$ photons *as they are incident upon the input slit.* This is independent of the df. Hence, what is the probability P_0 that n_1 slit photons will enter cell 1, and ..., and n_M photons will enter cell M? One possibility is that P_0 is a constant, independent of $\{n_m\}$. But that actually assumes the unknown spectrum to be equal-energy white (see Section VII), quite a restrictive assumption. We find next an expression for P_0 that allows for a more general state of prior knowledge about the spectrum, in fact the most general.

Let us regard the "fate" of each photon incident upon the entrance slit. Let q_m be the probability that such a photon will occupy general cell x_m of the object (its fate). The probabilities $\{q_m\}$ define an object that (temporarily) will be assumed to be known. It is the ideal, underlying, signal† spectrum that ultimately gives rise to the unknown number-count spectrum $\{n_m\}$. This ideal spectrum is not necessarily flat (the conventional assumption). It could, for example, be the Balmer line series.

The $\{q_m\}$ may be called the prior object or prior spectrum. We temporarily assume it to be known aside from the image data. Indeed, in the ML principle given later, the $\{q_m\}$ supplement the data and are inserted independently of them. We shall show in later sections how the $\{q_m\}$, or at least a probability law for $\{q_m\}$, may be estimated.

With definite values for $\{q_m\}$, probability P_0 obeys a simple chain rule

$$q_1^{n_1} q_2^{n_2} \cdots q_M^{n_M}.$$

This assumes that each photon placement is independent of all others, as would be true for an incoherently radiating source. Stated as an equality,

$$P_0(n_1, \ldots, n_M | q_1, \ldots, q_M) = q_1^{n_1} q_2^{n_2} \cdots q_M^{n_M}, \tag{14}$$

because this is the probability of the $\{n_m\}$ *conditional upon* definite numbers $\{q_m\}$. (In standard probabilistic notation, the vertical bar denotes "conditional upon.")

However, more generally the user has uncertainty about *what* prior spectrum $\{q_m\}$ may be present. That is, the user can prescribe only a probability law

$$p(q_1, \ldots, q_M) \tag{15}$$

on possible spectra.‡ This law defines the body of statistical prior knowledge at hand defining the "object class." It includes all biases and prejudgments that the user may have about the unknown object. We shall assume different forms for this object-class law later. For now, we shall see how it fits into the overall ML approach.

The existence of a probability law for $\{q_m\}$ affects expression (14) for P_0. By the total probability law of statistics, Eqs. (14) and (15) may be combined to yield a net

$$P_0(n_1, \ldots, n_M) = \int_{\Sigma q_m = 1} \cdots \int q_1^{n_1} \cdots q_M^{n_M} p(q_1, \ldots, q_M) \, dq_1 \cdots dq_M. \tag{16}$$

---

† By the law of large numbers (see, e.g., Frieden, 1983), $q_m = n_m/N$ as $N \to \infty$, q_m describes the relative *intensity* of resolution cell x_m in the limit of an infinite number of photons. This intensity is called the signal.

‡ Readers who wish thoroughly to confuse themselves can dwell on the fact that expression (15) is also a "probability of probabilities." This concept is in fact philosophically sound, and the basis of all Bayesian estimation methods. However, it will be more fruitful to regard expression (15) as simply a probability of signal *intensities*.

Again, this is the a priori probability of a number-count object $\{n_m\}$ if the unknown spectrum $\{q_m\}$ is chosen randomly from a class of spectra defined by $p(q_1, \ldots, q_M)$.

According to statistical mechanics the total probability P of a number-count object $\{n_m\}$ is the simple product of Eq. (16) with the degeneracy law $W(n_1, \ldots, n_M)$ of Eq. (12). Thus

$$P(n_1, \ldots, n_M)$$

$$= \prod_{m=1}^{M} \frac{(n_m + z_m - 1)!}{n_m!(z_m - 1)!} \int_{\Sigma q_m = 1} \cdots \int q_1^{n_1} \cdots q_M^{n_M} p(q_1, \ldots, q_M) \, dq_1 \cdots dq_M$$

$$= \text{maximum} \tag{17}$$

is the ML principle for the unknown spectrum $\{n_m\}$. The maximum is sought through choice of $\{n_m\}$. This defines the ML answer in the absence of image data, but with knowledge of object class (15).†

V.   MOST PROBABLE OBJECT IN THE PRESENCE OF DATA

The data $\{i_m\}$ are the measured output spectrum at the exit slit. This is simply the convolution of the number-count object at the entrance slit with the point spread function $s(x)$ of the intervening optics and following electronics,

$$i_m = \sum_{n=1}^{M} o_n s_{mn} = h \sum n_n v_n s_{mn}, \qquad m = 1, \ldots, M,$$

$$s_{mn} \equiv s(x_m - x_n). \tag{18}$$

The presence of noise in $\{i_m\}$ has been temporarily ignored. Note that the convolution has a discrete form because of the discrete size Δx of the resolution cells in object and image space. Also, it is assumed for simplicity that the form of $s(x)$ does not vary with abscissa x_m.

Hence, the object $\{o_m\}$ or $\{n_m\}$ that satisfies the ML principle (17) must also obey the M constraint equations (18). In addition, there is the normalization constraint (3). These may be tagged onto Eq. (17) by the conventional use of

† Bayesian estimation in general permits estimation in the absence of data. However, we are not advocating the abandonment of spectrometers; see Section V.

Lagrange multipliers. In addition, it turns out to be more convenient to maximize ln P instead of P directly. Then the ML principle becomes

$$\ln P(n_1,\ldots,n_M) + \mu(\sum n_m - N) + \sum_{m=1}^{M} \lambda_m(h \sum_n n_n v_n s_{mn} - i_m) = \text{maximum},$$

(19)

for P given by Eq. (17). This states that the estimated object is the most probable member of object class $p(q_1,\ldots,q_M)$ that also obeys the image data. *The estimate now combines the known physics of photons, prior knowledge $\{z_m\}$ of degrees of freedom, prior knowledge $p(q_1,\ldots,q_M)$ of object class, and posterior knowledge of image data.* The remaining gap concerns how to establish probability law $p(q_1,\ldots,q_M)$.

VI. OBJECT-CLASS LAW

Probability law (15) describes the user's judgment as to what signal object is present. More specifically, it displays the user's judgment as to the probable values of the prior signal $\{q_m\}$ *and* the user's assessment of the strength of this judgment. This is clarified as follows.

Suppose the user thinks that the prior $\{q_m\}$ ought to have values near a definite set of number $\{Q_m\}$, called biases. Suppose also that the user wants to express this suspicion with the *highest possible conviction.* Then the user asserts that there is *no spread* possible about the $\{Q_m\}$, or

$$p(q_1,\ldots,q_M) = \delta(q_1 - Q_1)\cdots\delta(q_M - Q_M).$$

(20)

The user's opinion is exceedingly strong: no other signal object can be present. Note that there is still an estimation problem, however. Even with one spectrum $\{Q_m\}$ present *any number-count object $\{n_m\}$* may result. If, for example, the signal object $\{Q_m\}$ is a true rect function, the *number-count* object $\{n_m\}$ may be a rectangle with one or more grooves and bumps on it. This is the object that emerges from the entrance slit and is later imaged; hence it is the one that is sought.

As to which object $\{n_m\}$ is present, this is where the image data in principle (19) enter in. Their effect is to force the estimate toward the true $\{n_m\}$ present. The effect of the $\{Q_m\}$ is to exert merely a beneficial† biasing effect on the estimate, as shown later.

---

† Conversely, if the user has the highest conviction *but is wrong,* the user's $\{Q_m\}$ will exert a deleterious effect on the estimate (see Section XI.B).

An analogy is the corresponding problem of two-dimensional object restoration. Suppose that the object really *is* a gray transparency with a dark letter A imprinted on it. The $\{Q_{mn}\}$ now describe the ideal signal A and its background. But because the number N of photons passing through the transparency is finite, the actual number-count object $\{n_m\}$ will randomly depart from the ideal A.

Alternatively, the user may have *less conviction* than is expressed by Eq. (20) about the unknown $\{q_m\}$. If so, the user should permit q_1 (for example) to have a finite range of values, perhaps with Q_1 as now the most likely value (formerly it was the *only* value permitted). An example is

$$p(q_1) = (1/\sqrt{2\pi})\sigma^{-1} \exp[-(q_1 - Q_1)^2/2\sigma^2],$$

the normal law. (The same goes for the other $\{q_m\}$.) The size of σ defines the user's conviction about Q_1 as the dominant value for q_1. The smaller it is made, the more conviction is expressed. In the limit $\sigma \to 0$, the state (20) of utmost conviction is achieved once again. Of course, forms for probability law $p(q_1)$ other than the normal will also express these tendencies. These are discussed in the next two sections.

VII.   CASE OF A "WHITE" OBJECT; MAXIMUM ENTROPY

Suppose that prior knowledge is of the maximum-conviction form (20), where in addition all

$$Q_m = 1/M.$$

This corresponds to prior knowledge of an equal-energy white spectrum. Combining Eqs. (16) and (20) then yields

$$P_0(n_1, \ldots, n_M) = (1/M)^{\sum n_m}.$$

But $\sum n_m$ is the constant N, by Eq. (3). Hence, P_0 is now a constant, independent of the $\{n_m\}$. This has an important ramification.

Principle (17) now consists of just the Bose–Einstein degeneracy factor, exactly Kikuchi and Soffer's form (1977). Also, as these authors showed (see also Section IX.B), in the case (11a) of sparsely occupied df it becomes Jaynes's maximum-entropy form (39) (Jaynes, 1968). Hence, both the Kikuchi–Soffer and Jaynes estimators are special cases of the ML approach, corresponding to the prior knowledge that the unknown spectrum is equal-energy white with the highest conviction.

This is a satisfying result. It has long been known that maximum entropy exerts a smoothing tendency on its estimate (Frieden, 1972). Now we see where this originates: in the high-conviction assertion of an equal-energy white object. If the underlying object is "known" to be flat, the most probable estimate will also tend to be flat.

It has also been observed in past empirical work with maximum entropy that the outputs are quite accurate estimates of the true objects, often exhibiting resolution beyond (it would seem) the realm of possibility; see Frieden (1972) or Wernecke and D'Addario (1977). Now we can see why this should be true. These high-resolution estimates occurred only when the object consisted of isolated impulses against basically a flat background. That is, the object was essentially "equal-energy white." In this case, as just discussed, the ML estimate *is* Kikuchi–Soffer or maximum entropy. These high-resolution outputs were therefore maximum likely as well!

The reader should be cautioned at this point, however, that not all ML estimates (19) will turn out to obey maximum entropy (see Section IX).

VIII.   SCENARIOS FOR KNOWING $p(q_1, \ldots, q_M)$

We shall next investigate a limited but interesting number of cases for which specific object-class laws $p(q_1, \ldots, q_M)$ are formed. Remarkably, in some cases the estimation principle (19) may itself be used to form p. In these cases we formally set all $\lambda_m = 0$, because object class is defined independent of knowing the data.

An interesting case that we shall *not* consider is where the user knows *power spectra* for the object and the noise and from these wants to infer $p(q_1, \ldots, q_M)$.† Indeed, the problem has not been solved yet, to our knowledge.

The problem of forming $p(q_1, \ldots, q_M)$ is central to our whole approach. It is also the most difficult step. Once known and substituted into Eq. (19), a solution $\{\hat{n}_m\}$ could always be at least *numerically* found—but not so with forming $p(q_1, \ldots, q_M)$. It often takes some ingenuity to frame the given prior knowledge in the form of a probability law. Some of the easier cases are taken up next. Luckily, the reader will also find them applicable in many cases.

A. Knowledge of Empirical Data

One scenario on which to build a $p(q_1, \ldots, q_M)$ is *empirical* evidence, that is, *actual observation of a number-count object* $\{n_m\}$ prior to estimating the object. The proviso is that object $\{n_m\}$ belongs to the "same class" as the

† Problem posed by P. A. Jansson.

unknown object. We shall assume this. There are two different assertions that the user may make, given the information $\{n_m\}$. These are that the data represent a ML estimate of the object or that the data only imply a fixed probability law (to be found) on possible objects. We consider these alternative assertions in Section VIII.A.1 and VIII.A.2.

1. Maximum Conviction That Empirical Data Represent the Object

First consider the assertion that, with strongest conviction, the $\{n_m\}$ represent the ML estimate of the unknown object in the absence of image data. We found that a state of strongest conviction is represented by form (20) with biases $\{Q_m\}$. Then what are the $\{Q_m\}$ in terms of observables $\{n_m\}$?

The ML principle (19) gives the ML solution $\{n_m\}$ either in the presence or in the absence of image data. In the latter case, the $\{\lambda_m\}$ are merely all set equal to zero. Although Eq. (19) is usually used to predict values $\{n_m\}$ for a known $p(q_1, \ldots, q_M)$, we shall instead use it in reverse, finding probabilities $\{Q_m\}$ from observed $\{n_m\}$. This use follows the assumption that the *observed* $\{n_m\}$ were *maximum likely* as well.

The principle (19) with $P(n_1, \ldots, n_M)$ given by Eq. (17) and the use of maximum conviction (20) results in a statement

$$\ln \prod_{m=1}^{M} \frac{(n_m + z_m - 1)!}{n_m!(z_m - 1)!} Q_m^{n_m} + \mu(\sum n_m - N) = \text{maximum}, \qquad (21)$$

through choice of the $\{n_m\}$. The $\{\lambda_m\}$ were set equal to zero because the situation is prior to observing the image.

The solution may be found by the use of Stirling's approximation to the factorial,

$$\ln n! \approx n \ln n. \qquad (22)$$

Expanding the logarithm of the product and using identity (22) yields

$$\sum (n_m + z_m - 1) \ln(n_m + z_m - 1) - \sum n_m \ln n_m$$
$$+ \sum n_m \ln Q_m + \mu(\sum n_m - N) = \text{maximum}. \qquad (23)$$

This estimator, without the first sum, was first derived by Hershel (1971). We have ignored a term in the $\{z_m\}$ because these do not affect the maximization. The solution is obtained merely be setting the derivative $\partial/\partial n_m$ of Eq. (23) equal to zero,

$$Q_m = Cn_m/(n_m + z_m - 1). \qquad (24)$$

Parameter C is a normalization constant incorporating μ. [That Eq. (24) attains a maximum and not a minimum for Eq. (23) may be verified by taking

$\partial^2/\partial n_m \partial n_k$ of Eq. (23) and noticing that this gives a diagonal matrix of numbers all of which are negative.]

As mentioned before, normally Eq. (24) would be used to express n_m in terms of a known Q_m. Here, we use it the other way around because it is assumed instead that n_m is known (observed in the prior object) and Q_m is to be estimated. Equation (24) thereby provides a means of knowing the prior spectrum $\{Q_m\}$. We see that each Q_m depends not only on the observed prior object value n_m but also on the number z_m of degrees of freedom. In particular, each Q_m is not simply linear in the corresponding prior object value n_m, which intuition might otherwise suggest. The essential reason is that n_m represents the number of photons in all possible arrangements over the z_m df, whereas Q_m represents simply the probability of occupation of cell m *prior to* the df (image) phenomenon. Note, in particular, that $Q_m \propto n_m$ when the df are sparsely occupied ($z_m \gg n_m$), as intuition suggests.

In summary, with the $\{n_m\}$ known as empirical observables, Eq. (24) permits the $\{Q_m\}$ to be known. Finally, by maximum conviction, $p(q_1, \ldots, q_M)$ is of the form (20).

2. Impartial Use of Empirical Data

Having considered maximum conviction, we now consider the opposite situation, called fair or impartial use of the data. Here, our conviction will only be as strong as the data permit. The motivation is as follows.

If a coin is flipped 100 times and 80 of the outcomes are heads, we cannot say that the probability q_1 of a head is 0.80 for sure. For example, even if $q_1 = 0.001$, there would still be a finite chance of obtaining the 80 heads. Hence, all q_1 between 0 and 1 are possible, and the evidence of $n_1 = 80$ heads *implies no more than a probability law* $p(q_1)$ describing the chance that any one value of q_1 is the true one. This law will have its peak at $q_1 = 0.80$, of course, so that value $q_1 = 0.80$ is maximum likely. However, it is not the only possibility.

So it is with knowledge of an empirical prior object $\{n_m\}$. If $N = 100$ photons come from the object, and $n_1 = 80$ are from cell 1, this does not imply that for sure $q_1 = 80/100$. All values of q_1 between 0 and 1 are still possible, according to a probability law. Hence, considering all the cells simultaneously, we seek the probability law

$$p(q_1, \ldots, q_M | n_1, \ldots, n_M), \tag{25}$$

that is, $p(q_1, \ldots, q_M)$ contingent upon the evidence $\{n_m\}$. As a matter of nomenclature, because we are allowing *all* q values to exist, we call this fair or impartial conviction about the $\{q_m\}$.

This problem may be solved using elementary probability theory. Bayer's rule states that the answer may be formed as

$$p(q_1, \ldots, q_M | n_1, \ldots, n_M) = \frac{P(n_1, \ldots, n_M | q_1, \ldots, q_M) p_0(q_1, \ldots, q_M)}{P(n_1, \ldots, n_M)} \quad (26)$$

(see, e.g., Frieden, 1983). The right-hand quantities have to be found.

Probability $P(n_1, \ldots, n_M | q_1, \ldots, q_M)$ describes Eq. (17) when one particular set $\{q_m\}$ is present,

$$P(n_1, \ldots, n_M | q_1, \ldots, q_M) = \prod_{m=1}^{M} \frac{(n_m + z_m - 1)!}{n_m!(z_m - 1)!} q_m^{n_m}. \quad (27)$$

Probability $p_0(q_1, \ldots, q_M)$ represents what the user supposes the probability of the $\{q_m\}$ would be prior to observation of the empirical $\{n_m\}$ (hence it is a "preprior" probability law). Prior to these observations, it is reasonable to assign all possible values of q_m the *same* probability of occurrence. This is the MacQueen–Marschak (1975) definition of maximum prior ignorance,

$$p_0(q_1, \ldots, q_M) = \prod_{m=1}^{M} \text{rect}(q_m - \tfrac{1}{2}),$$

$$\text{rect}(x) \equiv \begin{cases} 1 & \text{for } |x| < \tfrac{1}{2} \\ \tfrac{1}{2} & \text{for } |x| = \tfrac{1}{2} \\ 0 & \text{for } |x| > \tfrac{1}{2}. \end{cases} \quad (28)$$

This says that the prior spectrum is *anything* with equal likelihood. Note that this assumption is consistent with the overall stance of fair conviction in this section.

The denominator $P(n_1, \ldots, n_M)$ in Eq. (26) is the integral of the numerator over all $\{q_m\}$. By Eqs. (27) and (28) the integral is Dirichlet's (see Gradshteyn and Ryzhik, 1965), with the result

$$P(n_1, \ldots, n_M) = \frac{(M-1)!}{(L+M-1)!} \prod_{m=1}^{M} \frac{(n_m + z_m - 1)!}{(z_m - 1)!}, \qquad L \equiv \sum n_m. \quad (29)$$

Note the distinction between L and N: the empirical object prior to data $\{i_m\}$ has L photons, whereas the unknown object that forms $\{i_m\}$ has N photons.

With all the right-hand members of Eq. (26) now known, it becomes

$$p(q_1, \ldots, q_M) = \frac{(L+M-1)!}{(M-1)!} \frac{q_1^{n_1} \cdots q_M^{n_M}}{n_1! \ldots n_M!}. \quad (30)$$

(The contingency upon $\{n_m\}$ is, for brevity, suppressed in the left-hand argument.) This is the prior-knowledge probability law that we sought. As a check, notice that in the limit of no prior data $\{n_m\} = 0$, $L = 0$, it becomes a constant. This is consistent with the maximum-prior-ignorance condition (28) previously adopted, which supposed no prior data present.

B. Maximum Conviction without Empirical Data

Perhaps the commonest circumstance is where *no empirical data* $\{n_m\}$ are at hand, but the user has an *opinion* as to what the object is. A way of phrasing this opinion is, "Before seing the image data, I highly suspect the object to be of such and such a shape $\{n_m\}$." If, in fact, the user thinks it *most likely* that the object has this shape, then the derivation of the preceding section again applies. [Note that the $\{n_m\}$ in principle (21) are mere assertions. They may be real data or, as in this case, imagined or anticipated data.] Hence the result (24) may again be used to find the prior spectrum $\{Q_m\}$.

A word of assurance is perhaps necessary here. The reader might suppose that because the particular values $\{Q_m\}$ are input into the principle (19), the answer $\{\hat{n}_m\}$ that results will necessarily be proportional to the $\{Q_m\}$. This is not true. The image data in Eq. (19) exert their own, independent influence in biasing the answer toward peaks and valleys that are consistent with *it*. Thus, the $\{Q_m\}$ exert a biasing tendency but *only* a tendency. The main influence on the solution $\{\hat{n}_m\}$ is exerted by the image data.

The reader may surmise that perhaps the image data themselves may be used as the prior spectrum $\{Q_m\}$. Of course the image is a blurred version of the object, but nevertheless it does bear a resemblance to it. In this case the estimated object $\{\hat{n}_m\}$ will be biased toward the image values, which actually is a helpful tendency because the image is relatively smooth. Empirically, this helps to keep down noise and artifact oscillations in the $\{\hat{n}_m\}$, as we find when testing out this idea in Section XI.

C. Zero Prior Knowledge; Maximum Ignorance

Probably the most common state of prior knowledge is a state of zero prior knowledge (sometimes also called maximum ignorance). In this state the $\{z_m\}$ are known, defining the coherence of the source and the geometry of the spectrometer (as previously found), but "nothing else" is known about the object. In this "nothing else" lies a mystery. What does it mean? Jaynes (1968) and Kikuchi and Soffer (1977) implicitly assumed it to mean an equal-energy white object

$$p(q_1,\ldots,q_M) = \delta(q_1 - 1/M)\cdots\delta(q_M - 1/M), \tag{31}$$

as previously discussed. We shall see that this is consistent with the assertion that

$$n_m \propto z_m - 1 \tag{32}$$

is the ML object in the absence of image data [see Eq. (38)]. This looks reasonable because this prior object is then simply proportional to the number of degrees of freedom.

However, consider the opposing argument. Because the $\{z_m\}$ are known, by Eq. (32) a *unique object* is ML in the absence of data. Can this represent a state of *maximum ignorance*?

In particular, suppose that all the $\{z_m\}$ are equal. Then by Eq. (32) the prior object is imagined to be uniform. Why should such a choice of object be preferred over any other shape? Would it not make more sense to allow for many possible ML objects to be present a priori, each with the *same* probability? (ML solutions do not have to be unique, as discussed later.)

Perhaps the greatest fault with the choice (31) for representing zero prior knowledge is its assertion, with maximum conviction, that the prior object is equal-energy white. *This is a definite statement about shape.* Again, can this represent a state of *zero* prior knowledge?

For these reasons, following MacQueen and Marschak (1975), we use as an alternative definition of maximum ignorance

$$p(q_1, \ldots, q_M) = \prod_{m=1}^{M} \text{rect}(q_m - \tfrac{1}{2}). \tag{33}$$

Now *any* distribution $\{q_m\}$ of prior objects can be present (so long as they obey normalization), and *each distribution* has the same probability of occurring. Further, the use of this definition in principle (19) (with all $\lambda_m = 0$ because there are no image data) yields a ML prior object obeying

$$\sum (n_m + z_m - 1) \ln(n_m + z_m - 1) + \lambda(\sum n_m - N) = \text{maximum} \tag{34}$$

[see Eq. (47b).] This equation has an interesting property. Suppose, in particular, that all the $\{z_m\}$ are equal. Then Eq. (34) is a *global criterion*; that is, it is blind to interchanges of n_i with n_j, for any i, j. For example, if the solution to Eq. (34) were $\{n_m\} = (4, 2, \ldots)$, then the different object $(2, 4, \ldots)$ would also be a solution. The only way such interchanges would not result in a new object is if all the $\{n_m\}$ were equal. But such an object is actually the solution to

$$\sum n_m \ln n_m + \lambda(\sum n_m - N) = \text{minimum} \tag{35}$$

and so cannot also satisfy Eq. (34) (compare right-hand sides).

The result is that the ML prior object that is consistent with definition (33) is actually an *ensemble* of objects, each occurring with the *same* likelihood.

This is more in the spirit of maximum ignorance than was the preceding definition (31).

But if this is true, how could its use in the ML principle (19) (including data) result in a *unique* object estimate $\{n_m\}$? The answer is that the image data will constrain the estimate toward the one object that is consistent with *its* shape. This will rule out the other members of the object ensemble.

In summary, it makes more sense for "maximum ignorance" about a spectrum to mean an infinity of equally likely possibilities (33) than to imply the unique, equal-energy white spectrum (31).

IX. ESTIMATORS

The aim of this chapter has been to show how ML estimators may be formed for spectroscope outputs. Now we are in a position to do this. Having established a general ML principle (19) and specific forms of prior knowledge (20), (24), (30), (31), and (33), we now can combine results. What does estimator (19) become under different forms of prior knowledge? And what is its solution, in terms of data $\{i_m\}$?

A. High-Conviction Cases

Here the prior probability law is of the general form (20). To review, this includes the case where the $\{Q_m\}$ are simply guessed at, based on the user's expectations, or where the image data $\{i_m\}$ are used to represent $\{Q_m\}$, or the case (24) of empirical data and high conviction, or the case (31) of an equal-energy white signal spectrum.

Substitution of (20) into principle (19) yields

$$\ln \prod_{m=1}^{M} \frac{(n_m + z_m - 1)!}{n_m!(z_m - 1)!} Q_m^{n_m} + \mu(\sum n_m - N)$$

$$+ \sum \lambda_m(h \sum n_n v_n s_{mn} - i_m) = \text{maximum}. \tag{36}$$

Expanding the logarithm of the product, using Stirling's approximation (22) and ignoring terms that are constant in n_m, yields

$$\sum (n_m + z_m - 1) \ln(n_m + z_m - 1) + \sum n_m \ln(Q_m/n_m)$$

$$+ \mu \sum n_m + \sum \lambda_m h \sum n_n v_n s_{mn} = \text{maximum}. \tag{37}$$

This is the restoring principle for this case.

Its solution is obtained by setting $\partial/\partial n_m$ of Eq. (37) equal to zero, resulting in

$$\hat{n}_m = \frac{(z_m - 1)Q_m}{\exp(-\mu - hv_m \sum_n \lambda_n s_{nm}) - Q_m}. \tag{38}$$

This resembles the Bose factor $[\exp(hv/kT) - 1]^{-1}$ of quantum optics. Of course, the resemblance is no coincidence, both deriving from similar physics.

In the case (24) of empirical data, substitution into Eq. (38) gives $\hat{n}_m$ in terms of the observed *prior* object $\{n_m\}$.

The unknowns in Eq. (38) are μ and $\{\lambda_n\}$. These are found by demanding $\hat{n}_m$ to obey the image constraint equations (18) and normalization (3). Because the unknowns enter in a nonlinear way, the resulting $M + 1$ equations were solved by an iterative technique—Newton–Raphson relaxation (see, e.g., Hildebrand, 1956). Empirical cases are studied in Section XI.

B. Entropylike Estimators

The entropy H of an object $\{n_m\}$ is defined as

$$H \equiv - \sum_{m=1}^{M} n_m \ln n_m. \tag{39}$$

A principle of estimation H = maximum has long been advocated by Jaynes (1968). We note that the estimator (37) is *not* of this simple form, although one sum (the second) is close to it. The conclusion is that Jaynes's principle is not generally ML.

In this vein, however, consider limiting cases (11a) of sparsely occupied df sites. With the approximation

$$\ln(n_m + z_m - 1) = \ln\left[(z_m - 1)\left(1 + \frac{n_m}{z_m - 1}\right)\right] \approx \ln(z_m - 1) + \frac{n_m}{z_m - 1},$$

it follows that

$$(n_m + z_m - 1)\ln(n_m + z_m - 1) \approx (n_m + z_m - 1)\ln(z_m - 1) + n_m \tag{40}$$

after ignoring the small term proportional to $n_m/(z_m - 1)$.

Substitution of Eq. (40) into principle (37) yields an estimator

$$\sum n_m \ln(Q_m/n_m) + \text{(constraint terms)} = \text{maximum}, \tag{41}$$

the constraint terms all linear in the $\{n_m\}$. We call this a principle of maximum cross entropy.† In the limit of all $Q_m = 1/M$ it becomes simply maximum entropy.

---

† As a matter of nomenclature, Shore (1981) calls it *minimum* cross-entropy.

We can summarize this situation by the statement that the ML object obeys Jaynes's maximum-entropy principle when the "white" object definition (31) of maximum ignorance is used and when the object is of such low intensity that the df sites are mostly unoccupied. The latter situation is obeyed by weak astronomical objects such as planets in the visible and IR regions and the sun in the visible region (see Kikuchi and Soffer, 1977).

The "Burg entropy" is of the form

$$H_1 \equiv + \sum_{m=1}^{N} \ln n_m.$$

A principle H = maximum was first proposed by J. P. Burg (1967) in the estimation of power spectra from autocorrelation data. There is one condition under which estimator (37) goes into the Burg form, described next.

Consider limiting cases (11b) of highly occupied df sites. With the approximation

$$\ln(n_m + z_m - 1) = \ln\left[n_m \left(1 + \frac{z_m + 1}{n_m} \right) \right] \approx \ln n_m + \frac{z_m - 1}{n_m},$$

it follows that

$$(n_m + z_m - 1)\ln(n_m + z_m - 1) \approx (n_m + z_m - 1)\ln n_m + z_m - 1 \quad (42)$$

after ignoring the small terms proportional to $(z_m - 1)/n_m$.

Substitution of (42) into principle (37) yields an estimator

$$\sum (z_m - 1)\ln n_m + \sum n_m \ln Q_m + (\text{constraint terms}) = \text{maximum}, \quad (43)$$

the constraint terms all linear in the $\{n_m\}$. Terms in purely $z_m - 1$ have been dropped because these do not affect the maximization.

In the limit of all $Q_m = 1/M$, because $\sum n_m$ obeys normalization, the net principle is maximization of the first term in Eq. (43). This is a weighted form of Burg's entropy, with weights $z_m - 1$.

We can summarize this section as follows. The ML object obeys maximum entropy of type H (Jaynes) or type H_1 (Burg) when the "white" object definition (31) of maximum ignorance is used and when the df sites are sparsely or highly populated, respectively.

C. Estimators for Empirical Data; Impartial Conviction

In a previous section we found the object class (30) that is implied by observation of an empirical object $\{m_m\}$,

$$p(q_1, \ldots, q_M) = \frac{(L + M - 1)!}{(M - 1)!} \frac{q_1^{m_1} \cdots q_M^{m_M}}{m_1! \ldots m_M!}, \qquad L = \sum m_n. \quad (44)$$

We now want the estimator implied by this form of prior knowledge. This is obtained by substituting Eq. (44) into principle (19). After expanding the logarithm of the product, evaluating the integral by the use of Dirichlet's equality, and using Stirling's approximation (22), we obtain

$$\sum (n_m + z_m - 1) \ln(n_m + z_m - 1)$$
$$+ \sum (n_m + m_m) \ln(n_m + m_m) - \sum n_m \ln n_m$$
$$+ \mu(\sum n_m - N) + \sum \lambda_m (h \sum n_n v_n s_{mn} - i_m) = \text{maximum.} \quad (45a)$$

Quantities constant in the $\{n_m\}$ have been ignored. This is the general estimator for this state of prior knowledge.

We note that the first sum in Eq. (45a) has the form of the negative of an entropy [see Eq. (39)]. This entropy will be called photon-site entropy because it involves the df sites $\{z_m\}$. Likewise the second sum in Eq. (45a) is the negative of "empirical" entropy, in that it involves the empirical data $\{m_m\}$. Hence, overall, estimator (45a) is one of *minimum* photon-site entropy plus minimum empirical entropy plus maximum Jaynes entropy (third sum).

It is interesting that the solution to Eq. (45a) may sometimes be obtained by differentiation and sometimes not. Taking the second derivative of Eq. (45a) shows that the first derivative yields a maximum if and only if

$$n_m/(z_m - 1) < m_m/n_m. \quad (45b)$$

This is an interesting requirement relating the relative sparsity of df sites (left side) to the "strength" of the prior data $\{m_m\}$ (right side). When differentiation does work, it leads to a quadratic equation for the $\{\hat{n}_m\}$,

$$\hat{n}_m^2 + \hat{n}_m(m_m + z_m - 1 - \alpha_m) + m_m(z_m - 1) = 0. \quad (46a)$$

Parameter α_m lumps together all the unknowns μ and $\{\lambda_m\}$, through

$$\alpha_m \equiv \exp(-1 - \mu - hv_m \sum \lambda_n s_{mn}). \quad (46b)$$

Of course, the roots of Eq. (46a) may easily be found. The unknown quantities μ and $\{\lambda_m\}$ may be found by substituting $\{\hat{n}_m\}$ from Eq. (46a) into the imaging equations (18) and normalization equation (3). Because of the nonlinear nature of these equations, a Newton–Raphson (or other iterative) method of solution would be necessary.

Limiting forms of principle (45a) are easily established, as before. If, for example, the exposure time is so short that there is no photon degeneracy, i.e., $z_m = 1$ for all m, it becomes

$$\sum (n_n + m_m) \ln(n_m + m_m) + \text{(constraint terms)} = \text{maximum.} \quad (47a)$$

This is a principle of *minimum* empirical entropy. The solution $\{\hat{n}_m\}$ to Eq. (47a) cannot be obtained by setting its derivative equal to zero. That would instead *minimize* Eq. (47a). In fact, the solution $\{\hat{n}_m\}$ tends to be extreme and bumpy, the opposite tendency of maximizing entropy (39) (as might be expected).

In the limiting case of no prior data, all $m_m = 0$. This describes zero prior knowledge by MacQueen and Marschak's criterion. Then Eq. (45a) becomes a principle of minimum photon-site entropy,

$$\sum (n_m + z_m - 1) \ln(n_m + z_m - 1) + (\text{constraint terms}) = \text{maximum}. \quad (47b)$$

Tendencies of this solution were discussed following Eq. (34). See also Section XI.B.

If condition (11a) of sparsely populated df holds true, as shown in Eq. (40) the first sum in Eq. (45a) becomes absorbed into the linear constraints so that the estimator is

$$\sum (n_m + m_n) \ln(n_m + m_m) - \sum n_m \ln n_m + (\text{constraint terms})$$
$$= \text{maximum}. \quad (48)$$

The solution may again be found by differentiation, because the second derivative of Eq. (48) is always negative, as is easily verified.

If there is high df occupancy (11b), estimator (45a) becomes

$$\sum (n_m + m_m) \ln(n_m + m_m) + \sum (z_m - 1) \ln n_m + (\text{constraint terms})$$
$$= \text{maximum}. \quad (49a)$$

This is a principle of minimum empirical entropy (first sum) plus maximum weighted-Burg entropy (second sum). Once again, differentiation does not always yield the solution. Taking the second derivative of Eq. (49a) shows that the first derivative attains a maximum if and only if

$$n_m/(z_m - 1) < m_m/n_m + 1. \quad (49b)$$

X. HOW TO HANDLE NOISE

The preceding ML estimators have assumed no noise to be present in the image data $\{i_m\}$. However, the incorporation of noise presents little problem. The estimators will simply be revised by having something added to them in all cases. They therefore will keep the forms previously derived. This is shown next.

Let the image data suffer from noise $\{\varepsilon_m\}$. Then the imaging equations are [compare Eq. (18)]

$$i_m = h \sum n_n v_n s_{mn} + \varepsilon_m, \quad (50)$$

where ε_m is the noise.

Now we have two sets of unknowns, object $\{n_m\}$ and noise $\{\varepsilon_m\}$. The proper ML principle to use is therefore that *the $\{n_m\}$ and $\{\varepsilon_m\}$ that occurred are both (jointly) most likely,*

$$\mathcal{P}(n_1,\ldots,n_M,\varepsilon_1,\ldots,\varepsilon_M) \equiv \mathcal{P}(\{n_m\},\{\varepsilon_m\}) = \text{maximum.} \qquad (51)$$

The definition of conditional probability gives

$$\mathcal{P}(\{n_m\},\{\varepsilon_m\}) = P(\{n_m\})P_1(\{\varepsilon_m\}\,|\,\{n_m\}), \qquad (52)$$

where P is the a priori probability of an object, as considered in previous sections, and P_1 is the probability of the noise conditional upon that object (thereby allowing for signal-dependent noise). Now $\mathcal{P}$ is a maximum if and only if $\ln \mathcal{P}$ is. Then Eqs. (51) and (52) yield an ML principle

$$\ln P(\{n_m\}) + \ln P_1(\{\varepsilon_m\}\,|\,\{n_m\}) = \text{maximum} \qquad (53)$$

through choice of $\{n_m\}$ and $\{\varepsilon_m\}$.

Of these two contributors to the maximum, the first is already given by Eq. (17). It takes on the particular forms previously given under different states of object class $p(q_1,\ldots,q_M)$. Hence, this part of the ML principle is already known.

The second contributor to Eq. (53) has to do with the type of noise assumed to be present. To keep things simple, we shall assume it to be additive and independent Gaussian but with generally position-dependent variance $\{e_m^2\}$. This describes, for example, Johnson noise and other noise types that enter into the output of the spectrometer. Thus, we have as the probability P_N of noise

$$P_N(\varepsilon_1,\ldots,\varepsilon_M) = \left[\prod_{m=1}^{M} (\sqrt{2\pi}\,e_m)^{-1} \right] \exp\left(-\sum_{m=1}^{M} \frac{\varepsilon_m^2}{2e_m^2} \right). \qquad (54)$$

The tie-in between this P_N and P_1 is found next.

Because the noise is given to be additive, it is independent of the underlying object $\{n_m\}$:

$$P_1(\{\varepsilon_m\}\,|\,\{n_m\}) = P_N(\{\varepsilon_m\}). \qquad (55)$$

Accordingly, the overall ML principle (53) becomes

$$\ln P(\{n_m\}) - \sum_{m=1}^{M} \frac{\varepsilon_m^2}{2e_m^2} = \text{maximum,} \qquad (56)$$

after using Eq. (54) and ignoring a sum in $\{e_m\}$ that does not affect the solution. As in preceding sections, the image data are added on as Lagrange constraints

$$\sum \lambda_m (i_m - h \sum_n n_n v_n s_{mn} - \varepsilon_m) + \mu(N - \sum n_m) \qquad (57)$$

to the left side of Eq. (56).

Equations (56) and (57) constitute the overall ML principle that we sought. This principle embodies all the physical and prior knowledge that we have about the object and noise. The specific $\ln P(\{n_m\})$ to be used will depend on the particular prior knowledge $p(q_1, \ldots, q_M)$ at hand. Specific $\ln P(\{n_m\})$ were derived in preceding sections. We shall carry through some particular cases next.

XI. TEST CASES

We shall find estimates $\{\hat{n}_m\}$ in numerical cases consisting of experimental data (Section XI.A) and simulated data (Section XI.B). In all cases, high-conviction form (20) will be assumed because the solutions $\{\hat{n}_m\}$ are relatively easy to form and because of the ease with which prior knowledge $\{Q_m\}$ may be varied. In Section XI.A low photon occupancy n_m/z_m will be assumed, whereas in Section XI.B both low and high occupancies will be allowed. Note that low occupancy is consistent with maximum entropy according to Section IX.B, that is, particlelike behavior for the photons, whereas high occupancy is consistent with wavelike behavior for the photons. How, then, does the wave or particle nature of photons affect the estimates?

A. Application to Experimental Data

Under the prior-knowledge conditions mentioned earlier, the principle given by Eqs. (56) and (57) is augmented by Eq. (41) to become

$$\sum_m n_m \ln(Q_m/n_m) - \sum_m \frac{\varepsilon_m^2}{2e_m^2} + \sum_m \lambda_m(i_m - h \sum_n n_n v_n s_{nm} - \varepsilon_m)$$

$$+ \mu(\sum_m n_m - N) = \text{maximum}. \tag{58}$$

By the form of the first sum, the object is being estimated by a principle of maximum cross entropy. By the second sum, the noise is being estimated by a principle of least squares (notice the minus sign). The df have dropped out because of the sparsity constraint (11a). This is convenient, because it avoids the need for their detailed calculation.

The solution $\{\hat{n}_m\}$, $\{\hat{\varepsilon}_m\}$ to Eq. (58) is obtained merely by setting $\partial/\partial n_m$ (all $\{\varepsilon_m\}$ fixed) of Eq. (58) equal to zero and by setting $\partial/\partial \varepsilon_m$ (all n_m fixed) equal to zero. These result, respectively, in solutions

$$\hat{n}_m = Q_m \exp(-1 - \mu - hv_m \sum_n \lambda_n s_{nm}) \tag{59a}$$

and

$$\hat{\varepsilon}_m = -\lambda_m e_m^2. \tag{59b}$$

Note from the form of Eq. (59a) that $\hat{n}_m$ can only be positive, regardless of parameters μ and $\{\lambda_n\}$. Hence, a positivity constraint is a *natural consequence* of the overall approach. For other cases of prior knowledge, such as the case of emission spectra, *boundedness* would result. By comparison, all other restoration methods that enforce positivity or boundedness do it in an ad hoc manner.

It only remains to find the unknown parameters μ and $\{\lambda_n\}$. These are found by substituting the forms (59a) and (59b) for $\{n_m\}$ and $\{\varepsilon_m\}$ into the constraint equations, resulting in conditions

$$i_m = h \sum_n v_n s_{mn} Q_n \exp(-1 - \mu - hv_n \sum_k \lambda_k s_{kn}) - \lambda_m e_m^2,$$

$$N = \sum_n Q_n \exp(-1 - \mu - hv_n \sum_k \lambda_k s_{kn}). \tag{60}$$

This is a system of $M + 1$ equations in the $M + 1$ unknowns μ and $\{\lambda_m\}$. It may be solved by regarding the known left-hand sides as target values in a relaxation program that iteratively homes in on the solution. The Newton–Raphson method works fine for this job. With an initial trial solution of all $\lambda_m = 0$ and $\mu = N$, the final solution is attained in usually about 10 iterations.

To proceed further, we have to make assumptions about the forms for $\{Q_n\}$ and $\{e_m\}$. These describe prior knowledge about the object and the noise. For simplicity, let all $\{Q_m\}$ be equal. This means that we have maximum conviction that the unknown spectrum is a flat one. Such a spectrum may be called "quasi-featureless." This prior knowledge should exert a smoothing influence on the solution. Sure enough, by Eq. (58) the object is now being restored by a principle of maximum entropy H, known to foster smoothness in its outputs (Frieden, 1972).

Let us assume the noise to be strict-sense stationary of order 1. This means that all the $\{e_m\}$ values are equal. Such a situation is very common.

Under these prior conditions, the restoring algorithm given by Eqs. (59) and (60) becomes very close to one previously used by Frieden (1972). That one was also a maximum-entropy restoring algorithm in the presence of additive noise. We present some results of the latter method applied to spectral data.

Using spectral image data provided by Jansson et al. (1970), we restored a portion of the Q branch of the v_3 band of CH_4. Results are shown in Fig. 4. Note the smoothness of the output, no doubt enforced by the prior conviction of a flat object. But, despite this smoothness, the right-hand line has been split into two components. These same data were also restored by Jansson, using his bound-constrained algorithm (see Chapters 4, 6, and 7

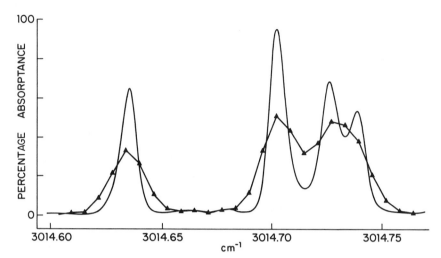

Fig. 4 Restoration by maximum entropy (solid curve) of a portion of the Q branch of the v_3 band of the infrared spectrum of CH_4 near 3014.7 cm^{-1}. Triangles indicate the experimental inputs $\{i_m\}$ used in the restoring scheme, some of which lie outside the plotted region and are not shown.

in this text). The restoration by Jansson is very similar to ours, with relative differences in the principal peaks amounting to a few percent.

It is important to realize, however, that this similarity is a coincidence of the case. Jansson's boundedness constraint *always* tends toward smoothness near the bounds, regardless of a priori knowledge of object class, whereas our ML estimator [Eqs. (17) and (19)] can be constrained toward bumpy solutions, the bumps consistent with the prior knowledge at hand. In the high-conviction case that we are treating here, the bumps would be manifest in the inputs $\{Q_m\}$. We next study the effect of varying the $\{Q_m\}$ in some computer simulations.

B. Effect of Choice of Prior Object $\{Q_m\}$ and of z

How do the form $\{Q_m\}$ of a prior object and the number z of degrees of freedom affect the estimate $\{\hat{n}_m\}$? This was the problem (37) augmented by data constraints (50). Suppose in addition that the image data $\{i_m\}$ suffer from Gaussian additive noise, so that Eq. (54) is true. Then the total estimator given by Eqs. (56) and (57) becomes

$$- \sum n_m \ln(n_m/Q_m) + \sum (n_m + z - 1) \ln(n_m + z - 1) - \frac{1}{2e^2} \sum \varepsilon_m^2$$

$$+ \mu(\sum n_m - N) + \sum \lambda_m(\sum n_n s_{nm} + \varepsilon_m - i_m) = \text{maximum.} \quad (61)$$

(For simplicity we have taken all $hv_m = 1$, $z_m = z$, and $e_m = e$.) This has a solution

$$\hat{n}_m = (z - 1)Q_m / [\exp(-\mu - \sum \lambda_n s_{nm}) - Q_m],\tag{62}$$

$$\hat{\varepsilon}_m = \lambda_m e^2\tag{63}$$

after setting the derivatives $\partial/\partial n_m$ and $\partial/\partial \varepsilon_m$ of Eq. (61) equal to zero. The unknown parameters μ and $\{\lambda_m\}$ number $M + 1$ and must satisfy the $M + 1$ constraint equations

$$\sum \hat{n}_n s_{nm} + \hat{\varepsilon}_m = i_m, \qquad m = 1, \ldots, M,\tag{64}$$

and (3). Solutions to these equations were obtained using the Newton–Raphson method.

The algorithm was applied to computer-simulated image data, represented by the dashed curve in Fig. 5. The corresponding object is represented by the shaded area. The latter was convolved with a sinc-squared kernel so as to form a diffraction-blurred signal image. Diffraction is particularly difficult to overcome in a restoration, because all spatial frequencies beyond that imposed by the finite lens pupil are *missing* from the signal image. Recovery of these frequencies is called superresolution. Until the pioneer work of Biraud (1969), Jansson (1968), Jansson *et al.* (1970) practical superresolution was thought to be impossible.

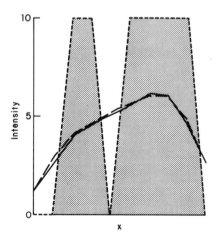

Fig. 5   Object (shaded) used in computer simulations. Its diffraction image is the solid curve. The data image (dashed curve) is the diffraction image plus 4% amplitude random noise. All plotted points are spaced by one-half the Nyquist interval. Hence to resolve the central dip in the object would require superresolution.

The image data were formed by adding 4% (of peak image value) Gaussian noise to the signal image. Sampling was at one-half the Nyquist interval so as to permit superresolution of the central dip in the object.

The outputs (62) are shown in Fig. 6(a–d), representing four assumed forms for $\{Q_m\}$, each form accompanied by either a low $z\,(z = 2)$ or a high $z\,(z = 100)$. Intermediate z values did not cause significantly different results.

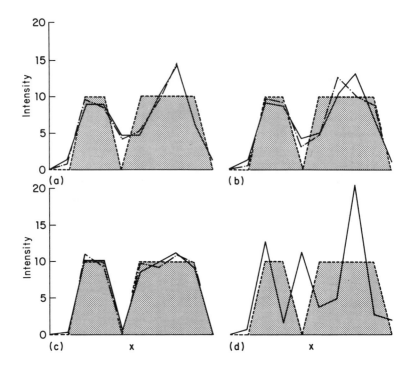

Fig. 6 Maximum-likelihood restorations (62) (solid and dot-dashed curves) of the data image in Fig. 5. Each solid curve is for $z = 100$, each dot-dashed curve for $z = 2$, indicating particlelike and highly Bose photons, respectively. The root-mean-square error RMSE of each restoration from the true object (shaded) is as follows. Each of (a)–(d) results from a different assumed form for biases $\{Q_m\}$, as indicated.

		RMSE	
Part	$\{Q_m\}$	$z = 100$	$z = 2$
(a)	constant	2.9	2.8
(b)	image	2.6	2.2
(c)	true object	0.9	0.6
(d)	true object, displaced	6.5	—

In Fig. 6(a), $\{Q_m\}$ was made constant, corresponding to prior knowledge of a flat spectrum. This is a rather conservative guess at the object.† Both restorations exhibit resolution of the central object dip, indicating super-resolution.

The restoration for low z suffers a bit more overshoot at the gradient edges. This is consistent with a "photon-bunching" phenomenon that occurs as the solution to estimator (34). Notice that our estimator (61) includes Eq. (34). If Eq. (34) per se is solved, the answer is that all N photons jam into the resolution cell x_m that has the largest degeneracy z_m over $m = 1, \ldots , M$. The other cells are empty, and the estimated spectrum is a spike. The presence of image data, and of noise, will compromise this extreme behavior. However, there is still a tendency to bunch photons in some cells and deplete them from others, as seen in Fig. 6(a–c).

By contrast, the solution for high z ($z = 100$) corresponds to maximum entropy [see the derivation of Eq. (41)], and hence is smoother in the figure.

In summary, then, ML estimates for particlelike photons are smoother than for wavelike photons. This perhaps goes against intuition, but it was true in other test cases as well [Fig. 6(b,c)].

With nothing else known about object class, one tactic is to make $\{Q_m\}$ proportional to the image data, on the basis that the object ought to look something like the image. Certainly the image should provide more "ground truth" about the object than did the complete grayness used previously. This was tried in Fig. 6(b). Indeed, the central dip is resolved better than in Fig. 6(a), and the plateau tops are better approximated as well, indicating that the tactic has some merit. The root-mean-square errors from the true object are lower than in Fig. 6(a) as well.

The *best* $\{Q_m\}$ to use ought to be the true object, of course, and we tried this out in Fig. 6(c). As expected, results are much better than in Fig. 6(a) or 6(b).

Finally, in Fig. 6(d) we asked what the effect would be of inputting for $\{Q_m\}$ the true object but displaced to the left by one point. This would test the sensitivity of the tactic to registration error. Results show that the output is indeed sensitive to such error. If a high-gradient $\{Q_m\}$ is used, it had better be in the correct registration. (The restoration for $z = 2$ is not shown because the algorithm would not converge under these contradictory conditions.)

The benchmark comparison with these results would be the ordinary least-squares restoration of the given image. This was carried through. However, the results cannot be plotted because, owing to the noise level in the data, the output was highly oscillatory, varying in size by $\pm 10^6$, and showing no resemblance to the true object.

---

† But only as far as resolution goes. By contrast, we showed in Section VIII.C that the "flat spectrum" hypothesis is actually a strong assumption about *shape*.

XII. CONCLUSION

We have shown that it is possible to combine the physics of the "taking" conditions with prior knowledge of "object class" to come up with a maximum-likelihood estimate of a spectrum. The resulting estimation principle depends strongly on what these quantities are in a given problem. No one algorithm can apply to all problems and still be called maximum likelihood. We have summarized some of the more important cases in Table I. It is up to the user, ultimately, to define his or her particular case of prior knowledge. The more descriptive of the unknown it is, the better will be the resulting ML estimate.

TABLE I

Estimators for Different Prior-Knowledge Scenarios

Prior knowledge	Maximum-likelihood estimator (equation)
High conviction of object $\{Q_m\}$, general occupancy ratio n_m/z_m	minimum photon-site entropy plus maximum cross entropy [(37)]
High conviction of object $\{Q_m\}$, low occupancy ratio n_m/z_m	maximum cross entropy [(41)]
High conviction of white object $Q_m = 1/M$ for all m, low occupancy ratio n_m/z_m	maximum Jaynes entropy [(39)]
High conviction of white object $Q_m = 1/M$ for all m, high occupancy ratio n_m/z_m	maximum weighted Burg entropy [(43)]
Maximum ignorance (MacQueen–Marschak), general occupancy ratio n_m/z_m	minimum photon-site entropy [(47b)]
Impartial conviction, empirical data, general occupancy ratio n_m/z_m	minimum photon-site entropy plus minimum empirical entropy plus maximum Jaynes entropy [(45a)]
Impartial conviction, empirical data, low occupancy ratio n_m/z_m	minimum empirical entropy plus maximum Jaynes entropy [(48)]
Impartial conviction, empirical data, high occupancy ratio n_m/z_m	minimum empirical entropy plus maximum weighted Burg entropy [(49a)]

REFERENCES

Biraud, Y. (1969). *J. Astron. Astrophys.* **1**, 124–127.
Born, M., and Wolf, E. (1959). "Principles of Optics." Macmillan, New York.
Burg, J. P. (1967). "Maximum entropy spectral analysis." Paper presented at the 37th Annual Society of Exploration Geophysicists Meeting, Oklahoma City.
Frieden, B. R. (1972). *J. Opt. Soc. Am.* **62**, 511–517.

Frieden, B. R. (1983). "Probability, Statistical Optics and Data Testing." Springer-Verlag, New York.

Gradshteyn, I. S., and Ryzhik, I. M. (1965). "Tables of Integrals, Series and Products." Academic Press, New York.

Hershel, R. S. (1971). Ph.D. Dissertation, University of Arizona, Tucson.

Hildebrand, F. B. (1956). "Introduction to Numerical Analysis." McGraw-Hill, New York.

Jansson, P. A. (1968). Ph.D. Dissertation, Florida State University, Tallahassee.

Jansson, P. A., Hunt, R. H., and Plyler, E. K. (1970). *J. Opt. Soc. Am.* **60,** 596–599.

Jaynes, E. T. (1968). (*IEEE Trans. Syst. Sci. Cybern.* **SSC-4,** 227–241.

Kikuchi, R., and Soffer, B. H. (1977). *J. Opt. Soc. Am.* **67,** 1656–1665.

MacQueen, J., and Marschak, J. (1975). *Proc. Natl. Acad. USA* **72,** 3819–3824.

Shore, J. E. (1981). *IEEE Trans. Acoust. Speech, Sig. Process.* **ASSP-29,** 230–237.

Wernecke, S. J., and D'Addario, L. R. (1977). *IEEE Trans. Comput.* **C-26,** 351–364

CHAPTER **9**

Fourier Spectrum Continuation

Samuel J. Howard

Physics Department, The Florida State University
Tallahassee, Florida

I.	Introduction	262
II.	Advantages of Fourier Analysis	264
III.	Restoration and Enhancement of Experimental Data	264
	A. Introductory Remarks	264
	B. Deconvolution	264
	C. Apodization	266
	D. Resolution and the Uncertainty Principle	267
	E. Continuation of the Fourier Spectrum	268
IV.	Discrete Fourier Transform and Discrete Function Continuation	271
	A. Discrete Fourier Transform	271
	B. Extrapolation of Discrete Fourier Spectra	277
V.	Constraint of Finite Extent	278
	A. Theory	278
	B. Application	280
VI.	Concluding Remarks	285
	References	286

LIST OF SYMBOLS

a_n	coefficients of cosine terms of Fourier series
A_n	coefficients of cosine terms of discrete Fourier series
b	number of unique coefficients in complex Fourier series expression for $u(k)$
b_n	coefficients of sine terms of Fourier series
B_n	coefficients of sine terms of discrete Fourier series
c	number of unique coefficients in complex Fourier series expression for $v(k)$
$C_n, F(n)$	discrete Fourier spectral components as given by discrete Fourier transform, which are the coefficients of a complex Fourier series
$\mathcal{F}$	forward Fourier transform operator
$\mathcal{F}^{-1}$	inverse Fourier transform operator
$f(k)$	general function of integer variable k
$f(x)$	general function
$F(w)$	Fourier transform of $f(x)$
$i(x), i$	"image" data recorded after degradation by convolution and noise

DECONVOLUTION:
WITH APPLICATIONS IN SPECTROSCOPY

$I(w)$, $O(w)$, $\tau(w)$, $N(w)$	Fourier transforms of $i(x)$, $o(x)$, $s(x)$, $n(x)$
j	imaginary number $\sqrt{-1}$
k	integer variable for spatial function, the index of sampled values
n	integer Fourier spectral variable
$n(x)$, n	noise, which together with the convolution of the object with the impulse response makes up the image
N	number of sample points, which is the same as the number of complex coefficients
$\hat{o}$	inverse-filtered estimate of object
$o(x)$, o	"object" or function sought by deconvolution
$\text{rect}(x)$	rectangular function with height and width of unity
$s(x)$, s	impulse response function
$u(k)$	discrete Fourier series composed of only low-frequency and dc terms
$U(n)$	discrete Fourier transform of $u(k)$
$v(k)$	discrete Fourier series composed of high-frequency terms such that when added to $u(k)$ a series of a complete Fourier spectral band is formed
$V(n)$	discrete Fourier transform of $v(k)$
w	Fourier spectral variable
x	generalized independent variable, in this work the wave number for spectroscopy data, but for other research areas time, mass number, wavelength, angle, etc.
x'	variable of integration
Δw	standard deviation of Fourier transform
Δx	standard deviation of spatial function
Ω	highest frequency present in band-limited data
Ω_p	cutoff frequency defining Frieden's optimum processing bandwidth
$\text{III}(x)$	Dirac "comb" $\sum_{n=-\infty}^{\infty} \delta(x - n)$

I. INTRODUCTION

Experimental data are not always displayed in a form that reveals the maximum available information. A modified form may make more of the inherent information manifest and accessible. A different form may also provide easier comparison or more convenient mathematical manipulation and a simpler theoretical treatment. For these reasons, further processing of experimental data may be desired.

Other than the operations that emphasize certain aspects of the data, there are destructive influences that reduce our knowledge of the features of the data under investigation. These influences introduce deviations from the true values of the quantity being measured called experimental error, or simply error. Many of these degrading operations are selective, in that their effects are more damaging to certain features of the data. For example, the errors that occur in most experimental data are most damaging to the fine detail. This selective effect is most clearly displayed under the Fourier transform, where it is seen that the lower frequencies of the Fourier spectrum of the data are little affected but that the higher frequencies are nearly all in error. Also, these errors render ineffective many possible mathematical operations or transformations by introducing much larger errors into the output.

The deviations due to some of these destructive influences are reversible. These are usually described as systematic errors. Many of the degradation processes that affect images and most recorded data are classified as systematic errors. For many of these cases the error may be expressed as a function known as the impulse response function. Much mathematical theory has been devoted to its description and correction of the degradation due to its influence. This has been discussed in some detail by Jansson in Chapter 1 of this volume. In that correction of this type of error usually involves increasing the higher frequencies of the Fourier spectrum relative to the lower frequencies, this operation (deconvolution) may also be classified as an example of "form alteration."

The remaining errors in the data are usually described as random, their properties ultimately attributable to the nature of our physical world. Random errors do not lend themselves easily to quantitative correction. However, certain aspects of random error exhibit a consistency of behavior in repeated trials under the same experimental conditions, which allows more probable values of the data elements to be obtained by "averaging" processes. The behavior of random phenomena is common to all experimental data and has given rise to the well-known branch of mathematical analysis known as statistics. Statistical quantities, unfortunately, cannot be assigned definite values. They can only be discussed in terms of probabilities. Because (random) uncertainties exist in all experimentally measured quantities, a restoration with all the possible constraints applied cannot yield an exact solution. The best that may be obtained in practice is the solution that is "most probable." Actually, whether an error is classified as systematic or random depends on the extent of our knowledge of the data and the influences on them. All unaccounted errors are generally classified as part of the "random" component. Further knowledge determines many errors to be systematic that were previously classified as random.

More generally, any additional knowledge of the data and influences on them counters the effects of the destructive influences on the data, both random and systematic, to increase our knowledge of the important parameters in the data. Prior knowledge of the physical limits to the values that the data may assume, for example, may be employed as constraints to recover additional information not given in the original data.

II. ADVANTAGES OF FOURIER ANALYSIS

One advantage of Fourier analysis is that it provides a way of exhibiting the data in another form, so that different aspects of the data become evident. An important example of this concerns the interferogram taken with a Michelson interferometer, for which only limited interpretation is allowed. One cannot locate or measure the electromagnetic spectral components until the Fourier transform is computed to separate the various components into distinct "lines." It is also noted that many of the processes that degrade experimental data are manifested largely in the higher-frequency portion of the Fourier spectrum. That much of the error may be localized in a narrow band of high frequencies, along with the fact that this band may be adequately represented by a small number of discrete Fourier spectral components, allow considerable reduction of the computational burden of restoration problems. Here we are making the not unreasonable assumption for most experimental data that the informational content of the very highest frequencies (as recorded) is nominal. An additional advantage of Fourier analysis is that many of the physical processes that affect optical images and many other types of experimental data have a simple mathematical form under Fourier transformation. Thus they lend themselves to convenient mathematical manipulation. This has given rise to a large body of mathematical theory relating to restoration and other enhancement of degraded data.

III. RESTORATION AND ENHANCEMENT OF EXPERIMENTAL DATA

A. Introductory Remarks

Given degraded experimental data, most researchers desire a restoration of the original appearance of the quantities being measured. Many, however, are satisfied with the results of almost any ad hoc method that improves the resolution. Another important consideration for many researchers is the unambiguous display of the quantities of interest in the data. Each of these items will be addressed in turn.

B. Deconvolution

For many data of interest, the degrading effects can be expressed as a convolution of the original appearance of the data with the appropriate impulse response function plus additive noise,

$$i(x) = \int_{-\infty}^{\infty} o(x')s(x - x')\,dx' + n(x), \qquad (1)$$

where o is the original appearance of the data, s the functional form of the impulse response function, n the noise term, and i the final appearance of the data recorded. This model for the degradation has been thoroughly discussed earlier by Jansson, and only a brief summary of the important results will be given here. There is an advantage to expressing Eq. (1) in terms of the Fourier transform because many of the restoration operations discussed in this chapter are most efficiently performed in the Fourier domain. This yields

$$I(w) = O(w)\tau(w) + N(w), \qquad (2)$$

where capital letters generally denote the transformed quantities. One exception is the transform of the impulse response, which is denoted by τ. The convolution becomes a simple multiplication in the Fourier domain. This suggests the equally simple inverse operation for recovering the original appearance of the data o by dividing this equation through by τ and solving for O:

$$O(w) = I(w)/\tau(w) - N(w)/\tau(w). \qquad (3)$$

In principle, inverse transformation would produce the restoration. The noise term in this equation, however, is predominantly random and increases with frequency for the deconvolution of nearly all experimental data. Random error is almost impossible to treat quantitatively and is the primary obstacle to an accurate restoration. The determination of optimum solutions would require the inclusion of statistical criteria (see Chapter 8). Nevertheless, for many data of interest, simply dividing the transform of the data by the transform of the impulse response function, $I(w)/\tau(w)$ and inverse transforming (often called inverse filtering), yields a good approximation to o when the noise level is low. For higher noise levels, the increasing noise term rapidly obscures all information in the deconvolved results, making truncating or filtering of the resulting spectrum necessary to achieve meaningful results.

There are a number of methods, nearly all iterative, that treat the inverse problem of recovery and successfully (some quite imaginatively) deal with the noise problem. The straightforward inverse-filtered estimate was adhered to in this research because of the possibility of saving computational time in the overall restoration. In practice, only discrete data are taken.

The discrete Fourier transform (DFT) of the data is evaluated to take advantage of the considerable speed and accuracy of the fast-Fourier-transform algorithm as calculated by modern digital computers. For most data (and most restorations) encountered in practice, it is found that only a relatively small number of Fourier spectral components are required to represent the data adequately, the higher frequencies being mostly noise. Because of the small number of Fourier spectral components involved, as well as the fast speed of the FFT for determining the DFT, inverse filtering is a computationally economical restoration precedure. However, the increasing noise prevents the most desirable restoration when the noise level is appreciable. In the restorations treated here, the most noise-free low-frequency components of the inverse-filtered spectrum will be used as the initial stage in the restoration because of the considerable saving of computer time. The Fourier spectrum will then be continued from the truncated inverse-filtered estimate to yield the complete spectrum. Very fast algorithms have been developed for recovering a band of frequencies beyond the given low-frequency band (Howard, 1981a,b) to yield an overall restoration procedure that is very economical computationally.

Function continuation procedures are applied to many other problems besides inverse-filtered Fourier spectral continuation and will be discussed in a separate section.

C. Apodization

To display unambiguously the quantities of interest in the data implies that some operation, or series of operations, must be performed to alter the form of the data. In most spectroscopy and chromatography data, the quantities of interest are discrete and are usually displayed as sharply peaked functions or "lines." However, all recording instruments broaden and alter the functional form to a certain extent, so that the data recorded for a monochromatic source would not be an infinitely sharp line, or Dirac δ function. The lines recorded would be broadened and generally have artifacts around them (such as sidelobes, or "ringing"). It is important to display each discrete component in the data unambiguously so that important parameters (such as wave number, intensity, or chemical compound) may be determined from them. For sharply peaked functions, this would generally involve improving the resolution and removing the artifacts, so that overlapping with adjacent peaked functions would be minimized.

It is found that multiplication of the Fourier transform of the data by a carefully chosen window function is very effective in removing the artifacts around peaked functions. This process is called apodization. Apodization with the triangular window function is often applied to Fourier transform spectroscopy interferograms to remove the ringing around the infrared spectral lines obtained from the transformation. However, other, more appropriate window functions for these data are currently enjoying more widespread use (Bell, 1972). Apodization is often done at the expense of resolution because the peaked functions are usually considerably broadened in the process. However, apodization usually produces a less ambiguous result on the whole, for artifacts such as ringing (characteristic of convolution with the sinc function) extend a considerable distance from each peak. This results in overlapping with adjacent peaks. However, we shall find later that both removal of artifacts *and* resolution improvement are possible within the context of the Fourier spectrum continuation techniques.

A general discussion of resolution is provided in the following section.

D. Resolution and the Uncertainty Principle

Deconvolution, the inverse operation of recovering the original function o from the convolution model as given in Eq. (1), employs procedures that almost always result in an increase in resolution of the various components of interest in the data. However, there are many broadening and degrading effects that cannot be explicitly expressed as a convolution integral. To consider resolution improvement alone, it is instructive to consider other viewpoints. The *uncertainty principle* of Fourier analysis provides an interesting perspective on this question.

Letting Δx denote the standard deviation of the spatial function and Δw the standard deviation of the spectral function, we must necessarily have

$$\Delta x \, \Delta w \geq 1/4\pi. \tag{4}$$

A proof of this relation may be found in Bracewell (1978). Note that the spectral variable used in this and the next chapter is the same as that defined in Eqs. (7) and (8). Now consider a spatial distribution $f(x)$ and its Fourier spectrum $F(w)$ that come close to satisfying the equality in Eq. (4). We may take Δx and Δw as measures of the width, and hence the resolution, of the respective functions. To see how this relates to more realistic data, such as infrared spectral lines, consider shifting the peak function $f(x)$ by various amounts and then superimposing all these shifted functions. This will give a reasonable approximation to a set of infrared lines. To discuss quantitatively what is occurring in the frequency domain, note that the Fourier spectrum of each shifted function by the shift theorem is given simply by the spectrum of the unshifted function multiplied by a constant phase factor. The superimposed spectrum would then be

$$[1 + \exp(j2\pi wx_1) + \exp(j2\pi wx_2) + \cdots]F(w),$$

where $x_1, x_2, \ldots$ are the amounts that the spatial functions are shifted. As Δx becomes smaller, each peak becomes narrower, and merged peaks become resolved. The Δw of each superimposed peak would necessarily have to increase all by the same amount to satisfy the uncertainty principle. With the aid of the foregoing relation, it is not too difficult to show that the superimposed Fourier spectrum is generally broadened also.

These arguments may be generalized to show that any overall increase in resolution is accompanied generally by a broadening of the Fourier spectrum. This seems to be borne out in practice because almost any operation that alters the magnitude of the high frequencies relative to the low frequencies affects the resolution. Broadening the Fourier spectrum or boosting the higher frequencies relative to the lower frequencies generally improves resolution and increases the detail, and narrowing the Fourier spectrum or reducing the highest frequencies almost always degrades resolution and causes smoothing. This is apparently the reason for the success of such ad hoc methods as zeroing the dc component and lowest frequencies of the Fourier spectrum, using high-pass filters, and the unconstrained Van Cittert method for the improvement of resolution. However, it needs to be pointed out that any boosting of the higher frequencies relative to the lower frequencies in an effort to improve resolution must be done in a meaningful way and not with random magnitudes and phases such as one would obtain by the addition of random noise. Actually, multiplication by the simple window functions discussed earlier is one example of a meaningful way. These window functions affect the magnitudes only, which are altered in some smooth way. This seems to have the effect of producing smooth results without violent oscillations, and without the introduction of extraneous detail (which random phase changes would introduce).

Incidentally, the uncertainty principle associated with the name of Heisenberg, well known in quantum mechanics, follows from the expression given here when de Broglie's relationship connecting the momentum of a particle with its wavelength is included.

E. Continuation of the Fourier Spectrum

This subject has been discussed by Jansson in Chapter 4, so the present discussion will be kept brief.

Because no unique extension of the Fourier spectrum exists in the absence of additional information, the impetus of the research described in this chapter has been to discover as much prior information as possible about the data and the influences on them, and to find ways to formulate this information as "constraints" to produce increasingly more probable values for the

Fourier spectrum. Ways of successfully applying six constraints to experimental data for spectral restoration have been developed. All of these constraints cannot generally be applied to any given set of data. The constraints must be appropriate to the data under consideration.

The first constraint developed by the author was based on prior knowledge of the correctly restored function. The first step in the restoration was the calculation of the inverse-filtered estimate with only the most noise-free low-frequency spectral components retained. Prior knowledge of the correctly restored function was used to construct an artificial function that, it was hoped, would closely resemble the correctly restored function. Specifically, it was hoped that this function's spectrum beyond the truncation frequency Ω_p would resemble the original spectrum of the undistorted function, so that it might be substituted into the spectrum of the inverse-filtered estimate from Ω_p on to yield a complete set of spectral components that were reasonably correct. This procedure was successful in significantly improving gas chromatographic data (Howard, 1978; Howard and Rayborn, 1980). The artificial function was created by measuring the heights, widths, and locations of the inverse-filtered peaks (which represented the various chemical components of the sample) and constructing Gaussian functions with these parameters. Superimposing these Gaussian functions, which of course had no sidelobes or other artifacts, yielded the artificial function, which at least crudely resembled the correctly restored function for most cases. The disadvantage of this procedure is that it is very difficult to create an artificial function for the general case. If the peaks in the data are not fairly well separated, it is very difficult to obtain the required parameters. This method thus has limited application.

The second constraint restricted the Fourier spectrum. This was an ad hoc filter that was applied to the entire inverse-filtered spectrum to bring the magnitudes of the high-frequency values of the spectrum (which were mostly noise) to values much closer to the correctly restored ones. This procedure resulted in observable improvement over the inverse-filtered estimate for infrared lines obtained from grating spectroscopy (Howard, 1982).

Many of the most effective constraints set well-defined limits to the data function (or its spectrum) beyond which the correct function is not allowed to go. An important example of this type of constraint is "nonnegativity," whereby the correctly restored function is not allowed to extend below the zero baseline and thereby take on nonphysical negative values. This is an appropriate constraint for spectroscopy and optical images. A further example of the constraints of fixed limits is that of an "upper bound" to the values of the restoration. Another important constraint of this type is that of "finite extent," for which no deviations from zero are allowed for the spatial function over those intervals on the spatial axis that lie outside the known extent of the original object. This constraint would be appropriate for spectroscopy in which the electromagnetic radiation in the spectral bands outside the band of interest is filtered out.

It is found that recovering only a band of spectral components beyond the truncation frequency of the inverse-filtered spectrum, along with the use of discrete components (as given by the DFT), considerably reduces the computational burden. Incidentally, recovering only a finite band of frequencies implies the additional constraint of restricting the highest frequencies to zero. This is not a serious restriction, because nearly all of the detail of interest resides in the relatively narrow band restored. It is usually very difficult to restore the highest-frequency spectral components accurately. Actually, recovery of only a narrow band of spectral components is often necessary for stability in the equations resulting from the procedures that apply many of these constraints. This is especially true for the constraint of finite extent. Also, recovering only a narrow band always results in a smoother restored (spatial) function.

Very efficient ways of recovering only a band of discrete spectral components with the application of constraints of well-defined limits were developed in this research. These constraints are generally applied in the spatial domain and begin with the function (in the spatial domain) formed from only the low-frequency band of Fourier spectral components. This function is held fixed, and the function formed from a band of high-frequency spectral components immediately adjacent to the low-frequency band is added to this (fixed spatial) function and allowed to vary until the total function is out of the nonallowed region, or as far out of it as the noise and other error will allow it to go. Within this very loose general framework, a variety of approaches may be developed. Mathematically, this scheme may be formulated as a minimization procedure in which the expression to be minimized is given as some function of the deviations into the nonallowed region. Such functions could be the familiar sum of squares, the sum of absolute values, or other expressions. (For the constraint of finite extent, one may wish to minimize the maximum deviation of the sum of the two functions corresponding to the two spectral bands.) Various techniques, both numerical and analytical, may then be applied to find the continued spectrum that minimizes one of these "measures" of the error. A very direct numerical technique may be used to minimize the largest deviation of the error. In this research all further calculations were done with the sum-of-squares expression because many of these calculations could be done analytically *once and for all.*

The explicit sum-of-squares expression for each constraint of well-defined limits will be more fully discussed in succeeding sections.

IV. DISCRETE FOURIER TRANSFORM
AND DISCRETE FUNCTION CONTINUATION

A. Discrete Fourier Transform

As with the continuous Fourier transform, we could treat the equations of the discrete Fourier transform (DFT) completely independently, derive all the required theorems for them, and work entirely within this "closed system." However, because the data from which the discrete samples are taken are usually continuous, some discussion of sampling error is warranted. Further, the DFT is inherently periodic, and the limitations and possible error associated with a periodic function should be discussed.

The DFT is defined as

$$C_n \equiv \frac{1}{N} \sum_{k=0}^{N-1} f(k) \exp\left(-j\frac{2\pi}{N}nk\right) = F(n). \tag{5}$$

Its companion, the inverse DFT, is defined as

$$f(k) \equiv \sum_{n=0}^{N-1} C_n \exp\left(j\frac{2\pi}{N}nk\right). \tag{6}$$

In these equations, N is the number of sample points, and n and k are integers.

To illustrate more appropriately the relationship between the (continuous) Fourier transform and the DFT, the alternative form given in Chapter 1 will be employed. Accordingly, we define this transform as

$$\mathcal{F}\{f(x)\} \equiv \int_{-\infty}^{\infty} f(x) \exp(-j2\pi xw)\, dx = F(w). \tag{7}$$

Then to recover the original function $f(x)$ we must define the inverse transform as

$$\mathcal{F}^{-1}\{F(w)\} \equiv \int_{-\infty}^{\infty} F(w) \exp(j2\pi xw)\, dw. \tag{8}$$

The forward and inverse Fourier transforms are denoted by $\mathcal{F}$ and $\mathcal{F}^{-1}$, respectively. In the foregoing equations, we identify w with ω used in Chapter 1 via the relation $w = \omega/2\pi$.

The Fourier series, which has a discrete spectrum but periodic spatial function, is actually a special case of the Fourier transform. (Note that an equally spaced discrete spectrum necessarily implies a periodic function having a finite period given by the wavelength of the lowest frequency.) See Bracewell (1978) to see how the explicit form of the Fourier series may be obtained from the Fourier transform. Taking discrete, equally spaced samples of the periodic function (or samples over one period because the series repeats indefinitely), we arrive at the final form of the DFT (see, e.g., Howard, 1978). We may even skip the intermediate step of obtaining the Fourier series and directly consider discrete values of *both* the function and its spectrum as special cases of the Fourier transform. With the DFT, both domains must obviously be periodic as well as discrete. See also Fig. 3 of Chapter 1 to see how the DFT may be acquired from the Fourier transform by sampling and/or periodicity.

Addressing first the limitations of a periodic representation, such as with the DFT or Fourier series, we see that it is evident that these forms are adequate only to represent either periodic functions or data over a finite interval. Because data can be taken only over a finite interval, this is not in itself a serious drawback. However, under convolution, because the function represented over the interval repeats indefinitely, serious overlapping with the adjacent periods could occur. This is generally true for deconvolution also, because it is simply convolution with the inverse filter $\mathscr{F}^{-1}\{1/\tau(w)\}$. If the data go to zero at the end points, one way of minimizing this type of error is simply to pad more zeros beyond one or both end points to minimize overlapping. Making the separation across the end points between the respective functions equal to the effective width of the impulse response function is usually sufficient for most practical purposes. See Stockham (1966) for further discussion of endpoint extension of the data in cyclic convolution.

Another problem with a discrete periodic representation is that it is very difficult to represent functions that are not equal at their end points. An abrupt discontinuity across the end points of a periodic representation causes considerable amplification of the high frequencies, which, in general, are very difficult to work with. Also, unless the sampling interval is taken very fine, this function cannot be adequately represented by the DFT. However, this problem can be dealt with by extending each end point in a smooth curve that joins smoothly with the next period so as to yield an overall periodic function that is reasonably smooth. This is a valid operation because it is only necessary that the function be adequately represented over the interval of interest. The advantage of a smooth curve is that the high frequencies are now much smaller, and the overall numerical problem much more tractable.

With convolution and deconvolution, one must be careful to avoid end-point error with this type of function. Convolution with the function beyond the end point of the data will extend inside the interval containing the data about half the length of the impulse response function, so the error will extend about half the length of the impulse response function also (assuming the impulse response function is approximately symmetrical). To minimize this error, the function extending beyond the end points should approximate the true function if anything is known about it. However, simply extending a very smooth curve across the end points is usually sufficient for most practical purposes. A good discussion of the end-point error involved in the deconvolution of molecular scattering data is provided by Sheen and Skofronick (1974). Incidentally, a discrete function for extending the data in a very satisfactory manner is provided in a versatile spline-fitting computer program developed by De Boor (1978). It minimizes the curvature by minimizing the second derivative of the discrete data function. Weights can be assigned to particular points, such as the end points, to assure small deviations there, and small weights can be assigned to the points beyond the end points so as to have sufficient flexibility to allow the discrete function there to form a very smooth curve. This program is also useful for smoothing and base-line fitting. This technique for end-point extension that we have just discussed provides an alternative to data windowing.

Next, we shall discuss sampling error. Fortunately, Fourier analysis is one of the few disciplines that provide a quantitative treatment of this. As discussed earlier, sharp corners, discontinuities, and fine detail require the highest frequencies to represent them adequately. Smooth data may usually be represented adequately by a lower-frequency band. On sampling with a coarse interval, it is evident that all fine detail is missed. This error enters into the Fourier expression of the data as an impersonation, or "aliasing," of the lower frequencies by the higher frequencies; that is, for every low frequency in the Fourier expression, there is a set of high-frequency components that exhibit exactly the same behavior under the Fourier transform. From the sampled data alone it would be impossible to determine how much of the lower frequency is present or how much of any of the other frequencies in the higher-frequency set is present. All one would know is their sum. However, if all of the higher frequencies were absent, that is, if the data were "band limited," then a complete recovery of the continuous function would be allowed by the discrete samples.

One procedure for recovering the continuous (band-limited) function exactly is provided by the *Whittaker–Shannon sampling theorem*, which is expressed by the equation

$$f(x) = \sum_{n=-\infty}^{\infty} f(n/2\Omega)\,\mathrm{sinc}(2\Omega x - n). \tag{9}$$

This formula tells us that when the data are band limited with $w = \Omega$ the highest frequency present, the spacing of the sampling interval may be as large as $\Delta x = 1/2\Omega$ and yet allow complete recovery of the original continuous function. A proof of the sampling theorem may be found in Hamming (1962) and Papoulis (1962). An interesting discussion of sampling and aliasing error in terms of the comb (III) function is provided by Bracewell

(1978). The DFT representation, however, allows a more convenient and faster interpolation of periodic data. To see how this can be, we shall write a few terms of the inverse DFT, which is a discrete Fourier series:

$$f(k) = \sum_{n=0}^{N-1} C_n \exp\left(j\,\frac{2\pi}{N}\,nk\right) = C_0 + C_1 \exp\left(j\,\frac{2\pi}{N}\,k\right) + \cdots$$

$$= A_0 + A_1 \cos\left(\frac{2\pi}{N}\,k\right) + B_1 \sin\left(\frac{2\pi}{N}\,k\right) + A_2 \cos\left(\frac{2\pi}{N}\,2k\right) + \cdots. \quad (10)$$

Note that we could let k be a continuous rather than a discrete variable, and we would then obtain a continuous function

$$f(x) = A_0 + A_1 \cos\left(\frac{2\pi}{N}\,x\right) + B_1 \sin\left(\frac{2\pi}{N}\,x\right) + A_2 \cos\left(\frac{2\pi}{N}\,2x\right) + \cdots, \quad (11)$$

where the continuous variable x has replaced k. This is the same functional form as the continuous Fourier series with the period $L = N$:

$$f(x) = a_0 + a_1 \cos\left(\frac{2\pi}{L}\,x\right) + b_1 \sin\left(\frac{2\pi}{L}\,x\right) + a_2 \cos\left(\frac{2\pi}{L}\,2x\right) + \cdots. \quad (12)$$

The continuous function consisting of only N terms can at best only approximate the most general (continuous) function, for, as we know, the Fourier series of the most general periodic function requires an infinite number of terms. However, consider the periodic function under examination made up of only N or fewer terms of its Fourier series, that is, a band-limited function. Then, because of the linear independence of the sinusoidal terms, the coefficients of the DFT are exactly the same as those of the Fourier series of the continuous function. That is, we find

$$A_0 = a_0,\ A_1 = a_1,\ B_1 = b_1,\ A_2 = a_2,\ \ldots, \quad (13)$$

and the continuous function with the DFT coefficients produces exactly the same continuous function as the Fourier series. No information is lost on taking the Fourier transform. (For an interesting proof of this and other important theorems for the DFT, see Bracewell, 1978.) Therefore the N sample points obtained by taking the inverse DFT of these coefficients completely represent the original DFT coefficients, and hence the continuous function formed from them. We may state this another way. If the continuous periodic function is band limited, having N or fewer discrete spectral components, then N samples equally spaced over one period are sufficient for perfect interpolation between the points. This may be done by constructing a continuous series from the DFT coefficients. Note that only N samples are necessary because one cycle is fully characteristic of the periodic function. For an interesting derivation of the relationship between the coefficients of the Fourier series and the DFT, see Hamming (1962), whose explicit expression shows exactly which set of coefficients alias a particular given coefficient.

Closely associated with sampling error is another question: to what extent is the information given in the DFT, a finite number of discrete spectral components, capable of representing the continuous function over an infinite interval? A periodic function is certainly capable of being represented by a finite number of discrete samples. In the preceding paragraphs we have discussed how this is possible. For the nonperiodic function, though, even if it is band limited and the sampling interval is sufficiently small, a finite number of samples is insufficient to represent the continuous function correctly. The sampling theorem tells us that an infinite number of samples is required. There is an aspect of a finite band, though, that we have not yet considered in these problems. In particular, the property of finite extent of the band may be used as a constraint, as discussed earlier, to extrapolate the sampled function. We would hope that an infinite number of sample points may be obtained in this way, and that they will be unique. Unfortunately, this is not the case. Fiddy and Hall (1981) and Schafer *et al.* (1981) have pointed out that, given a finite number of samples of a function, the constraints of square integrability and band limitation are not sufficient to determine that function uniquely. A family of functions is the best that may be determined.

With experimentally obtained data, which always contain random noise, such questions as uniqueness are academic. With noise, an exact solution is impossible to obtain, and the best that one may do, in a statistical sense, is to find the solution that is most probable. Even a solution that is most probable (overall) is not always what is desired. A unique solution is forced by the criterion or method of solution adopted, such as minimizing the sum of squares or maximum deviation. Each of these criteria produces a solution that is optimum in some sense. Various aspects of the error will be minimized in each of these slightly differing solutions. We may want to choose the criterion that minimizes some aspect of this error (or emphasizes some other aspect of the restoration). However, we may want to choose some particular criterion simply because it leads to more convenient mathematical manipulation or faster numerical calculation.

Except for special cases, the differences among the restorations produced by the various criteria are usually small and inconsequential. In a gross sense, the restorations produced by the methods mentioned in this section that apply many of the important constraints are usually much closer to the original undistorted function, even in the presence of moderate to high noise levels. The primary objective in initiating these restoration procedures is thus largely accomplished.

For the representation of almost all experimental data over an interval, the DFT has been found to be very adequate and useful. The errors peculiar to the DFT representation, such as those due to periodicity and a finite number of samples, are usually small compared with the errors due to noise. Restoration operations, such as deconvolution, are usually adequately performed within the DFT formulation.

As discussed earlier, the information-containing components of the Fourier spectrum of experimental data (even after restoration) become smaller with increasing frequency and merge into the noise in the high-frequency portion of the spectrum. After taking the DFT of experimental data, it is found that the useful information is represented in most cases by a quite small number of discrete Fourier components. The information-bearing components that are aliased would necessarily be very small, and would contribute negligible error. However, the noise spectrum decays very slowly (if at all) and is still significantly large at the higher frequencies, and the highest of the noise frequencies would be aliased (for any sampling interval). Taking a finer sampling interval minimizes this aliasing. With a finer sampling interval, more of the high-frequency noise spectral components that were previously aliased now show up in the Fourier spectrum. (That they are now present in the Fourier spectrum implies that they are certainly not aliased.) Taking increasingly finer sampling intervals therefore further minimizes the aliasing of the low-frequency informational components by the high-frequency noise components. The reader should be aware that this has little effect on the low-frequency noise, however. Note that this reduction of the aliasing with a finer sampling interval closely corresponds to minimization of the error in the data in a statistical sense in that simply "getting more points" on the data allows a reduction of the error by least-squares calculations.

Straightforward inverse filtering with the spectral components given by the DFT would thus involve very few spectral components with the type of experimental data discussed here.

When using the fast-Fourier-transform algorithm to calculate the DFT, inverse filtering can be very fast indeed. By keeping the most noise-free inverse-filtered spectral components, and adding to these an additional band of restored spectral components, it is usually found that only a small number of components are needed to produce a result that closely approximates the original function. This is an additional reason for the efficiency of the method developed in this research.

The preceding discussion also suggests a way of "tightly packing" the data, that is, a way of representing the significant information in the data by a much smaller number of points. Truncating the Fourier spectrum at some point before the noise assumes a significant fraction of the magnitude and saving this much smaller number of discrete components is a procedure that usually preserves all the important information in the data.

Extrapolation of discrete Fourier spectra is accomplished by applying the constraints to the discrete Fourier series given by the inverse DFT.

B. Extrapolation of Discrete Fourier Spectra

The discrete Fourier spectrum as given by the DFT is represented by the coefficients of a complex Fourier series. The inverse DFT is, of course, this series:

$$f(k) = \sum_{n=0}^{N-1} C_n \exp\left(j \frac{2\pi}{N} nk\right) = C_0 + C_1 \exp\left(j \frac{2\pi}{N} k\right) + \cdots$$

$$= A_0 + A_1 \cos\left(\frac{2\pi}{N} k\right) + B_1 \sin\left(\frac{2\pi}{N} k\right) + A_2 \cos\left(\frac{2\pi}{N} 2k\right)$$

$$+ B_2 \sin\left(\frac{2\pi}{N} 2k\right) + \cdots + A_{N-1} \cos\left(\frac{2\pi}{N}\right)(N/2)k. \tag{14}$$

The coefficients of the sines and cosines will be real for real data. Restoring a high-frequency band of c (unique complex) discrete spectral components to a low-frequency band of b (unique complex) spectral components will be the same (when transformed) as forming the discrete Fourier series from the high-frequency band and adding this function to the series formed from the low-frequency band. When applying the constraints in the spatial domain, the Fourier series representation will be used.

In all restorations treated in this research, the series formed from the low-frequency band is held fixed and the series formed from the high-frequency band altered by allowing the coefficients to vary until the particular constraint or constraints are satisfied. Let $u(k)$ denote the series for the low-frequency band and $v(k)$ the series for the high-frequency band. The explicit Fourier series expression for each would be

$$u(k) = A_0 + A_1 \cos\left(\frac{2\pi}{N} k\right) + B_1 \sin\left(\frac{2\pi}{N} k\right) + A_2 \cos\left(\frac{2\pi}{N} 2k\right)$$

$$+ \cdots + B_{b-1} \sin\left(\frac{2\pi}{N}\right)(b-1)k, \tag{15}$$

$$v(k) = A_b \cos\left(\frac{2\pi}{N} bk\right) + B_b \sin\left(\frac{2\pi}{N} bk\right) + A_{b+1} \cos\left(\frac{2\pi}{N}(b+1)k\right)$$

$$+ \cdots + B_{b+c-1} \sin\left(\frac{2\pi}{N}(b+c-1)k\right). \tag{16}$$

The sum $u(k) + v(k)$ yields the restored function. The coefficients in $v(k)$ are varied to satisfy the constraints. Because $u(k)$ is constant for all function continuations discussed here, no useful purpose is served by writing out its Fourier series expression, and so the series representation will always be suppressed.

Let the discrete spectrum, which consists of the coefficients of $u(k)$ and $v(k)$, be denoted by $U(n)$ and $V(n)$, respectively. The low-frequency spectral components $U(n)$ are most often given by the most noise-free Fourier spectral components that have undergone inverse filtering. For these cases $V(n)$ would then be the restored spectrum. However, for Fourier transform spectroscopy data, $U(n)$ would be the finite number of samples that make up the interferogram. For these cases $V(n)$ would then represent the interferogram extension.

As mentioned earlier, the sum of the squared error is found to be the most convenient measure of the error because much of the calculation may be done analytically. In the following sections the sum of the squared error will be formulated for each of the constraints, and the form of the equations in the unknown Fourier coefficients for each constraint will be determined. Values of both artificial and experimental data will then be substituted in these equations to determine these unknown Fourier spectral components of the extended spectrum. From these, the completely restored function may be determined.

V. CONSTRAINT OF FINITE EXTENT

A. Theory

The constraint of finite extent applies to data that exist only over a finite interval and have zero values elsewhere. Let N1 and N2 denote the nonzero extent of the original undistorted data. The restored function $u(k) + v(k)$, then, should have no deviations from zero outside the known extent of the data. To find the coefficients in $v(k)$ that best satisfy this constraint, we should minimize the sum of the squared points outside the known extent of the object. Actually, recovering only a band of frequencies in $v(k)$ implies the additional constraint of holding all higher frequencies above this band equal to zero. This is necessary for stability and is an example of one of the smoothing constraints discussed earlier. We minimize the expression

$$
\sum_{\substack{k < \text{N}1, \\ k > \text{N}2}} [u(k) + v(k)]^2 = \sum_{\substack{k < \text{N}1, \\ k > \text{N}2}} \left\{ u(k) + A_b \cos\left(\frac{2\pi}{N} bk\right) \right.
$$

$$
+ B_b \sin\left(\frac{2\pi}{N} bk\right) + A_{b+1} \cos\left[\frac{2\pi}{N}(b+1)k\right]
$$

$$
\left. + \cdots + B_{b+c-1} \sin\left[\frac{2\pi}{N}(b+c-1)k\right] \right\}^2 .
$$

The standard procedure for finding the minimum of a function of several variables is to take the partial derivative with respect to each variable and set the result equal to zero. Taking the derivative with respect to each unknown coefficient and setting the result equal to zero gives (summation interval suppressed)

$$\frac{\partial}{\partial A_b} \sum [u(k) + v(k)]^2 = 0,$$

$$\frac{\partial}{\partial B_b} \sum [u(k) + v(k)]^2 = 0,$$

$$\frac{\partial}{\partial A_{b+1}} \sum [u(k) + v(k)]^2 = 0, \tag{17}$$

$$\vdots$$

$$\frac{\partial}{\partial B_{b+c-1}} \sum [u(k) + v(k)]^2 = 0.$$

Considering in detail the derivative with respect to A_b, we have

$$\sum \frac{\partial}{\partial A_b} \left[u(k) + A_b \cos\left(\frac{2\pi}{N} bk\right) + \cdots \right]^2 = 2 \sum [u(k) + v(k)] \cos\left(\frac{2\pi}{N} bk\right) = 0. \tag{18}$$

The other derivatives are calculated in a similar manner, and we find for the complete set of equations

$$\sum [u(k) + v(k)] \cos\left(\frac{2\pi}{N} bk\right) = 0,$$

$$\sum [u(k) + v(k)] \sin\left(\frac{2\pi}{N} bk\right) = 0,$$

$$\sum [u(k) + v(k)] \cos\left[\frac{2\pi}{N} (b + 1)k\right] = 0, \tag{19}$$

$$\vdots$$

$$\sum [u(k) + v(k)] \sin\left[\frac{2\pi}{N} (b + c - 1)k\right] = 0.$$

Because of aliasing, the total number of coefficients obtained should not be greater than N. We have a set of $2c$ linear equations for the $2c$ unknown coefficients. A number of standard methods are available for solving a set of linear equations. We used the Gauss–Jordan matrix reduction method.

However, there are important advantages to iterative methods when the number of equations to be solved is large. Once the coefficients have been obtained, they may be converted to complex form and added to the original spectrum. Taking the inverse DFT would then yield the restored function. However, if the number of solved coefficients is small, it may be quicker simply to substitute the coefficients into the series representation for $v(k)$ and add this series to $u(k)$.

B. Application

1. Artificial Data

To illustrate the effectiveness of the finite-extent constraint, we show the restoration of two merged, almost completely noise-free Gaussian peaks in Fig. 1. These data, like all data discussed in this chapter, make up a 256-point data field. The original peaks are shown in Fig. 1(a). The Fourier spectrum truncated after the seventh (complex) coefficient is shown in Fig. 1(b). The restoration of Fig. 1(c) was accomplished by restoring 16 complex coefficients to the Fourier spectrum with the summation region over the last 128 points of the data field of Fig. 1(b). Note that an almost perfect restoration was accomplished. The Fourier spectrum of a Gaussian function is also a gaussian and dies out very quickly, so restoring only 32 (16 unique complex) coefficients produced a very good approximation to the original data. It was found that when the minimization procedure involved only the last quarter of the data field (64 data points), a good restoration was nevertheless obtained. The solved coefficients were almost the same as in the previous case. Even

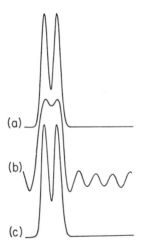

(a)

(b)

(c)

Fig. 1 Restoration of a Fourier spectrum of almost completely noise-free data. (a) Original two Gaussian peaks. (b) Same peaks with the spectrum truncated after the seventh (complex) coefficient. (c) Peaks in (b) with 16 (complex) coefficients restored.

when the summation was computed over only one-eighth of the data field (32 points), a crude approximation to the first few coefficients after the truncation frequency was obtained. However, the higher-frequency components were much larger than their true size.

It is important to note that these data are not completely error-free, and that some approximations are involved. There are several sources of small error, computational roundoff probably being dominant. Also, a Gaussian function never dies out exactly to zero. Its values, however, are very small even two or three standard deviations away from its peak. If the noise and other error are very much larger than this, however, severe computational difficulties ensue. The following figures illustrate the effect of moderate amounts of noise on the restoration.

The effects of a measured amount of Gaussian noise on the deconvolution of a single Gaussian peak are shown in Fig. 2 to enable a more quantitative discussion of the results. The original Gaussian function is shown in Fig. 2(a). The deconvolution of this function (with another gaussian) is shown in Fig. 2(b). This shows that the correctly restored result should be Gaussian in form and have a width about half of the original function. The restoration from the noisy data will be compared with this result. See Chapter 1 for further discussion of the relationship among the widths of the various functions under convolution.

Note that inverse filtering is sufficient for the restoration if the data are largely noise-free. We shall soon see, however, that even relatively small

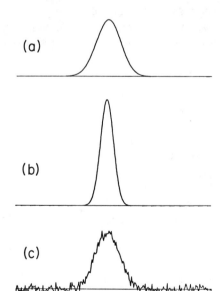

Fig. 2 Deconvolution of a single Gaussian peak. (a) Noise-free Gaussian peak. (b) Inverse filtering of the peak in (a) with another Gaussian function. The resulting peak is also Gaussian in form. (c) Peak in (a) with Gaussian noise of rms amplitude $\frac{1}{20}$ of the amplitude of the Gaussian peak superimposed.

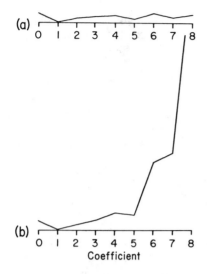

Fig. 3 Fourier spectrum of the noise that was superimposed on the Gaussian peak in Fig. 2(c). (a) Spectrum of the noise in the original peak. (b) Spectrum of the noise after inverse filtering. It is evident that the noise error increases considerably after the sixth complex coefficient.

amounts of noise necessitate truncation of all but a few coefficients of the Fourier spectrum. Figure 2(c) shows the same function shown in Fig. 2(a), with Gaussian noise of rms amplitude $\frac{1}{20}$ of the amplitude of the peak superimposed, a rather high level. On deconvolution of this noisy function, it was found that meaningful results could not be obtained unless the spectrum of the inverse-filtered result was truncated after the sixth (complex) coefficient. The reason for this is quite apparent when the spectrum of the noise in the inverse-filtered result is examined. Figure 3 illustrates this problem. Figure 3(a) shows the magnitude of the first nine Fourier spectral components of

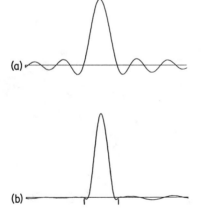

Fig. 4 Restoration by inverse filtering of the low-frequency band followed by spectral restoration of the high-frequency band with the constraint of finite extent. (a) Function produced by inverse filtering of the peak in Fig. 2(c) with six (complex) coefficients retained in the Fourier spectrum. (b) Improved function resulting from the restoration of 16 (complex) coefficients to the spectrum with the constraint of finite extent applied to the region indicated by the tick marks.

the noise in the original data [Fig. 2(c)]. Figure 3(b) shows the magnitude of these same spectral components after inverse filtering. From examination of Fig. 3(b) it is evident that keeping more than six (complex) coefficients in the inverse-filtered result would contribute unacceptably large noise error to the restoration. Even when only six or fewer coefficients are kept, the high noise level contributes considerable instability to the solution. This is especially apparent when the summation interval is decreased. In Figs. 4 and 5, the extent of the summation interval, the region to which the constraint is applied, is indicated by the tick marks. Figure 4(a) shows the inverse-filtered function with only six coefficients kept in its Fourier spectrum. Recall that peak broadening and ringing around the peak are characteristics of a function with a truncated spectrum. They come from the theoretical interpretation of convolution of the original function with the sinc function.

Letting the summation interval include the first negative sidelobe on each side of this function and solving for 32 (16 complex) coefficients produced the result shown in Fig. 4(b), a reasonably good result. However, taking the summation interval farther away from the peak produced a highly erroneous result, as shown in Fig. 5. For the summation interval defined by the tick marks, restoring 16 (complex) coefficients produced the result shown in Fig. 5(a), a very unsatisfactory result. To achieve improvement for this summation interval, no more than six (three complex) coefficients were sought. The result is shown in Fig. 5(b). As stated earlier, recovering only a small band of frequencies is an additional constraint that is necessary in many cases of heavily error-laden data.

(a)

Fig. 5 Restoration of the inverse-filtered result shown in Fig. 4(a) with the required summation for the constraint of finite extent taken over the interval indicated by the tick marks. (a) Restoration of 16 (complex) coefficients to the inverse-filtered estimate. (b) Restored function produced by restoring only three (complex) coefficients to the inverse-filtered estimate.

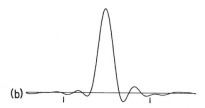

(b)

2. Experimental Data

The experimental data chosen for improvement were two merged infrared spectral lines, shown in Fig. 6(a). These were methane spectral lines taken with a two-pass Littrow-type diffraction grating spectrometer. Like all the grating spectrometer data discussed here, these data show isolated sets of infrared lines excerpted from data recorded and described by Hunt *et al.* (1978). To improve the resolution, inverse filtering was first performed. A Gaussian function was chosen as the impulse response function, which reasonably approximates the true impulse response for these lines (Jansson, 1968). The Fourier spectrum of a gaussian is usually an adequate approximation to most impulse response functions at their lower frequencies. However, because this spectrum dies out quickly, the approximation is usually poor at higher frequencies. If inverse filtering is not followed by spectral restoration and more high frequencies are retained, a better approximation to the correct impulse response will probably be needed. To minimize the error due to the noise in these methane lines, the deconvolution was truncated after the 10th (complex) coefficient, as shown in Fig. 6(b). To restore the spectrum after the 10th coefficient, the summation interval was chosen to cover the first 65 and the last 64 data points. This interval was chosen because it is evident from an examination of the original data that no information exists in these

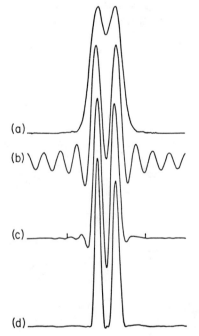

Fig. 6 Restoration of a Fourier spectrum of inverse-filtered noisy infrared peaks with the constraint of finite extent. (a) Two merged infrared peaks. (b) Inverse-filtered infrared peaks with the spectrum truncated after the 10th (complex) coefficient. (c) Spectrum restored by applying the constraint outside the marked region. Five (complex) coefficients were restored. (d) Spectrum restored with the constrained region including the first negative sidelobes and the dip between the peaks as well as all other regions outside the peaks. Sixteen (complex) coefficients were recovered.

regions. Attempting to restore 32 coefficients produced highly erroneous results. A much narrower band of frequencies must be attempted if a reasonable result is to be obtained. Solving for 10 (five unique complex) coefficients produced the restored peaks shown in Fig. 6(c). A considerable improvement in resolution was obtained. The constraint assured that the base line was straight in the summation region, as it should be. However, the first negative sidelobes around the peaks were still rather large and the function dipped strongly negative between the peaks. More of the function will have to be constrained for additional improvement. Extending the summation region to include the first negative sidelobe on each side of the peaks and most of the negative dip between the peaks produced the restoration shown in Fig. 6(d), which has a much-improved appearance. However, this operation is of questionable validity, because we suspect that information may exist in these included regions. For further improvement, without assumptions, additional contraints are needed. We shall see in the next chapter that the constraint of minimum negativity yields a better result for the restoration of these infrared lines.

In data-point units, the original infrared peaks were about 34 units wide (full width at half maximum). This corresponds to an actual width of approximately 0.024 cm^{-1}. The impulse response function was about 25 units wide. After inverse filtering and restoration of the Fourier spectrum, the resolved peaks were 11 and 14 units wide, respectively. This is close to the Doppler width of these lines.

VI. CONCLUDING REMARKS

Most of the original work of this chapter involves extrapolation of discrete Fourier spectra from some "region of support." This region of support may be the interferogram from Fourier transform spectroscopy. It may also be the inverse-filtered Fourier spectrum determined according to the convolution model, with the high-frequency spectrum truncated because of the increasing noise. A somewhat detailed discussion of the convolution model was included, because both the inverse filtering and the subsequent extension of the Fourier spectrum for restoration may be quantitatively discussed in terms of this model.

Even though restoration in two distinct spectral bands leads to very fast algorithms, it is still not optimum because of the residual error in the low-frequency band of spectral components used as a region of support. Perhaps the requirement that the inverse-filtered low-frequency spectrum (or, equivalently, its corresponding spatial function) be held constant for the restoration could be relaxed to allow small variations in the low-frequency spectral components closest to the truncation point. This could effect an improved solution that better satisfies the relevant constraints. Ways of weighting some spectral components more than others are currently being explored.

We shall end this chapter with a few practical remarks concerning the calculation of the inverse-filtered spectrum. In this research the Fourier transform of the data is divided by the Fourier transform of the impulse response function for the low frequencies. Letting $\hat{o}$ denote the inverse-filtered estimate and n the discrete integral spectral variable, we would have for the inverse-filtered Fourier spectrum

$$\hat{O}(n) = I(n)/\tau(n). \tag{20}$$

For the cases where τ is very small or zero at certain frequencies, the indeterminate discrete Fourier spectral components corresponding to these values of n may be determined along with the high-frequency band when the Fourier spectrum is continued. It is necessary to make sure that these coefficients, along with their associated sine and cosine terms, are included in the resulting equations that enforce the constraints.

Data are often normalized so that the area under the curve is preserved. This area is given by the dc spectral term, that is, for $n = 0$. To preserve the area in the discrete inverse-filtered result, every term should be multiplied by the dc spectral components of the impulse response function (if the impulse response function has not been normalized earlier). We would then have for $\hat{o}$

$$\hat{O}(n) = [I(n)/\tau(n)]\tau(0). \tag{21}$$

If unity is to be preserved for the area throughout, then each term of the inverse-filtered result should be multiplied by $\tau(0)/I(0)$. Note that spectrum continuation, because it does not involve the dc term, does not change the area under the data curve.

REFERENCES

Bell, R. J. (1972). "Introductory Fourier Transform Spectroscopy." Academic Press, New York.

Bracewell, R. N. (1978). "The Fourier Transform and Its Applications." McGraw-Hill, New York.

De Boor, C. (1978). "A Practical Guide to Splines." Springer-Verlag, New York.

Fiddy, M. A., and Hall, T. J. (1981). J. Opt. Soc. Am. 71, 1406.

Hamming, R. W. (1962). "Numerical Methods for Scientists and Engineers." McGraw-Hill, New York.

Howard, S. J. (1978). M.S. Thesis, Univ. of Southern Mississippi.

Howard, S. J. (1981a). *J. Opt. Soc. Am.* **71,** 95.

Howard, S. J. (1981b). *J. Opt. Soc. Am.* **71,** 819.

Howard, S. J. (1982). Ph.D. Dissertation, Florida State University.

Howard, S., and Rayborn, G. H. (1980). Deconvolution of Gas Chromatographic Data. NASA Contractor Report *CR*-3229.

Hunt, R. B., Brown, L. R., and Toth, R. A. (1978). *J. Mol. Spectrosc.* **69,** 482.

Jansson, P. A. (1968). Ph.D. Dissertation, Florida State University.

Papoulis, A. (1962). "The Fourier Integral and Its Applications." McGraw-Hill, New York.

Schafer, R. W., Mersereau, R. M., and Richards, M. A. (1981). *Proc. IEEE* **69,** 432.

Sheen, S. H., and Skofronick, J. G. (1974). *J. Chem. Phys.* **61,** 1430.

Stockham, T. G., Jr. (1966). *1966 Spring Joint Computer Conf. Proc.* **28,** 229.

CHAPTER 10

Minimum-Negativity-Constrained Fourier Spectrum Continuation

Samuel J. Howard

Physics Department, The Florida State University
Tallahassee, Florida

I.	Theory	290
II.	Application	295
	A. Artificial Data	295
	B. Experimental Grating Spectroscopy Data	297
	C. Fourier Transform Spectroscopy Data	302
III.	Concluding Remarks	323
	Appendix A	324
	Appendix B	325
	References	330

LIST OF SYMBOLS

A_n coefficients of cosine terms of discrete Fourier series

b Number of unique coefficients in complex Fourier series expression for $u(k)$

B_n coefficients of sine terms of discrete Fourier series

c number of unique coefficients in complex Fourier series expression for $v(k)$

$g(k)$ function of $u(k) + v(k)$ defined as $g(k) = 1/(1 + \exp\{K[u(k) + (k)]\})$

H Heaviside step function: $H(x) = 1$ when $x > 0$, 0 when $x \leq 0$

k integer variable for spatial function, the index of sampled values

K variable parameter in expression for $w(k)$

n integer Fourier spectral variable

N number of sample points, which is the same as the number of complex coefficients

T_m measured transmittance for absorption spectroscopy

$u(k)$ discrete Fourier series composed of only low-frequency and dc terms

U observed flux after absorption

U_0 incident flux on sample gas

$v(k)$ discrete Fourier series composed of high-frequency terms such that when added to $u(k)$ a series of a complete Fourier spectral band is formed

x_c denotes the original extent of the interferogram

I. THEORY

This chapter continues the development of restoration with the application of constraints that began in the preceding chapter. See Chapter 9 for an explanation of terms.

We discovered in Chapter 9 that the spatial function as given by the discrete Fourier transform (DFT) is a discrete Fourier series. Letting $u(k)$ denote the (known) series consisting of only low-frequency terms and $v(k)$ the series consisting of only high-frequency terms, we want to determine the unknown coefficients in $v(k)$ that best satisfy the constraints. Expressing deviations of the total function forbidden by the constraints as some function of $u(k) + v(k)$, we shall try to determine the coefficients of $v(k)$ that minimize these deviations. Sum-of-squares expressions for these measures of the error have been found to result in the most efficient computational schemes.

Minimizing the sum-of-squares expression for the constraint of finite extent resulted in a set of linear equations in the unknown coefficients. Chapter 9 closes with applications of the procedure that enforce this constraint. In this chapter we deal primarily with the constraint of minimum negativity, although we shall see later how other constraints may be included in the resulting formulation. The resulting much more general procedure has proved far more effective in correctly restoring the Fourier spectrum than the earlier methods mentioned. This observation is the motivation for reserving a separate chapter to discuss this method and its applications.

The constraint of minimum negativity (Howard, 1981) applies to data for which it is known that the correctly restored function should be all positive. For our formulation, we want to find the coefficients of $v(k)$ that best satisfy this constraint. These coefficients will be those that minimize the negative deviations in the total function $u(k) + v(k)$. The sum of the squared values of the negative deviations is given by

$$\sum_{k=0}^{N-1} \{H[-u(k) - v(k)][u(k) + v(k)]\}^2,$$

where H is the unit step function such that $H(x) = 1$ when $x > 0$ and $H(x) = 0$ when $x \leq 0$. The set of coefficients in $v(k)$ that minimizes this expression is the desired solution. However, we cannot carry out the minimization procedure with this expression. We cannot take the partial derivatives with respect to this function because it is not continuous. However, consider the alternative expression

$$\sum_{k=0}^{N-1} \left(\frac{1}{1 + \exp\{K[u(k) + v(k)]\}} [u(k) + v(k)] \right)^2,$$

where K is a variable parameter. Note that $1/[1 + \exp(Ky)]$ approaches $H(-y)$ as $K \to \infty$ for arbitrary y. Thus, as K in this summation approaches infinity, the entire expression approaches arbitrarily close to the original expression with the step function. Now we have an expression in closed form that consists entirely of analytic functions. We may carry out the minimization procedure by taking the derivatives with respect to the unknown coefficients and setting the results equal to zero. For simplification let

$$g(k) = 1/(1 + \exp\{K[u(k) + v(k)]\}).$$

Then we may express the resulting set of equations as follows:

$$\frac{\partial}{\partial A_b} \sum_{k=0}^{N-1} \{g(k)[u(k) + v(k)]\}^2 = 0,$$

$$\frac{\partial}{\partial B_b} \sum_{k=0}^{N-1} \{g(k)[u(k) + v(k)]\}^2 = 0,$$

$$\frac{\partial}{\partial A_{b+1}} \sum_{k=0}^{N-1} \{g(k)[u(k) + v(k)]\}^2 = 0, \tag{1}$$

$$\vdots$$

$$\frac{\partial}{\partial B_{b+c-1}} \sum_{k=0}^{N-1} \{g(k)[u(k) + v(k)]\}^2 = 0.$$

Considering in detail the derivative with respect to A_b, we have

$$\sum_{k=0}^{N-1} \frac{\partial}{\partial A_b} \left(\frac{1}{1 + \exp\{K[u(k) + v(k)]\}} [u(k) + v(k)] \right)^2$$

$$= 2 \sum_{k=0}^{N-1} \cos\left(\frac{2\pi}{N} bk\right) \frac{u(k) + v(k)}{1 + \exp\{K[u(k) + v(k)]\}}$$

$$\times \left(\frac{1}{1 + \exp\{K[u(k) + v(k)]\}} - K \exp\{K[u(k) + v(k)]\} \right.$$

$$\left. \times \frac{u(k) + v(k)}{1 + \exp\{K[u(k) + v(k)]\}} \right) = 0. \tag{2}$$

The second term in the rightmost set of large parentheses approaches zero as $K \to \infty$, so that for K large we have a good approximation

$$\sum_{k=0}^{N-1} \frac{u(k) + v(k)}{(1 + \exp\{K[u(k) + v(k)]\})^2} \cos\left(\frac{2\pi}{N} bk\right) = 0. \tag{3}$$

The other derivatives are calculated in a similar manner, and we have for the complete set of equations

$$\sum_{k=0}^{N-1} g^2(k)[u(k) + v(k)] \cos\left(\frac{2\pi}{N} bk\right) = 0,$$

$$\sum_{k=0}^{N-1} g^2(k)[u(k) + v(k)] \sin\left(\frac{2\pi}{N} bk\right) = 0,$$

$$\sum_{k=0}^{N-1} g^2(k)[u(k) + v(k)] \cos\left[\frac{2\pi}{N} (b + 1)k\right] = 0, \qquad (4)$$

$$\vdots$$

$$\sum_{k=0}^{N-1} g^2(k)[u(k) + v(k)] \sin\left[\frac{2\pi}{N} (b + c - 1)k\right] = 0,$$

where $b + c < N/2$. Because the unknown coefficients are also in the exponents, these equations are nonlinear in the coefficients and would ordinarily be difficult to solve. However, certain iterative techniques may be applied that not only yield a converging solution but also bring about considerable simplification in the equations. A form of the method of successive substitutions was developed for this research. The Newton–Raphson method also shows promise. However, because a matrix square in the number of equations, $2c$, must be solved for each iteration, the Newton–Raphson method may not be appropriate when c is very large. It does, however, converge faster than the method of successive substitutions. (See Hildebrand, 1965, for a discussion of the method of successive substitutions and the Newton–Raphson method.)

We now discuss in detail the particular iterative method chosen for this research. First, each of the equations is solved for one of the unknown coefficients within the first set of brackets. There are a number of ways of doing this. However, a reduction in the number of calculations results from solving for the coefficient associated with the multiplicative sinusoidal factor in each equation. Only the first equation of the resulting set will be written, because it illustrates all the salient features:

$$A_b = -\sum_{k=0}^{N-1} \frac{u(k) + v(k) - A_b \cos[(2\pi/N)bk]}{(1 + \exp\{K[u(k) + v(k)]\})^2}$$

$$\div \sum_{k=0}^{N-1} \frac{\cos^2[(2\pi/N)bk]}{(1 + \exp\{K[u(k) + v(k)]\})^2}. \qquad (5)$$

Next, the initial values of the coefficients are substituted in the right-hand sides of the equations. Zeros are substituted if no better initial values are known. The substantial advantage of this approach now becomes apparent.

For K sufficiently large, if the function $u(k) + v(k)$ is negative by even the smallest amount, the exponential will be extremely small and the factor $1/(1 + \exp\{K[u(k) + v(k)]\})$ will be approximately unity. If $u(k) + v(k)$ is positive by even the smallest amount, the factor $1/(1 + \exp\{K[u(k) + v(k)]\})$ essentially vanishes. These approximations approach exactness as $K \to \infty$. Thus the equations may be expressed essentially as follows:

$$
\begin{aligned}
A_b &= -\sum_{k=0}^{N-1} \left[u(k) + v(k) - A_b \cos\left(\frac{2\pi}{N} bk\right) \right] \\
&\quad \times \cos\left(\frac{2\pi}{N} bk\right) \left[\sum_{k=0}^{N-1} \cos^2\left(\frac{2\pi}{N} bk\right) \right]^{-1}, \\[2mm]
B_b &= -\sum_{k=0}^{N-1} \left[u(k) + v(k) - B_b \sin\left(\frac{2\pi}{N} bk\right) \right] \\
&\quad \times \sin\left(\frac{2\pi}{N} bk\right) \left[\sum_{k=0}^{N-1} \sin^2\left(\frac{2\pi}{N} bk\right) \right]^{-1}, \\[2mm]
A_{b+1} &= -\sum_{k=0}^{N-1} \left\{ u(k) + v(k) - A_{b+1} \cos\left[\frac{2\pi}{N} (b+1)\right] \right\} \\
&\quad \times \cos\left[\frac{2\pi}{N} (b+1)k\right] \left\{ \sum_{k=0}^{N-1} \cos^2\left[\frac{2\pi}{N} (b+1)k\right] \right\}^{-1}, \\[2mm]
&\ \ \vdots \\[2mm]
B_{b+c-1} &= -\sum_{k=0}^{N-1} \left\{ u(k) + v(k) - B_{b+c-1} \sin\left[\frac{2\pi}{N} (b+c-1)k\right] \right\} \\
&\quad \times \sin\left[\frac{2\pi}{N} (b+c-1)k\right] \left\{ \sum_{k=0}^{N-1} \sin^2\left[\frac{2\pi}{N} (b+c-1)k\right] \right\}^{-1}
\end{aligned}
\tag{6}
$$

for $u(k) + v(k) < 0$. That is, take the summation only over those data points for which the function $u(k) + v(k)$ is negative. It is evident that considerable computation is saved by avoiding calculation of the exponentials. Note that the negative values of $u(k) + v(k)$ change in each iteration as the new values of the coefficients are determined, so that a different summation is made in each iteration. For this reason it is necessary to test $u(k) + v(k)$ for its negative values before summing over them.

Although this procedure was developed from the constraint of minimum negativity, it will easily accommodate other constraints also, with only slight modification. Note that if the summation is not over a different set of data points for each iteration, but over a fixed set of points, the summation need be computed only once, because $u(k)$ and the sinusoids are constant for each value of the variable k. This is true for the finite-extent constraint, in which the summation is taken over the fixed interval outside the known extent of the data. The form of Eqs. (6) would be the same as that taken by the equations of finite extent when the method of successive substitutions is used. The constant summation interval of finite extent yields the constant coefficients (not to be confused with the sought coefficients) of the set of linear equations in the unknowns. The form of the method of successive substitutions for a set of linear equations is known as the Gauss–Seidel or the Jacobi method. In the former method each new unknown determined is used in improving the unknown in the succeeding equation. In the latter method every new unknown is determined from all the preceding ones before another iteration is begun. That both constraints result in very similar equations under the method of successive substitutions strongly suggests that both constraints could probably be included in one formulation, so as to yield the optimum resolution for a given set of data. It is found in practice that all that is necessary to implement successfully both constraints on the same set of data is to take the summation over all data points outside the known extent of the data (for the constraint of finite extent) while summing the negative values only over all other intervals (for the constraint of minimum negativity).

Further, if the data have an upper bound as well as a lower bound, this additional constraint may also be applied by summing over the data points above the upper bound [after subtracting the upper-bound value from $u(k) + v(k)$] as well as those below the lower bound. Although this additional constraint has been successful in minimizing values above the upper bound, it has resulted in little overall improvement in the restoration for the small number of data sets attempted here.

The modified procedure discussed here can successfully accommodate any of the constraints mentioned or any combination of them in a relatively uncomplicated way. Applying all the constraints simultaneously produces a result much more likely to be correct than would be the result of applying each constraint separately. The general procedure developed permits great versatility.

There are two different procedures by which the iteration may be carried out. One may substitute each improved coefficient, along with all the others, into the succeeding equation. This is known as the *point successive* procedure. Alternatively, one may determine all the unknown coefficients in all the equations using only the set of coefficients determined from the previous iteration. This is known as the *point simultaneous* procedure. Undesirable convergence properties of the point simultaneous procedure motivated its abandonment early in the research, however. The former procedure converges almost always when the tolerance is not chosen exceedingly small. The tolerance determines when the iteration is stopped. The iterations continue until the maximum change in any coefficient over its value in the previous iteration is less than the tolerance. With this procedure, convergence is quick to a large tolerance and much slower to a smaller tolerance. In this research the tolerance was usually taken to be about $\frac{1}{100}$ of the values close to the truncation point. With this tolerance, most of the restorations shown in this work usually required 10–50 iterations. However, convergence for some data was very slow. Occasionally a data run would take several hundred iterations. If the processing and restoration of data becomes a routine procedure in the laboratory, tolerances larger than that discussed here would probably be preferable. Tolerances several times larger have produced very satisfactory results.

II.   APPLICATION

A. Artificial Data

The procedure for implementing the constraint of minimum negativity as given in the preceding paragraphs will first be applied to the deconvolution of the artificial data shown in Fig. 2 of Chapter 9, to permit a better comparison of the effectiveness of this constraint with that of the constraint of finite extent. Figure 1(a) shows the result of inverse-filtering the Gaussian peak shown in Fig. 2(c) of Chapter 9 with six (complex) coefficients retained in its spectrum, and is the same function as that shown in Fig. 4(a) of Chapter 9. Restoration of 32 (16 complex) coefficients produced the result shown in

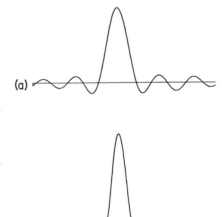

(a)

(b)

Fig. 1   Restoration by inverse filtering of the low-frequency band followed by spectral restoration of the high-frequency band with the constraint of minimum negativity. (a) Inverse-filtered result shown in Fig. 4(a) of Chapter 9. Six coefficients are retained. (b) Restored function produced by restoring 16 (complex) coefficients to the Fourier spectrum by applying the constraint of minimum negativity.

Fig. 1(b). This is slightly better than the result obtained with the constraint of finite extent in which the summation interval included the first negative sidelobes around the peak, which is shown in Fig. 4(b) of Chapter 9. It is vastly better, however, than the result obtained with the summation interval taken farther away from the peak as shown in Fig. 5 of Chapter 9.

We have found it to be generally true that the constraint of minimum negativity produces results much superior to those obtained with the constraint of finite extent. The minimum-negativity procedure has also been found to be extraordinarily insensitive to noise and other error. This is contrasted with the equations resulting from the constraint of finite extent, for which usually only a narrow band of coefficients may be permitted restoration to achieve a stable solution. The best overall results, however, are obtained with a combination of the two constraints.

Figure 2 shows the constraint of minimum negativity applied to the same deconvolution as that shown in Fig. 1 but with different truncation points for the Fourier spectrum. Figure 2(a) shows restoration to the inverse-filtered estimate with seven (complex) coefficients retained, and illustrates the distortion occurring when too many noise-laden coefficients are retained in the Fourier spectrum. From Fig. 3(b) of Chapter 9 it is evident that the seventh coefficient contains a large amount of noise error. Figure 2(b) shows restoration to the inverse-filtered result with only five complex coefficients retained in the Fourier spectrum. It differs little from the restoration with only six coefficients retained in the inverse-filtered estimate shown in Fig. 1. For both cases shown in Fig. 2, 16 complex coefficients were restored.

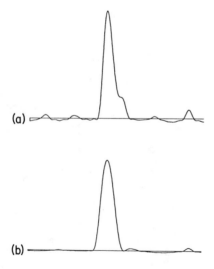

(a)

(b)

Fig. 2 Effect on spectral restoration of choosing a different truncation point for inverse filtering of the peak shown in Fig. 2(c) of Chapter 9. (a) Restored function produced by restoring 16 coefficients to the spectrum. Seven (complex) coefficients were retained in the inverse-filtered result. The effect of retaining too many error-laden coefficients in the inverse-filtered result is illustrated. (b) Restored function produced by restoring 16 coefficients to the spectrum. Five (complex) coefficients were retained in the inverse-filtered result.

B. Experimental Grating Spectroscopy Data

Figures 3 and 4 illustrate the application of the constraint of minimum negativity for restoring to inverse-filtered infrared spectral lines. A Gaussian function was chosen as an approximation to the impulse response function and 32 (16 unique complex) coefficients were restored in each case. The plots in Figs. 3(a) and 3(b) are the same as the plots in Figs. 6(a) and 6(b) of Chapter 9 and permit a comparison of the constraints of minimum negativity and finite extent for experimental data. Figure 3(c) shows the restoration using the constraint of minimum negativity. It is obviously far superior to the restoration shown in Fig. 6(c), which was obtained with the constraint of finite extent. In Fig. 3(c) the base line is quite straight in the regions outside the immediate vicinity of the peaks and is correct there, because examination of Fig. 3(a) reveals that no information exists in these regions. If, however, too many high-noise-contaminated coefficients are retained in the inverse-filtered spectrum, false detail will show up in these regions. Examination of the base line away from the peaks in the original data thus provides a crude measure of the error in the restoration, especially that caused by retaining too many error-laden high-frequency coefficients.

The data illustrated in Fig. 4(a) are methane absorption lines (0.02 cm^{-1} wide) observed with a four-pass Littrow-type diffraction grating spectrometer. For these data also, 256 points were taken. The data were obtained at low pressure, so that Doppler broadening is the major contributor to the true width of the lines. The straightforward inverse-filtered estimate with 15 (complex) coefficients retained is shown in Fig. 4(b). Figure 4(c) shows the restored function. The positions and intensities of the restored absorption

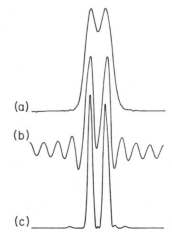

Fig. 3 Restoration of Fourier spectrum to the inverse filtering of two noisy infrared peaks using the constraint of minimum negativity. (a) Two merged infrared peaks. (b) Inverse filtering of the infrared peaks with the spectrum truncated after the 10th coefficient. (c) Spectrum restored by minimizing the sum of the squares of the negative regions of the inverse-filtered result. Sixteen (unique complex) coefficients were restored.

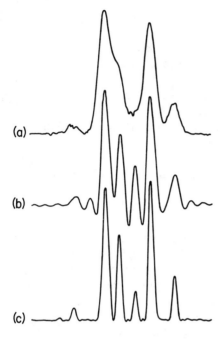

(a)

(b)

(c)

Fig. 4 Restoration of Fourier spectrum to the inverse filtering of strongly merged infrared peaks using the constraint of minimum negativity. (a) Noisy infrared data. (b) Inverse filtering of the infrared data with the spectrum truncated after the 15th coefficient. (c) Spectrum restored by minimizing the sum of the square of the negative regions of the inverse-filtered result. Sixteen (unique complex) coefficients were restored.

lines agree well with other independent studies. In data-point units, the original peaks were 19 units wide (full width at half maximum). The restored peaks are about 8–9 units wide, or close to the expected Doppler width. The Gaussian impulse response is approximately 16.5 units wide.

Note that in both Figs. 3 and 4 the negative values of the restored functions are very small.

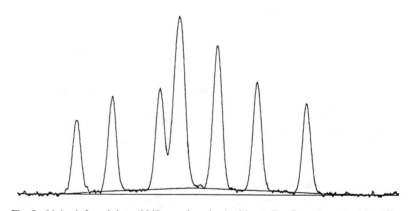

Fig. 5 Noisy infrared data (2048 sample points) with a spline fit to the curved base line.

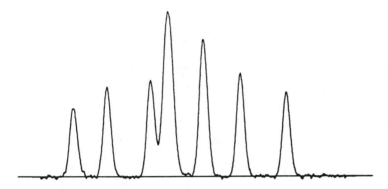

Fig. 6   Data of Fig. 5 adjusted to a function with zero base line according to the transmittance relation.

Figures 5–11 illustrate the restoration process in the presence of a drifting base line. These data are methane absorption lines taken with a four-pass Littrow-type diffraction grating spectrometer. For these data 2048 data points were taken. The impulse response function was approximated by a gaussian. The true width of these lines is approximately 0.02 cm^{-1}.

The original data are shown in Fig. 5. A spline fit (De Boor, 1978) has been made to the curved base line. These data are adjusted to a zero base line in accordance with the transmittance relation $T_m = U(x)/U_0(x)$, where T_m is the measured transmittance, $U_0(x)$ the unabsorbed flux, and $U(x)$ the observed flux after absorption by the gas in the sample. The relation between

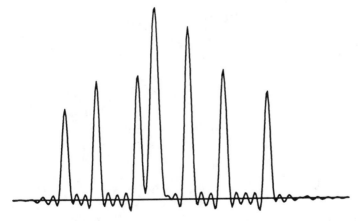

Fig. 7   Result of inverse-filtering the corrected data of Fig. 6 with a Gaussian impulse response function having a FWHM of 39 units. The Fourier spectrum was truncated after the 35th (complex) coefficient.

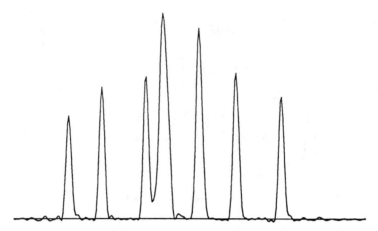

Fig. 8 Result of restoring 32 (16 complex) coefficients to the inverse-filtered spectrum of the function shown in Fig. 7 by minimizing the sum of the squares of the negative deviations.

U and U_0 is given by the Bouguer–Lambert law (see Chapter 2). The corrected data are shown in Fig. 6. Figure 7 illustrates the deconvolution of these corrected data with the Fourier spectrum truncated after the 35th coefficient. The Gaussian impulse response function had a full width at half maximum (FWHM) of 39 (data-point) units. Restoring 32 (16 complex) coefficients to the truncated spectrum by minimizing the sum of the squares of the negative values of the function of Fig. 7 produces the improved infrared lines shown in Fig. 8.

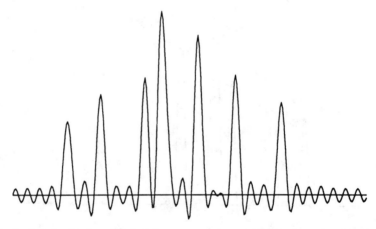

Fig. 9 Result of inverse-filtering the corrected data of Fig. 6 with a Gaussian impulse response function having a FWHM of 46 units. The Fourier spectrum was truncated after the 30th (complex) coefficient.

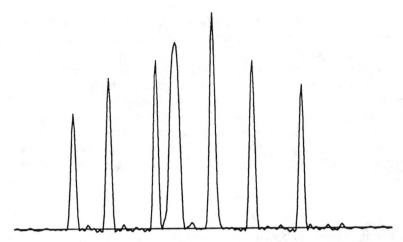

Fig. 10   Result of restoring 62 (31 complex) coefficients to the inverse-filtered spectrum of the function shown in Fig. 9 with the constraint of minimum negativity. The broadened spectrum of these narrower peaks necessitates recovering more spectral components to obtain reasonable results.

Figure 9 shows the result of inverse filtering with a Gaussian impulse response function having a FWHM of 46 units. The Fourier spectrum was truncated after the 30th coefficient. Note that the broader impulse response function should result in narrower restored peaks. Restoring 62 (31 complex) coefficients to the Fourier spectrum of the inverse-filtered result of Fig. 9 by minimizing the sum of the squares of the negative deviations produces the result shown in Fig. 10. Note that these peaks are narrower than those

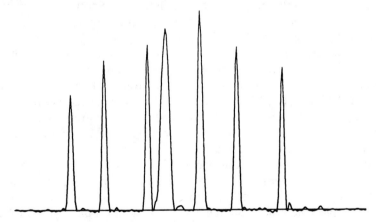

Fig. 11   Result of restoring 74 (37 complex) coefficients to the inverse-filtered spectrum of the function shown in Fig. 9 by minimizing the sum of the squares of the negative deviations.

of Fig. 8, which implies a broader Fourier spectrum. This necessitates recovering more spectral components to obtain reasonable results. Restoring 74 (37 complex) coefficients to the Fourier spectrum of the inverse-filtered result of Fig. 9 with the application of the constraint of minimum negativity produces the result shown in Fig. 11. Increased resolution is noted for all the peaks except the fourth one. A more detailed analysis has shown that this peak may be further resolved into three lines. The infrared lines obtained in the detailed restoration are closer to the true Doppler width than those obtained with the impulse response function having a FWHM of 39 units. We may physically interpret the restoration with the impulse response function having a 39-unit FWHM as a removal of only part of the distortion.

C. Fourier Transform Spectroscopy Data

1. Introductory Remarks

The treatment of Fourier transform spectroscopy (FTS) infrared data involves application of essentially the same techniques that apply to the other infrared data. However, sufficient difference exists in practice to warrant a separate discussion. In grating spectroscopy, the data recorded are the infrared lines, with random noise superimposed. The data recorded in Fourier transform spectroscopy form an interferogram. The Fourier transform of the interferogram produces the infrared spectral lines. It is evident, then, that for FTS data the Fourier spectrum of the interferogram and the electromagnetic spectrum are the same. The restoration problem is slightly different for each of these cases. It is seldom possible to obtain the entire nonzero extent of the interferogram experimentally because this generally requires a very large optical path difference. So for FTS data the purpose is to restore the interferogram function by applying constraints to its transform, the infrared spectral lines. For grating spectroscopy, restoration results from the deconvolution operation. In that case it is the inverse-filtered transform that is extended, extrapolated from a function that is truncated because of the increasing noise. In practice, FTS Fourier frequencies are seldom boosted to amplitudes greater than those in the original interferogram unless the lines are considerably broadened.

In the operating mode customarily used, which is to determine the existence, location, and intensity of the spectral lines, the interferometer produces an interferogram that is symmetric about the zero displacement position. If the zero displacement position (the maximum point on the "central fringe") is taken as the origin of the interferogram function, the Fourier transform of this will produce an infrared spectrum that is real and symmetric about its origin. Note that for this case it does not matter whether the forward or inverse Fourier transform is used, because the same result will be obtained. If a one-sided interferogram is given, it is usually symmetrically extended about the origin. An improvement in the efficiency of the required calculations is afforded by a real and symmetric function. See Chapter 17 in Bell (1972) for a detailed discussion of the use of the fast Fourier transform (FFT) in efficient computational schemes of interferometric data.

However, for illustration, only one side of the interferogram and its spectrum will be shown, usually the function of the positive spatial and spectral variable. In other operating modes of the interferometer, asymmetric interferograms are produced that have a complex Fourier transform. Asymmetric interferograms will not be treated in this work. For a more complete discussion of Fourier transform spectroscopy, the reader should consult Bell (1972), Vanasse and Strong (1958), Vanasse and Sakai (1967), Steel (1967), Mertz (1965), the *Aspen International Conference on Fourier Spectroscopy* (Vanasse *et al.*, 1971), and the two volumes of *Spectrometric Techniques* (Vanasse, 1977, 1981). A review of early work, which includes several major contributions of his own, is given by Connes (1969). Another interesting paper on the earlier historical development of Fourier transform spectroscopy is that by Loewenstein (1966).

The incomplete interferogram that is recorded could be represented by the complete interferogram function multiplied by the rectangle function. The transform, of course, is the convolution of the infrared lines with the sinc function, which is the origin of the oscillatory artifacts that alternate positively and negatively about the base line around the peaks. The complete interferogram should yield a set of infrared spectral lines that are all positive, so that the constraint of minimum negativity could appropriately be applied here. If all infrared frequencies outside the band of interest are filtered out before reaching the interferogram, then the constraint of finite extent could also be applied. In practice this constraint is often awkward to apply, however, because the filters never terminate a spectrum abruptly.

For FTS data, artifact removal is a consideration that is as important as resolution improvement for most researchers in this field. Interferogram continuation methods are not as yet widely known in this area. Methods currently in widespread use that are effective in artifact removal involve the multiplication of the interferogram by various window functions, an operation called apodization. A carefully chosen window function can be very effective in suppressing the artifacts. However, the peaks are almost always broadened in the process. This can be understood from the uncertainty principle. A window that reduces the function most strongly closest to the end points will yield a transform for the modified function that must be broader than it was originally. Alternatively we may employ the convolution theorem, which has it that the multiplication of the interferogram by the window function is the same as convolution of the infrared lines with the Fourier transform of the window function. One window function widely used for apodization is the triangular window function, which is very effective in removing negative artifacts. However, because its transform is the sinc-squared function, the small positive sidelobes characteristic of this function show up around sharp peaks in the spectrum.

For most data it is seldom possible to effect a complete restoration of the interferogram. On recovering even a relatively few additional data points, though, a considerable improvement in resolution is usually obtained. However, the artifacts in most cases are still appreciable. It has been found that multiplying the interferogram by an appropriate window function, one that does not go to zero at the end points, will effect, on extending the interferogram, a removal of almost all of the artifacts as well as an improvement in resolution over that of the lines of the original interferogram. However, the ultimate resolution obtained with this window function is not as good as that obtained with the straightforwardly extended interferogram in the absence of the window function. The use of these window functions essentially involves a trade-off between artifact removal and resolution improvement.

For FTS data, then, the general procedure for improvement is first to determine the number of additional points on the interferogram that can be recovered, then to multiply the (unrestored) interferogram by one of a class of window functions appropriate to the number of points to be recovered, and finally to restore the additional points to the interferogram using as many of the applicable constraints as possible. The Fourier transform of the interferogram is almost always computed with the DFT to take advantage of the speed of the FFT algorithm. We may therefore borrow almost all of the DFT formulation of the restoration problem for application to FTS data. Actually, deconvolution with a symmetric impulse response function is the same as multiplying the interferogram by a high-pass filter, because the Fourier transform of a real, symmetric impulse response function is also a real, symmetric function, and would produce no phase shifts. So convolution and inverse filtering would then be only special cases of multiplying the interferogram by a certain class of window functions before its continuation.

2. Simulated Fourier Transform Spectroscopy Data

For a more quantitative analysis of the errors involved, artificial data will be addressed first. To see the effects of truncation and restoration of a single spectral line, we shall first consider a monochromatic electromagnetic

source. Figure 12 could represent such a source. Of course, an entirely monochromatic source is physically impossible, but there are many wave trains
in nature whose total length is very long compared with its wavelength and
would therefore yield a very sharp spectral line. The case that we have
illustrated could closely approximate many aspects of such a source. Figure
12(a) could be a segment of an interferogram of a monochromatic source. Its
Fourier transform is shown in Fig. 12(b) and is a single sharp spectral line
(two lines actually, that are symmetric about the origin). Theoretically, the
Fourier transform of a monochromatic source of finite intensity is two Dirac
δ functions. They would be infinitely high and have zero width. Figure 13
illustrates a more realistic case, that of a finite interferogram. Figure 13(a)
could be the finite interferogram of an approximately monochromatic
source. Its Fourier transform, the sinc function, is shown in Fig. 13(b). This
much-broadened spectral line with the ringing around it closely approximates the lines obtained in practice from the interferograms of approximately monochromatic sources. For improvement of these spectral lines we
would like to remove the artifacts as well as increase the resolution. Most
presently used methods only remove the artifacts (by the process of apodization). Nearly all of the window functions employed for apodization have the
property of going smoothly to zero at the end point of the interferogram.
    One window function widely used is the triangular window function. This
function is shown in Fig. 14(d). Multiplying the interferogram of Fig. 13(a)
by this window function produces the altered interferogram shown in Fig.

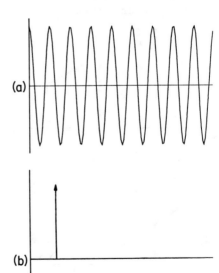

Fig. 12   (a) Interferogram and (b) spectrum
of a monochromatic source. The interferogram would be a cosine function that extends
indefinitely, its Fourier transform the Dirac δ
function.

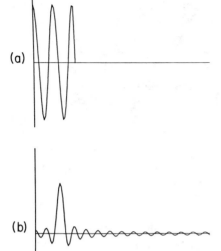

(a)

(b)

Fig. 13 Simulation of a monochromatic source with a finite arm displacement of the interferometer. (a) Truncated cosine function (30 discrete data points). (b) Its Fourier transform, the sinc function, which simulates the infrared spectral line.

15(a). The Fourier transform of this is shown in Fig. 15(b). Notice, importantly, that the negative artifacts have been completely removed. The small positive sidelobes remaining are not nearly so prominent as those in the original spectral line. This spectral function is the sinc-squared function, the Fourier transform of the triangle function. Use of the triangular window function has been very effective in reducing artifacts, but a heavy

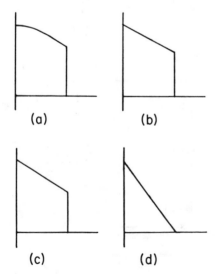

(a)   (b)

(c)   (d)

Fig. 14 Four window functions used to multiply the interferogram. (a) Gaussian window. (b) Triangular window. (c) Triangular window of greater slope than (b). (d) Triangular window tapering to zero at the end point of the interferogram (used for apodization).

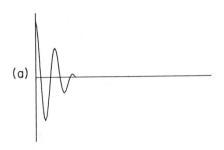

(a)

Fig. 15 Apodization, or the reduction of artifacts in the spectral line by the multiplication of the interferogram by a window function that tapers to zero at the end point of the interferogram. (a) Cosine interferogram of Fig. 13(a) multiplied by the triangular window function of Fig. 14(d). (b) Resulting spectral line, the sinc-squared function.

(b)

price has been paid. The spectral line has been considerably broadened, as is evident from a comparison of the widths of the lines in Figs. 13(b) and 15(b).

If increasing the resolution is the only consideration, then we may simply extend the interferogram function by the methods discussed earlier. Only the constraint of minimum negativity will be used in the following illustrations. The straightforward extension of the interferogram of Fig. 13(a) obtained by minimizing the negative values of the function shown in Fig. 13(b) produces the restored spectral line shown in Fig. 16(b), if the interferogram is extended by the amount shown in Fig. 16(a). The extent of the original interferogram is denoted by x_c (for cut-off point) on the plot. A considerable improvement in the resolution of the spectral line has been achieved. However, as is evident from examining the remaining function around the peak of Fig. 16(b), the artifacts are still rather large.

To increase the resolution *and* simultaneously reduce the artifacts, the interferogram must be multiplied by the proper window function before extension, as discussed earlier. A variety of window functions have been found that will bring about, to some degree, the simultaneous achievement of both goals. However, they all share the common property that they do *not* taper to zero at the end point of the interferogram. Consider the triangular window function shown in Fig. 14(b). Multiplying the interferogram of Fig. 13(a) by this window function and extending the interferogram to the point where the triangle function would cross the spatial axis yields the result shown in Fig. 17(a). The considerably improved spectral line with negative values

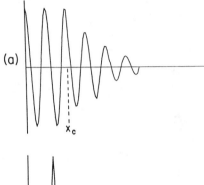

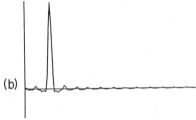

Fig. 16 Straightforward extension of the interferogram by minimizing the negative values of the spectrum of Fig. 13(b). (a) Interferogram (of 30 data points) extended by 50 data points. (b) Restored spectral line, which exhibits a considerable increase in resolution.

almost nonexistent is shown in Fig. 17(b). Note that the resolution is improved over that of the original spectral line of Fig. 13(b) and that the artifacts have been almost completely removed, with only small positive sidelobes remaining. However, note also that the resolution is not quite as good as that of the extended interferogram of Fig. 16(b), which has not been multiplied by a window function. This comparison illustrates the trade-off

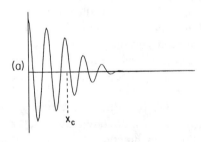

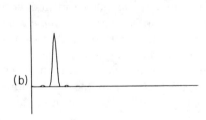

Fig. 17 Effectiveness in removing the artifacts from the spectrum of multiplying the interferogram by the proper window function before extending the interferogram by a finite number of points. (a) Cosine interferogram of Fig. 13(a) premultiplied by the triangular window function of Fig. 14(b) before extending by 50 data points. (b) Restored spectral line.

between resolution and artifact removal brought about by the use of a
window function. When restoring a given number of points to the interferogram, the researcher should choose the appropriate window function to
emphasize whichever of the two aspects he or she deems most important.

Thus far, the discussion has been restricted to triangular window functions.
However, it has been discovered that windows of many other functional
forms are capable of bringing about improvement in the spectral lines. In
this research the author has found that the window of Gaussian shape has
produced the best overall results. With the same interferogram and extension
by the same amount as in the previous example, premultiplication by the
Gaussian window function shown in Fig. 14(a) produced the restored interferogram shown in Fig. 18(a). The restored spectral line shown in Fig. 18(b)
has a resolution much improved over that of Fig. 17(b), where the triangular
window function was used, yet the artifacts are no worse. The researcher
should explore the various functional forms of the window function to find
the one best suited for his or her particular data.

To see the effects that noise would have on a single spectral line, consider
the interferogram of unity amplitude shown in Fig. 19(a), where Gaussian
noise of rms amplitude 0.1 is added. It is evident that this high noise level
considerably distorts the interferogram function. Figure 19(b) is the resulting
spectral line. Note that it is quite smooth, which is not quite what we would
expect considering the high noise level in the interferogram. The smoothness
comes about because the noise extends over only part of the data field—the
part over the finite interferogram. By minimizing the negative values of this
function, we obtain the restored interferogram and spectral line of Fig. 20.

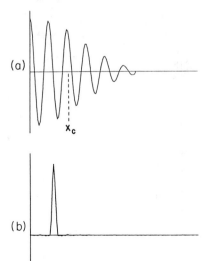

Fig. 18   Interferogram of Fig. 13(a) multiplied by the Gaussian window function of Fig.
14(a) to effect artifact removal. (a) Interferogram extended by 50 data points. (b) Restored
spectral line.

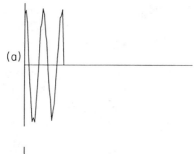

(a)

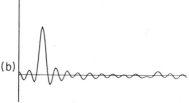

(b)

Fig. 19 Interferogram of Fig. 13(a) with Gaussian noise of rms amplitude 0.1 added. (a) Cosine interferogram of unity amplitude with random noise of rms amplitude 0.1 superimposed. (b) Single spectral line with the oscillatory artifacts.

Note that all cases shown of restoring this interferogram will involve extension by the same number of points. The resolution of this line is very much the same as that of the noise-free case. However, let us remove the debris from the base line to see what remains after reducing the artifacts. Multiplying this interferogram by the triangular window function of Fig. 14(b) and then extending, we obtain the interferogram shown in Fig. 21(a). The restored spectral line is shown in Fig. 21(b). Surprisingly, the resolution of the spectral line is almost the same as that of the noise-free case. However,

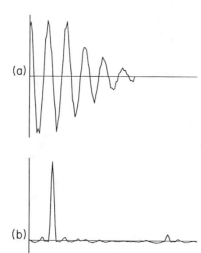

(a)

(b)

Fig. 20 Straightforward extension of the noisy interferogram of Fig. 19(a) by minimizing the negative values of the distorted spectral line of Fig. 19(b). (a) Interferogram extended by 50 data points. (b) Restored spectral line.

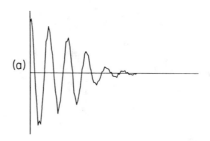

(a)

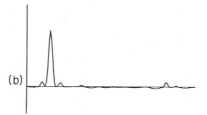

(b)

Fig. 21 Interferogram of Fig. 19(a) multiplied by a triangular window function before extension in an attempt to remove the artifacts. (a) Interferogram premultiplied by the triangular window function of Fig. 14(b) before extending by 50 data points. (b) Restored spectral line.

many low-level artifacts have been introduced. Multiplying the interferogram by the Gaussian window function and extending yields the interferogram of Fig. 22(a) and the restored spectral line of Fig. 22(b). The resolution of the spectral line has surprisingly changed very little as compared with the noise-free case. The main effect of the noise seems to be the introduction of many low-amplitude features scattered randomly about the base line.

Considering more complicated spectra, we show in Fig. 23 the (noise-free) interferogram and spectral lines for the superposition of two closely spaced

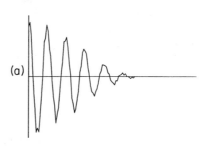

(a)

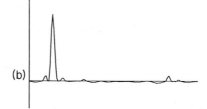

(b)

Fig. 22 Interferogram of Fig. 19(a) multiplied by a Gaussian window function before extension in an attempt to remove the artifacts. (a) Interferogram premultiplied by the Gaussian window function of Fig. 14(a) before extending by 50 data points. (b) Restored spectral line.

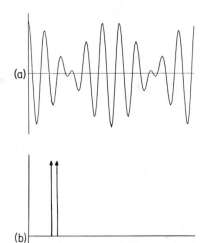

Fig. 23 Interferogram and spectrum of two monochromatic sources that differ slightly in wave number. (a) Noise-free interferogram, understood to repeat indefinitely. (b) The two closely spaced spectral lines.

frequencies. This example is shown to demonstrate the ability of the constraint of minimum negativity to separate these two closely spaced spectral lines. Figure 24(a) is the finite interferogram. Figure 24(b) is the spectrum. Note that the lines for this finite interferogram are completely merged. Extending the interferogram by the amount shown in Fig. 25(a) by minimizing the sum of the squares of the negative values of the merged spectral lines effectively separates these two lines, as Fig. 25(b) demonstrates.

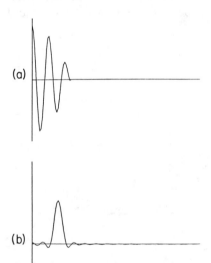

Fig. 24 Interferogram of the two monochromatic sources that would be obtained for a finite maximum path difference of the interferometer. (a) Finite interferogram. (b) Recorded spectrum. The two lines are completely merged into one.

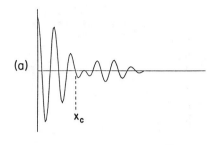

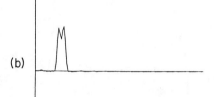

Fig. 25 Interferogram of Fig. 24(a) extended by minimizing the negative values of the spectrum of Fig. 24(b). (a) Interferogram extended by 50 data points. (b) Restored spectrum. The lines have been effectively resolved.

Considering increasingly more realistic examples, we show in Fig. 26(a) the interferogram corresponding to the superposition of four monochromatic sources of varying intensity. The largest cosine in the interferogram has an amplitude of 1.6 units. The four sharp spectral lines of this spectrum are shown in Fig. 26(b). A physically more realistic case, that of a finite interferogram with Gaussian noise of rms amplitude 0.1 added, is shown in Fig. 27(a). The degraded spectral lines are shown in Fig. 27(b), which are now

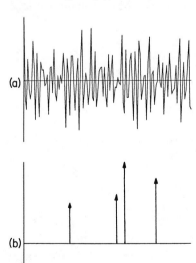

Fig. 26 Interferogram and spectrum of the superposition of four monochromatic sources. (a) Noise-free interferogram. The magnitude of the largest of the four cosine functions is 1.6 units. (b) Spectrum composed of four Dirac δ functions.

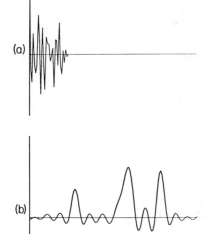

Fig. 27 Finite interferogram of the four monochromatic sources of Fig. 26 with Gaussian noise of rms amplitude 0.1 superimposed and the resulting degraded spectral lines. (a) Interferogram of 30 data points. (b) Merged and distorted spectral lines.

broadened and merged. Minimizing the negative values of this spectrum produces the extended interferogram and restored spectral lines of Fig. 28. As noted earlier, the noise seems to have little effect on the resolution. All the lines are completely resolved. This result also suggests that, even with the noise, there is not much interaction among the peaks on restoration. That is, the restored lines are very much like the result that one would obtain by restoring each line separately and then superimposing them. Even though most peaks show little interaction, occasionally a restoration will show wide

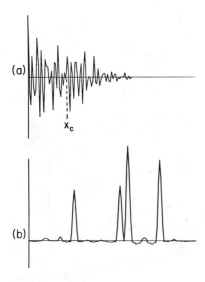

Fig. 28 Extended interferogram and restored spectrum produced by minimizing the negative values of the spectrum of Fig. 27(b). (a) Interferogram extended by 50 data points. (b) Restored spectral lines.

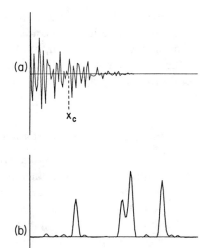

Fig. 29   Interferogram of Fig. 27(a) multiplied by the triangular window function of Fig. 14(b) before extension to remove the artifacts. (a) Interferogram extended by 50 data points. (b) Restored spectral lines.

divergence from this, especially when the noise level is high. Although the resolution of these lines is good, large artifacts still remain in the restoration.

If we multiply the interferogram by the triangular window function of Fig. 14(b) before extension, we obtain the interferogram and spectral lines shown in Fig. 29. The artifacts have been considerably reduced, but a slight loss of resolution has occurred. The best overall results are obtained by premultiplying the interferogram by the Gaussian window function of Fig. 14(a) before extension. These results are shown in Fig. 30. The spectral lines of

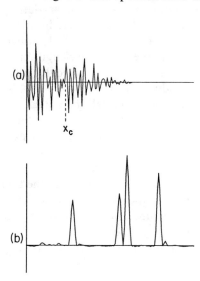

Fig. 30   Interferogram of Fig. 27(a) multiplied by the Gaussian window function of Fig. 14(a) before extension to remove the artifacts due to a finite interferogram. (a) Interferogram extended by 50 data points. (b) Restored spectral lines.

Fig. 30(b) are better resolved than those of Fig. 29(b), yet the artifacts are no worse. The major effect of the noise for all cases seems to be the introduction of randomly located low-amplitude artifacts (over and above those obtained for the noise-free case). Note also that the degradation and restoration operations affect each spectral line the same way, regardless of its wavelength.

3. *Experimental Fourier Transform Spectroscopy*

Finally, experimental FTS data will be addressed. For further discussion and description of the experimental data, and the experimental conditions under which the data were taken, see Toth *et al.* (1981). Given the interferogram, the Fourier transform must be computed to obtain the spectral lines, because the constraints are applied to the spectrum. We may attempt to restore the entire spectrum or isolated sets of spectral lines. The iterative method developed here seems to be adequate to restore a large spectrum consisting of many data points. The method of successive substitutions used is capable of accommodating a large set of equations. Attention here, however, will be restricted to isolated sets of lines. Because we are not sure at what spectral frequencies the information exists for these cases, only the constraint of minimum negativity can be validly applied for improvement. For a small set of spectral lines with only a few data points, it is usually unnecessary to multiply the interferogram function by a window function before extending, because the entire interferogram is recovered in most cases. The reason for this is that real spectral lines have a nonzero width, unlike the Dirac δ-function lines used in the artificial data, and a small number of data points implies a small number of discrete Fourier spectral components.

The experimental data illustrated here are methane absorption lines observed in the 3-μm-wavelength infrared spectral region. The narrow Lorentzian lines due to the molecular transitions are broadened by Doppler effects to produce a final line shape that is approximately Gaussian. See Chapter 1 for a more thorough theoretical investigation of the broadening factors for these lines and a better approximation to the true line shape. A two-sided interferogram is acquired that extends 50 cm on each side of the central fringe. Because 50 cm does not extend to the "washout" point on the interferogram (in practice, the washout point is usually considered to be the point where the rms amplitude of the noise exceeds the magnitude of the signal), the line shape recorded will be the convolution of the original line shape with the sinc function. Minimizing the negative values of this final spectral function is expected to produce a reasonable approximation to the complete interferogram and to the true line shape.

A very-low-frequency sinusoid was superimposed on these spectral lines owing to channeling. This comes about by reflections from the window surfaces that contain the sample gas. These often result in a spike on the interferogram, which produces a superimposed sinusoid on transforming. Rather than removing the sinusoid from the entire data set, we fitted a smooth curve to the base line of each isolated set of lines treated.

There are three major sources of error in these particular data. There is, of course, the noise. Then there is the inevitable error due to the base-line fit of this strongly varying function. Third, it is impossible to obtain completely isolated lines. Because the data are convolved with the sinc function, the peaks die out slowly, and an "isolated" set of lines never dies out exactly to zero. Some adjustment of the end points of this type of data is almost always necessary to make the data go smoothly to zero there. In spite of these errors, reasonably good restorations were obtained for all three of the data sets treated here.

The simplest case, that of two large infrared lines, is shown in Fig. 31(a). A smooth curve was fitted to the base line as shown. A spline-fitting computer program developed by De Boor (1978) was used to obtain this fit very conveniently. After the fit was obtained, the data were adjusted to a flat base line, as shown in Fig. 31(b), and the data field was extended by padding with zeros to yield an overall data field of $2^8 = 256$ points. Taking the Fourier transform, we obtained the interferogram function shown in Fig. 32. (Even though it was not obtained directly from the interferometer as recorded

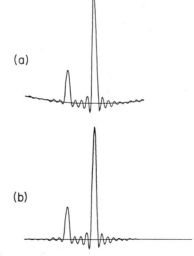

(a)

Fig. 31 Experimentally obtained methane absorption lines in the infrared spectral region. (a) Two isolated methane spectral lines. A smooth curve is fitted to the base line. (b) Replotted lines. All points on the smooth curve were assigned the value of zero, and all values on the spectral lines are adjusted accordingly.

(b)

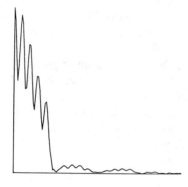

Fig. 32 Interferogram function for the set of spectral lines of Fig. 31(b).

data, we shall refer to this function as the interferogram function to avoid the introduction of new and possibly confusing terminology.) Note that, as expected, the interferogram function does have a rather sharp cutoff point. The residual low-amplitude interferogram extended beyond the cutoff point is due to the errors discussed in the preceding paragraph. The amplitude of this extended interferogram function provides a crude measure of this error. This residual extended interferogram function is cleanly truncated so as to yield a smooth noise-free set of infrared spectral lines to which the constraints may be applied.

Note that by the definition of the Fourier spectrum presented here, which we have given as the Fourier transform of the function or data under consideration, we may consider the interferogram function to be the "Fourier

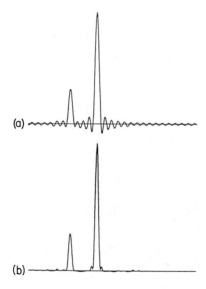

Fig. 33 (a) Smooth set of infrared spectral lines produced by truncating the interferogram function of Fig. 32 just before its sharp dropoff, keeping 27 data points. (b) Restored infrared spectral lines produced by minimizing the negative values of (a) and extending the interferogram by 60 data points.

spectrum" of the infrared lines under consideration. Although the term "Fourier spectrum" is usually reserved for the forward transform, this designation would make little difference on application to the symmetric FTS data treated here. We had earlier referred to the Fourier transform of the experimentally obtained interferogram as the Fourier spectrum.

The infrared spectral lines given by the truncated interferogram function are shown in Fig. 33(a). This last step is not absolutely necessary because the computer program written for restoration will restore a band of Fourier spectral components for any given spatial function, regardless of whether the spectrum has been truncated or not. So, if we prefer, we may use the program simply to "improve" the spectrum rather than restore the Fourier spectrum of the spatial function after spectral truncation. However, error will occur for an interferogram that is not truncated if the index of the furthermost nonzero point is greater than the highest index in the band of points of the restored interferogram function. Minimizing the negative values of the lines shown in Fig. 33(a) produces the restored lines shown in Fig. 33(b). Both lines show an increase in resolution. The origin of the two small "wings" on each side of the larger peak is unexplained. However, it occurs in the restoration of many lines of this data run.

A more complicated set of spectral lines is shown in Fig. 34(a). A spline fit is made to the base line as shown, which allows adjustment to a flat base line, shown in Fig. 34(b). Taking the Fourier transform of this yields the interferogram function shown in Fig. 35. Truncating this function after the sharp dropoff that signals the end of the interferogram, we take the Fourier

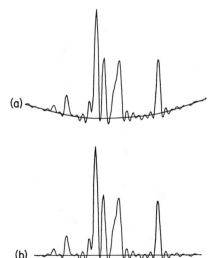

Fig. 34 Methane absorption lines in the infrared spectral region. (a) Isolated set of infrared spectral lines. A smooth curve is fitted to the base line. (b) Result of taking the differences from the curved base line drawn and replotting.

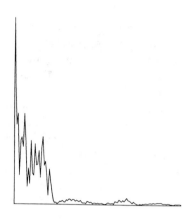

Fig. 35 Interferogram function obtained by Fourier-transforming the spectral lines of Fig. 34(b).

transform to obtain the much smoother set of spectral lines shown in Fig. 36(a). Applying the constraint of minimum negativity to these lines yields the restored lines of Fig. 36(b). All the lines show an improvement in resolution. Note that the merged peaks in the original data separate into two distinct lines in the restoration. As mentioned in the discussion of the previous example, two of the Fourier transforms involved here are not absolutely necessary. We could directly apply the constraint of minimum negativity to the data of Fig. 34(b) if they were relatively error-free.

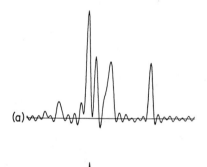

(a)

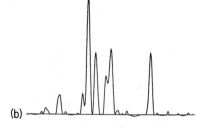

(b)

Fig. 36 (a) Smooth spectrum produced by truncating the interferogram function of Fig. 35 just before its sharp dropoff, keeping 28 data points. (b) Restored infrared lines resulting from minimizing the negative values of (a) and extending the interferogram by 60 data points.

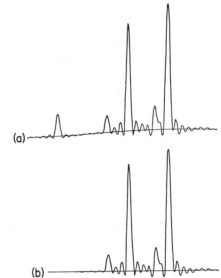

(a)

(b)

Fig. 37   Isolated set of infrared absorption lines in the infrared spectral region (a) with a smooth curve fitted to the base line and (b) adjusted to zero base line.

For the final example, we have the set of spectral lines shown in Fig. 37(a). Fitting a smooth curve to the base line and adjusting it to zero gives Fig. 37(b). Taking the Fourier transform produces the interferogram function shown in Fig. 38. Choosing a truncation point in the sharp dropoff of this function and transforming produces the smooth set of spectral lines shown in Fig. 39(a). Minimizing the negative values of these spectral lines produces the restored infrared spectrum shown in Fig. 39(b). Increased resolution is observed. Figure 40 is included to emphasize that a good restoration is not obtained everytime with this restoration procedure. By truncating the interferogram function of Fig. 38 and retaining one point less than for the previous figure, the Fourier transform shown in Fig. 40(a) is produced. Restoration from this function produces the restored lines of Fig. 40(b). This restoration

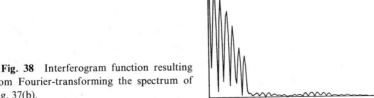

Fig. 38   Interferogram function resulting from Fourier-transforming the spectrum of Fig. 37(b).

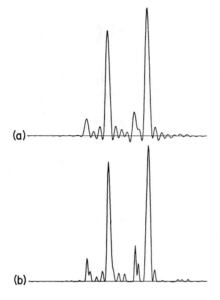

(a)

(b)

Fig. 39 (a) Smooth set of infrared spectral lines produced by the Fourier transform of the truncated interferogram of Fig. 38. (b) Restored infrared spectral lines resulting from minimizing the negative values of (a).

is not exactly what is expected from the original data, and is generally inferior to the restoration of Fig. 39(b).

I shall conclude this section with a few remarks on practical restoration with the DFT. To take the fullest advantage of the computational speed afforded by the FFT, the overall data field is usually composed of a number of data points given by 2 raised to some integral power. The data are usually

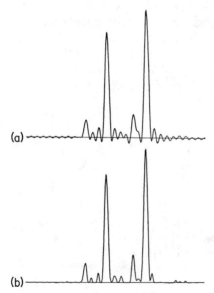

(a)

(b)

Fig. 40 (a) Smooth set of infrared spectral lines resulting from Fourier-transforming the truncated interferogram of Fig. 38 but retaining one datum point less in the interferogram than in Fig. 39. (b) Restored infrared spectral lines resulting from minimizing the negative values of (a).

padded with zeros to bring them up to this number. Algorithms for computing the FFT for bases other than 2, however, are becoming increasingly more popular with improvements in their efficiency.

Often FTS data are obtained that are "tightly packed"; that is, only the minimum number of points needed to represent them are given. If the spectral lines are given, the density of points representing them will be very sparse. The finite interferogram function for this case will extend over all (or nearly all) of the discrete points of the DFT. For the restoration, however, we find that there is little or no interferogram function to extend with this number of points. For further improvement, it is necessary to pad more zeros beyond the end of the interferogram function (or beyond both sides) to double the total number of data points (which brings the number up to the next integral power of 2). On transforming, it will be found that the density of points on the spectral lines has doubled. Applying the constraints will then restore the spectrum and extend the interferogram function. Greater resolution will be obtained for the spectral lines. This implies that more information is now carried by the data points. Many find this result surprising: that the additional data points included actually bring more information with them. Yet this property may be amply verified in practice with both artificial and experimental data. The additional information, of course, is brought in by the constraints. Also, if further extension of the interferogram is needed, more zeros may be simply padded onto it.

III. CONCLUDING REMARKS

The number of additional Fourier spectral components to recover is the option of the researcher. The number of iterations to execute with the most general computer program written is also the option of the researcher. A tolerance is presently used to determine the number of iterations performed. However, it is found in practice that only 5 or 10 iterations yield sufficiently accurate results for nearly all experimental data of interest. With the presently used computer program, restoration is to the spatial function, and the improved spatial function and the improved values of the coefficients are both generated with each iteration. If the improved Fourier spectrum is not desired, then additional computational time could be saved by neither reading nor writing the Fourier coefficients. When M data points are treated, the computer memory requirements are seldom more than $7M$ words. If it is not necessary to determine the extended Fourier spectrum, then more than $5M$ words are seldom needed in computer memory.

Furthermore, the iterative algorithm is very versatile in that the four constraints of finite extent, minimum negativity, minimization of values above an upper bound, and recovery of only a finite Fourier spectral band may be simultaneously applied. The formulation could also accommodate other constraints of definite limits. The method of successive substitutions is also able to accommodate a large number of equations, making it possible to solve for a large number of spectral components.

The researcher may want to combine the computer program used for inverse filtering with that used for spectral continuation so as to perform the complete restoration in one step. The truncation frequency of the inverse-filtered spectrum could be automatically determined from the rms of the noise and the signal, and the amplitude of the spectrum of the impulse response function.

Some of the most valuable recent advances in the field of deconvolution have come from the field of digital image processing. It is therefore appropriate to note in conclusion that all the techniques developed here for spectroscopy may easily be generalized to two dimensions. The three constraints of finite extent, minimum negativity, and recovery of a finite spectral band, for example, are appropriate to image processing. For medical images, some of these constraints would apply. However, ways of applying other constraints appropriate to these data would have to be developed. Also, in two dimensions the memory requirements for the iterative algorithm are considerably less than for the one dimensional case (in proportion to the number of data points). See Appendix A for the two-dimensional formulation of the iterative equations.

See Appendix B for recent improvements in the numerical efficiency of the iterative equations.

APPENDIX A

The equations for the coefficients of the sines and cosines in the frequency spectrum to be restored have also been derived for the two-dimensional case. The general expression for these coefficients for a square array is illustrated in the following two equations:

$$
A_{p,q} = -\left(\sum_{x=0}^{N-1} \sum_{y=0}^{N-1} \left\{ u_{x,y} + v_{x,y} - A_{p,q} \cos\left[\frac{2\pi}{N}(px + qy) \right] \right\} \right.
$$
$$
\left. \times \cos\left[\frac{2\pi}{N}(px + qy) \right] \right) \left\{ \sum_{x=0}^{N-1} \sum_{y=0}^{N-1} \cos^2\left[\frac{2\pi}{N}(px + qy) \right] \right\}^{-1} \quad (A.1)
$$

and

$$
B_{p,q} = -\left(\sum_{x=0}^{N-1} \sum_{y=0}^{N-1} \left\{ u_{x,y} + v_{x,y} - B_{p,q} \sin\left[\frac{2\pi}{N}(px+qy) \right] \right\} \right.
$$

$$
\left. \times \sin\left[\frac{2\pi}{N}(px+qy) \right] \right) \left\{ \sum_{x=0}^{N-1} \sum_{y=0}^{N-1} \sin^2\left[\frac{2\pi}{N}(px+qy) \right] \right\}^{-1} \quad \text{(A.2)}
$$

for $u_{x,y} + v_{x,y} < 0$, where $x, y, p, q = 0, 1, 2, \ldots, N - 1$, and N is the number of rows or columns. The discrete spatial variables are denoted by x and y, and the corresponding spectral variables by p and q. The functions $A_{p,q}$, $B_{p,q}$, $u_{x,y}$, and $v_{x,y}$ are the two-dimensional counterparts of the one-dimensional functions discussed in the text.

If one is looking at the descrete Fourier spectrum of real data with the dc component centered at $(N/2, N/2)$, one wishes to recover a band of frequencies within the desired spectral range in the first quadrant including the zero frequencies (corresponding to sinusoids oriented along the x or y axes). In addition, a corresponding band with one negative spectral variable should be included (exclusive of the zero frequencies). This is necessary to obtain a complete set of waveforms. This additional spectral band could either be in the second or fourth quadrant. The choice is completely arbitrary. No other frequencies are required. The reader will recall that the complex coefficients in the other two quadrants are the complex conjugates of those in the two quadrants chosen.

APPENDIX B

Recent investigations have been made into the computational efficiency of the restoration methods discussed in Chapters 9 and 10. Rewriting Eqs. (6), (A.1), and (A.2) in different forms to perhaps find a form that involved fewer computations was explored. Means to bring about a more rapid convergence of these equations were also investigated. The two-dimensional equations will be addressed in the following discussion, although most of the techniques mentioned will apply to the one-dimensional situation, also.

The summation over the squared sines and cosines, respectively, in the denominator of Eqs. (A.1) and (A.2) essentially acts as a weighting function applied to each Fourier coefficient. Calculating these summations for the coefficients restored over several iterations, the author observed that this weighting function, generally, varied very slowly over the Fourier frequencies and iterations. This strongly suggested that for the numerical calculations involved, this function could be replaced by a constant value.

$$1/\sum_x \sum_y \cos^2\left[\frac{2\pi}{N}(px + qy)\right] \approx 1/\sum_x \sum_y \sin^2\left[\frac{2\pi}{N}(px + qy)\right]$$

$$\approx \text{const} = S. \tag{B.1}$$

Replacing all the weighting functions by a single constant value in the numerical calculations for several test data sets resulted in little significant change in the image over the first few iterations. A considerable reduction in the number of computations results from the use of the constant factor.

To further reduce the computational burden, an attempt was made to separate the variables. To see how this may be implemented, let us consider Eq. (A.1), which enforces the minimum negativity constraint. Note that it may also be written:

$$A_{p,q} = -\frac{\sum_x \sum_y (u_{x,y} + v_{x,y}) \cos[(2\pi/N)(px + qy)]}{\sum_x \sum_y \cos^2[(2\pi/N)(px + qy)]} + A_{p,q} \tag{B.2}$$

for $u_{x,y} + v_{x,y} < 0$. Replacing the summation by the constant discussed in the preceding paragraph and denoting $u_{x,y} + v_{x,y}$ by $h(x, y)$, we will have

$$A_{p,q} = A_{p,q} - S\sum_x \sum_y h(x, y) \cos\left[\frac{2\pi}{N}(px + qy)\right] \tag{B.3}$$

for $u_{x,y} + v_{x,y} < 0$. The $A_{p,q}$ on the right-hand side of this equation, as for Eqs. (6), is understood to be the old value of this coefficient. The $A_{p,q}$ on the left-hand side is the new (hopefully improved) value of this coefficient after the current iteration. Finally, the summation over the negative values is equivalent to multiplying $h(x, y)$ by a mask that zeros all but the negative values of this band-limited function. This mask is unity over the negative values and is zero elsewhere. It would be a rect function in one dimension. In two dimensions, though, it could have many varied patterns. Denoting this mask by $m(x, y)$ we now have

$$A_{p,q} = A_{p,q} - S\sum_x \sum_y h(x, y)m(x, y) \cos\left[\frac{2\pi}{N}(px + qy)\right]. \tag{B.4}$$

The finite extent constraint may easily be included in this formulation by simply letting the nonzero extent of the mask also include all values of the data field outside of the limited extent of the object.

We may include the coefficient for the sine corresponding to any given frequency by writing the coefficients in complex form:

$$A_{p,q} + jB_{p,q} = A_{p,q} - S \sum_x \sum_y h(x, y)m(x, y) \cos\left[\frac{2\pi}{N}(px + qy)\right]$$

$$+ jB_{p,q} - jS \sum_x \sum_y h(x, y)m(x, y) \sin\left[\frac{2\pi}{N}(px + qy)\right]$$

$$= A_{p,q} + jB_{p,q} - S\left(\sum_x \sum_y h(x, y)m(x, y) \right.$$

$$\left. \times \left\{ \cos\left[\frac{2\pi}{N}(px + qy)\right] + j\sin\left[\frac{2\pi}{N}(px + qy)\right] \right\} \right)$$

$$= A_{p,q} + jB_{p,q} - S\{\sum_x \sum_y h(x, y)m(x, y)e^{[j(2\pi/N)(px+qy)]}\}. \quad \text{(B.5)}$$

The complex exponential separates into factors involving only a single variable. Lumping the constant S with the mask, we now have for the correction term:

$$A_{p,q} + jB_{p,q} = A_{p,q} + jB_{p,q} - \sum_x e^{[j(2\pi/N)px]} \sum_y h(x, y)m(x, y)e^{[j(2\pi/N)qy]}. \quad \text{(B.6)}$$

This separation of the variables allows a vast reduction in the number of calculations. This savings becomes more significant as the number of data points increases. Further, note that the final summation in this equation is in the form of a one-dimensional Fourier transform. This implies that the considerable calculational advantage of the fast Fourier transform (FFT) algorithm may be used here. The entire summation may be performed by repeated application of the one-dimensional FFT. This implies that for any data set that it is practical to apply the FFT, it would be also practical to apply the nonlinear constraints for improvement.

There is an important disadvantage stemming from variable separation and use of the FFT, however. One is limited to the use of the "point simultaneous" procedure in the determination of the coefficients. That is, all of the coefficients must be computed in an iteration before they can be re-substituted back in the iterative equations. With the "point successive" method, the improved coefficient determined by one equation is substituted in the succeeding equation to render additional improvement. With the test data treated in this work, the point simultaneous procedure converged much more slowly than the point successive method.

The application of the point simultaneous procedure to two-dimensional equations yields adequate restorations for many interesting two-dimensional data. However, faster convergence is clearly desirable. Two techniques have been developed by the author that have proven successful in achieving more rapid convergence. One method involves the use of the finite extent constraint applied in the spectral domain. The other procedure involves the use of more or less ad hoc multiplicative factors that weight the spectrum corrections determined from Eq. (B.6), thereby producing spectral values closer to the original.

To understand how the application of the finite extent constraint in the spectral domain produces improved results, a brief review of Fourier theory is necessary. From the first iteration, we would desire the spatial function produced only by the *high frequencies* (beyond cutoff), which cancels the negative values of the given band-limited spatial function (produced by the low-frequency spectral band). Then multiplying this function by the mask would produce the spatial function that is the *negative* of the given (low-frequency) band-limited function multiplied by the mask. The convolution theorem tells us that multiplication by the mask in the spatial domain results in convolution with the Fourier transform of the mask in the frequency domain. This gives us another interpretation for the Fourier spectrum produced by the iteration. It would be the convolution of the Fourier transform of the mask with the (original) low-frequency spectrum. Alternatively, it could be considered as the convolution of the Fourier transform of the mask with the *negative* of the high-frequency spectrum that produces the spatial function that cancels the negative values of the band-limited function.

Interpreting the degradation as a convolution of the desired high-frequency spectrum with the mask transform suggests the use of deconvolution procedures to restore the spectrum. Inverse-filtering would not apply with this type of response function. The constraint of minimum negativity could not appropriately be applied as the spectral components are generally both positive and negative. The constraint of finite extent could appropriately be applied because all values in the low-frequency range could be interpreted as lying outside the extent of the high-frequency range. Because the spectrum we are considering is the Fourier transform of the negative values of the original spatial function, the correct spatial function that lies outside the range of these negative values should produce a Fourier transform that zeros all frequencies lying in the low-frequency spectral range. To implement this constraint numerically, we now apply the mask in the *spectral* domain, the reverse of the usual procedure, to remove those frequencies that lie outside the low-frequency range. (The high-frequency band is saved, of course.) Taking the inverse Fourier transform of this low-frequency band, we multiply the resulting spatial function by the negative image of the original spatial mask, which passes only those values outside the extent of the original negative region and eliminates all others. After taking the Fourier transform of this spatial function, the resulting spectral values are added to the retained high-frequency spectrum to yield the corrected high-frequency band. Finally, the *negative* of this high-frequency band is joined to the original correct low-frequency band to produce the entire improved Fourier spectrum. Taking the inverse Fourier transform of this, we are ready to begin another iteration.

The correction to the spectrum that we discussed in the foregoing may also be considered as a type of "reconvolving factor" because it may be obtained by "reconvolving", or convolving the high-frequency spectrum again with the Fourier transform of the mask that passed only the negative values. This reconvolution was accomplished in the transformed domain because the inverse transform of the high frequencies was multiplied by the mask. The resulting function was then transformed to the frequency domain. These spectral values were then subtracted from the original high-frequency band in an attempt to produce some approximation to the original spectrum before the first convolution. The motivation for applying these corrections, of course, was that as the high-frequency band obtained from the transform of the negative regions is considered a reasonable first approximation to the restored high-frequency spectrum, a reconvolution should yield at least a crude approximation to the changes produced by the first convolution. Adding the negative of these changes to the spectrum should produce a high-frequency band closer, overall, to the original. Reconvolution produces essentially the same corrections as applying the finite extent constraint in the frequency domain, for if one follows the development closely, it is discovered that one set of corrections is the negative of the other. Regardless of the interpretation given these corrections, it has been proved to bring about significant improvement in the resulting images for a variety of test data. However, because two additional FFTs are involved in the generation of these corrections, very little overall reduction in computational time is presently being achieved.

It has also been discovered that a variety of simple multiplicative factors applied to the high-frequency spectrum during an iteration was successful in bringing about more rapid convergence of the equations. The primary motivation for applying these was the observation that the largest spectral components immediately beyond cutoff that were restored in the first few iterations were usually considerably smaller than the original components, but that their signs were nearly all correct. Any simple positive factor that rises slightly above unity for the values immediately beyond cutoff should produce spectral values closer to the original. In practice, a wide variety of factors has been shown to produce improvement. In the test data sets the author used, values from unity to slightly above two could be applied without serious distortion in the image. For the larger factors, however, tapering window functions applied to the Fourier spectrum were needed to reduce the spurious high-frequency components that were occasionally generated. In the test data treated by the author, a window of Lorentzian form that tapered gradually from values of two (for the high-frequency spectrum immediately beyond cutoff) to the higher frequencies produced the best combination of rapid and distortionless restorations. This window was gradually lifted as more of the high frequencies were restored. The factor of two for the lower frequencies was altered only slightly, however, as it was noted that the correction to the spectrum generated on each iteration by the method was always underestimated.

ACKNOWLEDGMENTS

There are several people I wish to thank for help in bringing this work to completion. I am deeply indebted to Dr. Robert H. Hunt in ways too numerous to mention. Many enlightening discussions were held with Bob Hunt and Pete Jansson on the subject of Chapters 9 and 10. Their knowledge and expertise were invaluable. I also appreciate their proofreading the initial drafts. To acknowledge others whose peripheral support was very helpful, I would certainly have to mention Clydeyne Nelson, Ken Ford, Beverly, my wife, as well as many others in the Physics Department at Florida State University.

Some illustrations and passages were borrowed from two papers of mine published in the *Journal of the Optical Society of America* (January 1981 and July 1981).

REFERENCES

Bell, R. J. (1972). "Introductory Fourier Transform Spectroscopy." Academic Press, New York.
Connes, P. (1969). "Lasers and Light." Freeman, San Francisco.
De Boor, C. (1978). "A Practical Guide to Splines." Springer-Verlag, New York.
Hildebrand, F. B. (1965). "Introduction to Numerical Analysis," pp. 450, 451. McGraw-Hill, New York.
Howard, S. J. (1981). *J. Opt. Soc. Am.* **71**, 819.
Loewenstein, E. V. (1966). *Appl. Opt.* **5**, 845.
Mertz, L. (1965). "Transformations in Optics." Wiley, New York.
Steel, W. H. (1967). "Interferometry." Cambridge University Press, New York.
Toth, R. A., Brown, L. R., Hunt, R. H., and Rothman, L. S. (1981). *Appl. Opt.* **20**, 932.
Vanasse, G. A., ed. (1977). "Spectrometric Techniques," Vol. 1. Academic Press, New York.
Vanasse, G. A., ed. (1981). "Spectrometric Techniques," Vol. 2. Academic Press, New York.
Vanasse, G. A., and Sakai, H. (1967). *In* "Progress in Optics," Vol. VI (E. Wolf, ed.), p. 261. North-Holland, Amsterdam.

Vanasse, G. A., and Strong, J. D. (1958). Application of Fourier transforms in Optics: Interferometric spectroscopy. *In* "Classical Concepts of Optics" (J. D. Strong, ed.), Appendix
F. Freeman, San Francisco.

Vanasse, G. A., Stair, A. T., Jr., and Baker, D. J., eds. (1971). *Aspen Int. Conf. Fourier Spectrosc.*
1970. AFCRL-71-0019, Spec. Rep. No. 114. A. F. Cambridge Res. Lab., Bedford, Massachusetts.

Index

A

Aberrations, optical, 230
Absorbance, 56
Absorptance, 54–57, 174
Absorption, apparent, 175
Absorption coefficient, 42–43, 175, 177
Absorption line, 41–43
   narrow, used in measuring response
     function, 59–61
   saturated, 59, 61, 105, 176
Absorption spectrum, 102, 171
Absorption spectroscopy, 41–44, 54–61,
   104–106
Aliasing, 23, 25, 88, 163, 170, 273, 276, 279
   of noise, 171
Alternating projection methods, 121–123
Alternatives to deconvolution, 30
Aluminum Kα doublet, 140, 145
Amplifier
   sample-and-hold, 169
   tuned, 167, 168
Analog application of constrained nonlinear
   deconvolution, 109
Analog signal processing, 166
Analog-to-digital converter, 169, 210
Analytic continuation, 97–98
Analytic function, 98, 102
Annealing, 126–128
Apodization, 64, 98, 212, 266, 303, 305
   triangular, 213
Area
   conservation of, 106
   under Fourier transform, 21
Artifacts, *see* Spurious components of solution
Artificial doublets in ESCA, 145

Associativity, 6, 40
Asymmetric deconvolution method, 76, 109
Autocorrelation function, 101–102, 165
Averaging of multiple scans, 179, 216

B

Background, 54, *see also* Baseline
Band limit, 78–81, 85, 113–114, *see also*
   Cutoff frequency
Bandwidth, optimum for processing, 80
Bandwidth extension, 97–98, 101–102, 106,
   123, 130, 261–331, *see also* Fourier
   spectrum extension
   in data communications, 111
Base line, 54, 115, 141–143, 181, 209–210
   determination of, 57
   in ESCA, 138
Bayes' rule, 243
Bayes' theorem, 129, 230, 236, 237
Beam experiment, 222
Benefits of nonlinear deconvolution methods,
   89–90, 96–97, 114, 130, 296, 323
Bias, user, incorporation of, 115
Bias circuit, 166
Bilevel signals, 111
Binding energy, 137
Biraud's method, 111–115
   in ESCA, 143
Bit-reversed sampling, 77, 127
Blackbody spectrum, 234
Blurring, 3
Boltzmann constant, 234
Boltzmann distribution, 126
Boltzmann factor, 116, 126, 128

Bose–Einstein statistics, 116, 229, 232, 235, 239
Bouguer–Lambert law, 42, 174–177, 300
Bound, 89, 97, 104, 108, 115, 125–126, 128, 131, 181, 229, 253, 269, *see also* Constraint
    upper, 104, 269
Bucket brigades, 110
Burg entropy, 230, 247, 250
Burg's method of maximum entropy, 118, 143

C

Calibration of spectra (wavelength and wavenumber), 171–173
Carbon-rod source, 157
Cauchy distribution, *see* Cauchy function
Cauchy function, 10, 33, 39, 53, *see also* Lorentzian line shape
Cautions in use of deconvolution, 90, 130, 155–156, 224
Cell, resolution, 230
Center of mass, *see* Centroid
Central fringe, in Fourier spectroscopy, 302
Central-limit theorem, 8–10
Centroid, 10, 21
Chain rule for probabilities, 236
Channeling, 317
Charge-coupled devices, 110
Charging of sample in ESCA, 137, 149–150
Chemical bonding, 136–137
Chemical shift, in ESCA, 137
Chopper, mechanical, 53, 164–166
Chromatography, 266, 269
Circulant matrix, 74
Circular buffer, 77
Clean algorithm, 32
Clipping, 103, 107–108, 122, 130, 184
Coherence area, 233
Coherence time, 233
Coherent radiation 18, 29, 45–49
Collimator, 45
Collision broadening, 39–42
Comb function, *see* Infinitely-replicated impulses
Combined Doppler, natural, and collision broadening, 40–41, 43, 179
Commutativity, 6, 40
Component probability, 119
Compton scattering, 222

Computer program, deconvolution, 106, 150
Condensed-phase infrared spectra, 99
Conditional probability, 236, 251
Confidence, user, incorporation of, 115–116
Conjugate-gradient method, 129
Constrained methods, 90, 93–134, 143–151, 183–331, *see also* Nonlinear method
Constraint, 79, 86, 90, 96–97, 108, 114, 144, 181, 268–270, 275, 323–324
    finite extent, 278–280, 293
    positivity, 33
Contraction-mapping theorem, 108
Convergence
    of alternating projection method, 123
    of Gold's method, 99, 100
    of Howard's method of minimum negativity, 292, 295, 327–330
    of iterative method, 78
    of Jansson's method, 183, 203–205
    of nonlinear overrelaxation, 108
    of overrelaxation, 85–86
    proof of, 130
Conviction
    highest possible, 238
    user, incorporation of, 115–116, 235–239
Convolution, discrete, 4–6, 73, 79
Convolution theorem, 22–24, 102, 303–304, 328
Coriolis interaction, 217
Cross entropy, 247, 252
Cubic equations, 112
Curve fitting, 90, *see also* Optimization
Curve-resolving instrument, 33
Cutoff frequency, 64, 78, 80, 83, 97–98, 269, 307, *see also* Band limit

D

Data
    communications, 111
    preparation of, 179–182
Data acquisition rate, enhanced, 90, 156
Decision rule, 125–126
Deconvolution
    defined, 28, 188–189
    methods,
        alternating-projection, 121–123
        analytic continuation, 97–98
        annealing, 126–128
        asymmetric, 76, 109

Biraud's, 111–115
Burg's, 118–143
conjugate-gradient, 129
constrained methods, 93–134
direct approach, 69–71, 141
error-reduction algorithm, 122
Fourier inverse filter, *see* inverse filter
Frieden's, of maximum information, 129
Frieden's, of maximum likelihood,
    115–121, 227–259
Frieden's, of maximum entropy, 117
Gauss–Seidel, 76
Gold's ratio, 99
Howard's, of finite extent, 123–124,
    261–287
Howard's, of minimum negativity, 124,
    289–331
inverse filter, 80–87, 98, 112, 118, 122,
    124, 265, 272, 295
iterative, based on direct approach,
    69–71, 141
Jansson's, 29, 144–151, 170, 172,
    187–225, 253
linear,
    defined, 68, 89
    limitations, 68, 71, 73–74, 78, 84–86,
        89–90, 96, 131
linear programming, 129
matrix inversion, 73–80
maximum a posteriori estimation, 230
maximum entropy, 117–118, 239–240
Monte Carlo, 77, 125–128
nonlinear iterative, 102–111
nonlinear, 68, 90, 93–331
objective function, *see* Deconvolution,
    methods, optimization
optimization, 30–33, 125–128
overrelaxation, 76
Phillips and Twomey, 88, 129
point-Jacobi, 76
point-simultaneous, 75–77, 103, 294, 327
point-successive, 76, 105, 145, 327
Poisson transform, 88
projection, 88
pseudoinverse, 88
real-time superresolving, 109
reblurring, 78, 86, 108–109, 179
regularization, 124, 129
Schell's, 101–102, 118
sharpness-constrained Wiener inverse
    filter, 82–83

simultaneous displacement, 76
singular value decomposition, 88
steepest-descent, 129
stepwise implementation of inverse filter,
    86–87, 143
successive substitution, 293–316
superresolving transversal filter, 109–111
Van Cittert's, 29, 71–72, 76, 79, 83–86,
    99–100, 142–143, 268
Wiener inverse filter, 82–83, 101
Degeneracy, 249, 257
    factor, 234–235, 237
Degrees of freedom, 30, 113, 116, 232–234,
    238
Delta function, *see* Dirac δ function
Detector, 48
    infrared, 158, 163
    photoconductive, 166
    photovoltaic, 166
    spectral response, 54
    visible, 164
Determinant, 100
DFT, *see* Discrete Fourier transform
Differentiation of convolution, 7–8
Diffraction, 62, 230, 255
    limit, 106
    pattern, 45–51
Digital filtering, 170
Digital image processing, *see* Image
    processing
Digital transmission, 111
Digitizing, signal, 169, 206
Dirac δ function, 7, 16, 19, 24, 28, 138, 266,
    305, 316
Direct approach to deconvolution, 69–71,
    141
Dirichlet's equality, 249
Dirichlet integral, 243
Discrete convolution, 23
Discrete Fourier transform, 24–27, 271–276,
    290, 304
Dispersion, 160, 172
    variation of, 178
Dispersive spectrometer, 230
Distortions in ESCA spectra, 138–141
Distributivity, 6
Doppler broadening, 40–41, 58, 213, 297,
    316
    emission line, 119
Doppler line shape, 61, 161, 176, 217, 222,
    302

Doublet, x ray, 140
Drift, detector and/or source, 55
Duty cycle of square wave, 146

E

Echelle grating, 157
Eigenvalue, 78, 100, 103
Electrical filtering, 51–53
  in ESCA, 140
Electron energy-analyzer broadening, 140
Electron microscopy, 121–122
Electron spectroscopy for chemical analysis,
  71, 99, 135–152
Electrostatic charging of sample, *see* Sample
  charging in ESCA
Elsasser band model, 118
Emission spectrum, 40, 119, 230
Empirical data, impartial use, 242
Empirical entropy, 230, 249–250
Empirical evidence for object-class law,
  240
End effects, 4–6
Ensemble average, 80
Entropy, 117–118, 129
  Burg, 230, 247, 250
  cross, 247, 252
  empirical, 230, 249–250
  Jaynes', 230, 239, 247, 249
  Kikuchi–Soffer, 230, 239
  maximum, 239–240
  photon-site, 230, 249–250
Entropy-like estimators, 247–248
Equal-energy spectrum, 239–240
Equivalent width, 7, 30, 44, 56–58, 162
Error, 31, 102, 263
  in direct method of deconvolution, 71
  roundoff, 74, 78
Error-reduction algorithm, 122
ESCA, *see* Electron spectroscopy for chemical
  analysis
Estimate of spectrum, 229
Estimators, maximum likelihood, 246–250
Even function, 18, 20
Exponential
  relaxation function, 107, 144
  truncated, 17
Extrapolation, *see also* Bandwidth extension,
  98, 106
  of Fourier spectra, 277–278

F

Fast Fourier transform, 23–27, 122, 124, 265,
  322–323, 327
Fermi–Dirac statistics, 116
FFT, *see* Fast Fourier transform
Filter, *see also* Smoothing
  digital, 170
  inverse, *see* inverse filter
  rolloff, 168
  superresolving, 109–111
Finite extent, 102, 112, 269, 278–280, 293,
  303
Firmware, 111
Fitting methods, 30–33
Flatness of spectral line at wing and peak,
  118–120
Flood-gun method of minimizing charging
  effects in ESCA, 150
Fourier analysis, advantages of, 264
Fourier inverse filter, *see* Inverse filter, 80
Fourier spectra, 4
Fourier spectrometer, resolution of, 62–63
Fourier spectroscopy, 25, 45, 90, 98, 164,
  211–213, 219–221, 264, 278, 302–323
Fourier spectrum continuation, *see* Fourier
  spectrum extension
Fourier spectrum extension, 102, 106,
  111–115, 261–331
Fourier transform, 11–27
  alternate convention, 11, 271
  in alternating project method, 122
Fourier's integral theorem, 18
Fredholm integral equation, 29, 73
Freezing, in annealing, 127
Frieden's method
  of maximum information, 129
  of maximum likelihood, 115–121, 227–259
  of maximum entropy, 117
Full width at half maximum 63, 161
Functional analysis, 108
Future of deconvolution, 131
FWHM, *see* Full width at half maximum

G

Gamma-ray spectrum, 222–223
Gas, spectrum of, 38–44
Gate function, 165
Gauss–Jordan method, 279

Gauss–Seidel method, 76, 294
Gaussian, *see also* Gaussian function
  form of Maxwell distribution, 40
Gaussians, convolved, 23
Gaussian absorption line, 59
Gaussian distributed error, 31
Gaussian elimination, 74
Gaussian function, 8, 13, 33, 239
Gaussian instrument function, 178, 189, 194, 216
Gaussian line, 171, 213
Gaussian line-broadening function, 87
Gaussian window function, 309, 315
Gerchberg–Saxton method and related methods, 121–123
Gold's ratio method, 99
Grain
  in annealing, 126
  in decision rule approach, 125
  negative, in annealing, 128
  as photon in Frieden's method of maximum likelihood, 115
Graphic display, 182
Grating,
  echelle, 157
  equation, 160

H

Half-width
  of absorption line, 59–61
  Doppler, 40
  at half-maximum, 8, 63
  Lorentzian, 39, 58
Hamiltonian, 127
Heaviside step function, 17, 52, 79, 124, 138, 168, 290
Heisenberg uncertainty principle, 38, 233, 267–268
High frequencies
  attenuation of, 194
  loss of, 30, 64, 194
High-conviction estimators, 246–247
Hilbert space, 123
Holography, 121
Howard's method of finite extent, 123–124, 261–287
Howard's method of minimum negativity, 124, 289–331
HWHM, *see* Half-width, at half maximum

I

Ignorance, maximum prior, 229
Ill-conditioned matrix, 74
Ill-posed problem, 124, 129
Image function, 28
Image processing, 4, 32, 88, 96, 109, 239, 324, *see also* Two-dimensional problems
Impartial conviction, 247–250
Impartial use of empirical data, 242
Impulse, *see also* Dirac δ function
  first derivative of, 7
Impulsive object, 240
Incoherent radiation, 18, 49–51, 236
Inelastic-scattering baseline correction, 141–143
Inelastic scattering of electrons, 138–139
Infinitely replicated impulses, 16, 273
Infrared detector, 158, 163
Infrared spectrum, 102, 105, 125, 171, 215–221, 297
  condensed phase, 99
  of liquid, 44
Inherent broadening, deconvolution of, 43–44, 176–177
Inherent line breadth, 38
Instrument, curve resolving, 33
Instrument function, 28, 29, 161, 172, 177–179, 206, *see also* Spread function
  effect of errors on, 207–208
Instrument response function, *see* Instrument function
Instrumental considerations, 153–185
Integral equation, 29, 188
Integrated absorption, 44
Integration of convolution, 7–8
Intensity statistics, 118–120
Interaction, human, 75, 129
Interferogram, 302
Interferometric spectroscopy, *see* Fourier spectroscopy
Interleaved sampling, 77
Interpolation, 215, 274
Inverse filter, 80–87, 98, 112, 118, 122, 124, 265, 272, 295
  stepwise implementation, 86–87
Isolated line, used to measure spread function, 58, 150, 161
Isotopic shift, 215
Interative deconvolution method based on direct approach, 69–71, 141

J

Jacobi method, *see* Point-Jacobi method
Jansson's method, 29, 170, 172, 183–184,
    187–225, 253
    applied in ESCA, 144–151
Jaynes' entropy, 230, 239, 247, 249

K

Kernel, convolution, 83
Kikuchi–Soffer entropy, 230, 239
Kirchoff's law, 52

L

Lagrange multiplier, 112, 116, 238
Least-squares criterion, 27, 31–32, 80–84,
    86, 101, 124, 129, 252, 257, 276, 278,
    290
Light pen, 182
Limitations of linear methods, *see*
    Deconvolution, methods, linear,
    limitations
Limiting resolution, 63
Line breadth, inherent, 38–44
Line pairs, optical, 63
Line strength, measurement of, 57
Linear combination, 4
Linear deconvolution method, defined, 68, 89
Linear differential correction method, 32
Linear equations, methods of solution, 76
Linear-programming method, 129
Linearized methods, 32
Linewidth, determination of, 31
Liquids, infrared spectra of, 44
Littrow spectrometer, 157, 178, 284, 297, 299
Lock-in detection, *see* Phase sensitive
    detection
Lorentzian line shape, 10, 17, 39, 40, 44, 58,
    140, 143, 213, 222
Lost frequencies, 79, 84–85, 89–90, 97–98,
    120, 123, 255

M

MacQueen–Marschak maximum prior
    ignorance, 243
MAP, *see* Maximum a posteriori estimation

Marquardt's method of optimization, 32
Matrix formulation of deconvolution, 73
Matrix inversion, 73–80
Maximum likelihood applied to empirical data,
    252–257
Maximum a posteriori estimation, 230
Maximum conviction, 239
    empirical data, 241
    without empirical data, 244
Maximum entropy, 117–118, 239–240
Maximum ignorance, 244–246
Maximum information, 129
Maximum likelihood, 115–121, 129
    estimate, 230
    estimators, table, 258
    methods, 227–259
Maximum prior ignorance, 229
    MacQueen–Marschak definition, 243
Maximum-information method, 129
Maxwell distribution, 40
Memory, efficient use of, 77, 108
Methods, of deconvolution, *see* Deconvolution
Microprocessor-based instrument, 111
Minimization, 30–33
Minimum mean-square-error, *see* Least-squares
    criterion
Minimum negativity, *see* Howard's method of
    minimum negativity
Model, of object or spectrum, 31
Modes, 232
Modulation, *see also* Chopper, mechanical
    to create artificial ESCA couplets, 145
Molecular scattering, 273
Molecular-beam scattering, 99
Moment, 21–22
    second, 10
Monochromatic radiation, 305
    coherent, 45–47
    x rays in ESCA, 137
Monochromator, *see* Spectrometer
Monte-Carlo methods, 77, 125–128
Most probable number-count set, 235
Most probable object
    definition, 235–237
    in presence of data, 237–238

N

Natural line broadening, 38–44
Natural line shape, 138, 141

NEP, *see* Noise equivalent power
Newton's method, 52, 58, 142
Newton–Raphson method, 117, 247, 249, 253, 255, 292
Noise, 20, 29–30, 33, 51, 80, 130
    additive, 80, 82, 112, 116–117, 196, 251
    before blurring, 111, 113
    broadband, 195
    effect on deconvolution, 173, 281, 296
    Gaussian-distributed, 82, 117, 251, 256, 281
    Johnson, 251
    and lost frequencies, 84–85
    in maximum likelihood estimate, 250–252
    in measuring equivalent width, 57
    in measuring response function, 60
    nonadditive, 33
    signal dependent, 33
    in Van Cittert's method, 72
    white, 165, 195
Noise-equivalent power, 164–165
Noiseless data, 130, 188–194, 191, 193, 203, 280, 304–309, 311–312
Nonlinear method of deconvolution, 68, 90, 93–134, 143–151, 183–331
    benefits, 89–90, 96–97, 114, 130, 296, 323
    error tolerance, 58
    iterative, 102–111
Nonlinear superresolving transversal filter, 110
Nonlinearity, 184
    defined, 97
Nonnegativity, 97, 100, 101, 184, 269, *see also* Positivity
Nonphysical solution, elimination of, 104–106, 201, 269
Norm, 129
Normal distribution, *see* Gaussian function
Normal equations, 124
Normal modes, 116
Normalization, 4, 6–8, 48, 241
    of photons in object, 231, 237, 247
    of response function, 48, 60, 145
Nuclear spectrum, 129, *see also* Gamma-ray spectrum
Number-count set, 116
Nyquist interval, 25, 106, 255, 256

O

Object class, prior knowledge of, 238
Object function, 28

Object-class law, 236, 238–239, 240
Objective function, 30–33, 125–126
Odd function, 18, 20
Offset
    additive, 54–56
    in base line, effect of, 210
Optical density, 56
Optimization, 30–33, 125–128
Orthogonal functions, 18
Overrelaxation, 76

P

Parabolic relaxation function, 107
Parameters, fitting of, *see* Fitting methods
Passes, of dispersion, 62, 158
Path difference, of Fourier spectrometer, 98, 302
Perfectly resolving instrument, 28, 38, 48, 50, 79
Phase problem, 121
Phase-sensitive detection, 53, 166
Phillips–Twomey method, 88, 129
Photoelectron spectroscopy, *see* Electron spectroscopy for chemical analysis
Photon bunching, 257
Photon statistics, 229, 232
Photon-site entropy, 230, 249, 250
Physical knowledge, 229
Physical model for object, 232–235
Physical realizability of solution, 72, 89–90, 96–97, 99, 101–103, 111, 124, 143, 229
Pipeline processor, 109
Planck's constant, 231, 234
Plasmon, 143
Point-Jacobi method, 76, 294
Point-simultaneous method, 75–77, 103, 294, 327
Point-spread function, *see also* Response function, 62
Point-successive method, 76, 105, 145, 294, 327
Poisson transform, 88
Polarization, 46
Polynomial, fitting to find derivatives, 20–21
Polynomial filter, *see* Polynomial smoothing
Polynomial smoothing, 77–79, 84–85, 105–106, 108, 181
Positive-definite matrix, 100

Positivity, 33, 97, 115, 126, 128–129, 144, 229, 253, 290, *see also* Nonnegativity
Posterior knowledge, of image data, 238
Power spectral density, 80, 89, 194, 240
Power spectrum, *see* Power spectral density
Power theorem, 19
Preamplifier, 166
Predisperser, 157
Preprior probability law, 243
Pressure broadening, *see also* Collision broadening, 42, 58, 61, 213
Prior knowledge, 30–33, 89, 115, 263, *see also* Constraint
   of object class, 238
   physical, 229
   statistical, 229
Prior object, 236
Prior spectrum, 116, 118–120, 236
Probability, 89, 263
   conditional, 236
Probability-based method, 115–121, 126–129, 227–259
Program, computer, 150
Projection method, 88
Prolate spheroidal wave function, 123
Proof of convergence, *see* Convergence
PSD, *see* Power spectral density
Pseudoinverse, 88

Q

Quadratic relaxation function, 107
Quantum optics, 247
Quasi-featureless spectrum, 253
Quasi-monochromatic source, 49

R

Raman spectrum, 214, 219–222
Ramp function, 17
Random-grain model, 115–120, 125–128
Rayleigh criterion, *see* Rayleigh limit
Rayleigh distance, *see* Rayleigh limit
Rayleigh limit, 46, 49, 62–64, 161
Rayleigh width, *see* Rayleigh limit
Rayleigh's theorem, 19
RC filter, 51–53, 140, 167–168, 180
Real-time superresolving deconvolution, 109
Reblurring, 78, 86, 108–109, 179

Reconvolving factor, 329
Rectangle function, 8, 12, 48, 85, 143, 303
Recurrence formula, 142
Regularization, 124, 129
Relaxation factor, 76, 78, 85–86, 103
   optimum, 78
Relaxation function, 103–108, 144, 183–184, 189, 194–195, 201–208, 223
Relaxation methods, 29, 75
   nonlinear, 102–111
Replication of impulses, *see* Infinitely replicated impulses
   in analyzing of discrete Fourier transform, 24–27
Resolution cell, 230
Resolution criteria, 61–64
Resolution target, 105
Resolution versus acquisition time trade-off, 156
Resonance contour, *see also* Lorentzian line shape, 17
Response function, 23, 48, 54–55, 304
   approximating, 178
   determining 58–61, 179
   Gaussian, 59–60
   see also Spread function
Response time, detector, 167
Restoration, 68
Restoring principle, 229
Rise time, 63
Root-mean-square error, 205, 208–209, *see also* Least-squares criterion
Roundoff error, 74, 78, 281

S

Sample charging in ESCA, 137, 149–150
Sample-and-hold amplifier, 169
Sampling, 4, 7, 24–27, 106–108, 170
   bit-reversed, 77, 109–127
   in data acquisition, 163
   density, 180, 276
   to enforce finite extent constraint, 112–114
   theorem, 25, 63, 112, 170–171, 273, 275
Saturated absorption line, 59, 105, 176
Saturation of residual, in Biraud's method, 113
Savitzky and Golay, *see also* Polynomial smoothing
   differentiation method, 21
   smoothing method, 79, 181, 197, 223

Scale factor
  in Fourier transform, 11, 19
  in Jansson's method, 145
Scanning
  continuous, 170, 179
  rate, 171
  step, 171, 180
Schell's method, 101–102, 118
Self-convolution, 112
sgn function, *see* Sign function, 17
Shannon entropy, 117
Sharpness-constrained Wiener inverse filter,
    82–83
Shift theorem, 20, 267
Shift-variant integral equation, 29
Shift-variant spread function, 74–75, 79, 100
Si function, 13, 46
Sidelobes, 212, 219–221, 266, 283, 296,
    305
Sign function, 17
Signal-processing function, 165
Signal-to-noise ratio
  constant, 53, 156
  and detection, 163–166
  required for deconvolution, 174
Signal-to-noise versus resolution trade-off, 86,
    130, 137
Similarity theorem, 12, 19
Simulated data, 90, 188, 255, 280–283,
    304–316
Simultaneous displacement method, 76
Sinc function, 12–13, 49, 85, 180, 212, 267,
    303, 305
Sinc-squared function, 25, 46, 62–64, 80,
    143, 255, 304–305
Singular matrix, 74
Singular-value decomposition, 88
Sinusoid, 16
Slit, curved, 158
Slit, spectrometer, 45, 49
Smoothing, 4–7, 78, 180–181, 195,
    197–200, 222
  by digital filtering, 170
  by electrical filtering, 51–53
  by maximum entropy, 117–118, 239
  by polynomial convolution, *see* Polynomial
      smoothing
  by reblurring, 86, 108–109
  by Wiener filter, 80–82
Source, 229
  carbon rod, 157

emission spectrum of, 54
  laser line, 58
  spectral line, 58
Sparrow's criterion, 62
Spectra, Fourier, 4
Spectral line of known shape, 29, 79
Spectral purity, 233
Spectral slit width, 161
Spectrometer,
  analytical characterization, 160
  dispersive, 97, 156–173, 215–216, 230
    broadening, 44–61
  Fourier, 90
    broadening, 44
  Litrow, 157
  multiplex, 90
  non-ideal, 51, 189
  rapid-scanning, 55
  single-beam, 54
  theoretical resolving power, 62–63
  Wadsworth-Littrow, 158
Spline fit, 273, 299, 317, 319
Splitting of lines
  spontaneous, 118
  torsional, 217
Spread function, *see also* Response function,
    28, 59
  shift variant, 74
Spreading, *see* Blurring
Spurious components of solution, 30, 115,
    117, 126, 128, 195, 201, 203, 297, 311
Square wave, 145–147, 166
Standard deviation, 8
Stationary process, 165
Statistical knowledge, 229
Statistics, 89
Steepest-descent method, 129
Steepness of solution, 83
Stepwise implementation of inverse filter,
    86–87, 143
Stepwise refinement, 18
Stirling's approximation, 241, 246, 249
Stray flux, 54
Structured flowchart
  alternating projection method, 122
  Biraud's method, 113
Successive substitution method, 293, 316
Superposition, 18, 22, 46, 49–50, 96
Superresolution, *see also* Bandwidth extension,
    64, 106, 114, 120, 122, 129, 130,
    255–257, 323

Superresolving transversal filter, 109–111
Support, region of, 285
Surface research, 136
Swapping domains, 19–20
Symmetry of spread function, matrix
    formulation, 73–74
Synchronous detection, *see* Phase sensitive
    detection

T

Tangent-arm grating drive, 160, 179
Taylor series, 98
TDL, *see* Tunable-diode laser
Telescope, 45
Television lines, 63
Thermal equilibrium, 127
Thermal physics, 117
Toeplitz matrix, 74
Torsional motion, molecular, 217
Total probability law, 236
Trade-off
    acquisition time versus resolution, 156
    signal-to-noise versus resolution, 86, 90,
        308
    smoothness versus consistency, 88
Transfer function, 62
    electrical filter, 52
Transmittance, 54–57, 102, 104
Transpose of matrix, 100
Trapezoidal relaxation function, 107
Trial grain, 127
Triangle function, 8, 13
Tunable-diode laser, 177–178, 217
Tunable laser, 41
Two-dimensional gel electrophoresis, 32
Two-dimensional problems, 4, 32, 79, 96,
    239, 324, 327
Two-point resolution criteria, 61–63

U

Uncertainty in data, *see* Noise

Uncertainty principle, *see* Heisenberg
    uncertainty principle
Uniqueness, 30, 115, 126, 128, 245, 275
Upper bound, 107, 294
User bias, 118
User conviction, 115–116, 238

V

Van Cittert's method, 29, 71–72, 76, 79,
    83–86, 99–100, 142–143, 268
Van Cittert–Zernicke theorem, 233
Variance,
    of convolution product, 23
    of a distribution, 22
    expressed in terms of Fourier transform, 22
Video, 109
Voigt function, 10, 40–41, 43, 179, 213

W

Wadsworth–Littrow prism monochromator,
    158
Washout point, 316
Weight, 4, 6
Weighting function, *see* Relaxation function
White spectrum, 239–240
Whittaker–Shannon sampling theorem, *see*
    Sampling, theorem
Wiener filter, 80–83, 85, 89, 122
    inverse, 82, 101
    sharpness-constrained inverse, 83
    smoothing, 80–82
Window function, 303, 309, *see also*
    Apodization
Wrap-around, 4–6

X

X ray, 136
    doublet, 140
    line-broadening, 144